Update on Nutrition and Food Allergy

Update on Nutrition and Food Allergy

Editors

Nicolette W. de Jong
Harry J. Wichers

MDPI • Basel • Beijing • Wuhan • Barcelona • Belgrade • Manchester • Tokyo • Cluj • Tianjin

Editors
Nicolette W. de Jong
Erasmus MC
The Netherlands

Harry J. Wichers
Wageningen UR
The Netherlands

Editorial Office
MDPI
St. Alban-Anlage 66
4052 Basel, Switzerland

This is a reprint of articles from the Special Issue published online in the open access journal *Nutrients* (ISSN 2072-6643) (available at: https://www.mdpi.com/journal/nutrients/special_issues/Nutrition_and_Food_Allergy).

For citation purposes, cite each article independently as indicated on the article page online and as indicated below:

LastName, A.A.; LastName, B.B.; LastName, C.C. Article Title. *Journal Name* **Year**, *Volume Number*, Page Range.

ISBN 978-3-0365-5347-4 (Hbk)
ISBN 978-3-0365-5348-1 (PDF)

Contents

About the Editors

Nicolette W. de Jong

Dr. Nicolette W. de Jong is an associate professor and staff member at the department of Internal medicine, section Allergology & Clinical Immunology, ErasmusMC Rotterdam, the Netherlands. She has a background in clinical chemistry, with a specialization in immunology. She earned a PhD degree in 2004 with a thesis on Occupational and Food Allergy. During the past 20 years, she has initiated and coordinated many clinical trials, mainly focused on the diagnosis and treatment of food allergy. She is Principle Investigator (PI) of a multi-center project on cashew allergy in children with three PhD students, which was granted €1m by NWO-TTW. In 2019, she received a second grant of €1m from NWO-TTW for a project on cow's milk tolerance induction in co-operation with WUR. Since 2021, she has been PI on an NWO-TKI-PPS project (granted €1.6m) focused on the sensitizing capacities of novel nutritional proteins, together with WUR and UIPS with three PhD students. Furthermore, Nicolette is a graduated biology teacher and highly interested in (digital) education and the development of new educational programs in medicine (e-learning). She is coordinator and lecturer at the Erasmus University Center, in BSc in "Immunology and infection" and in MSc in "Clinical reasoning in Medicine". She is an associate member of the Institute of Medical Education Research (iMERR), INDANA, and NVvAKI. Internationally, she is a member of AAAAI and EAACI and a member of several EAACI Task Forces.

Harry J. Wichers

Prof. Dr. Harry J. Wichers is a biochemist/immunologist by training. He obtained his PhD degree from the University of Groningen, on the subject "Biotechnological production of pharmaceuticals via cultured plant cells". He moved in 1990 to Wageningen University and Research Centre, where he was involved in research on the biochemical characterization of quality-related parameters that determine food quality (taste, texture, and notably colour).

Currently, Harry Wichers is working on the relationship between (the digestion of) food and its components and (in particular) the gastro-intestinal immune system, including allergies. In this research, data on the immunomodulatory effects of food components, or intermediates that are formed during digestion, are integrated with data on their characteristics in raw materials and data on the impact of processing. The ultimate objective is to develop sensorily attractive foods that can contribute to maintaining a balanced and active immune system.

Preface to "Update on Nutrition and Food Allergy"

Food allergy is increasingly recognized as a growing public health burden and has been referred to as the "second wave" of the allergy epidemic, following asthma. Recent research shows a prevalence of 10% in young children in Europe. Causes may include pollution, dietary changes, and less exposure to microbes. Recently, processing operations, e.g., heating and drying, also appear to have an effect on the allergenicity and toxicity of food proteins. So far, the molecular and cellular mechanisms involved in sensitization to food allergens are not fully understood. A recent hypothesis is based on the dual allergen exposure of food allergens to the skin and gut. Observations that children with atopic dermatitis are more often sensitized to food allergens led to the hypothesis that exposure on the skin leads to sensitization, whereas early oral ingestion leads to tolerance. The scope of this Special Issue is to describe causes of sensitization to food allergens and food allergy, thereby focusing on effects of processing operations of food proteins, extraction of food proteins for diagnostic tools and allergenicity of rather new allergens, e.g., seaweed. Furthermore, this Special Issue pays attention to other subjects focusing on food allergy. One example describes the huge role of the amount of allergens (the dose) in different situations. How much allergen should be used in, e.g., food challenges, and introduction of foods, is very important, but also the exact amount of ingredients in packed food is indispensable.

This Special Issue is addressed to researchers and health care professionals with an interest in food allergy. The ten articles are written by groups of professionals throughout the Netherlands, with a wide experience in food allergy and a huge variety of expertise, e.g., dieticians, allergists, pediatric allergists, biochemical analysists, immunologists, and research nurses.

Nicolette W. de Jong and Harry J. Wichers
Editors

nutrients

MDPI

ditorial

Update on Nutrition and Food Allergy

Nicolette W. de Jong [1,2,*] and Harry J. Wichers [3]

1 Department of Internal Medicine, Section of Allergology & Clinical Immunology, Erasmus MC, University Medical Centre Rotterdam, 3015 GD Rotterdam, The Netherlands
2 Department of Paediatric Allergology, Sophia Children Hospital, Erasmus MC, University Medical Centre Rotterdam, 3015 GD Rotterdam, The Netherlands
3 Food & Biobased Research, Wageningen University & Research Centre, 6708 WG Wageningen, The Netherlands; harry.wichers@wur.nl
* Correspondence: n.w.dejong@erasmusmc.nl

Citation: de Jong, N.W.; Wichers, H.J. Update on Nutrition and Food Allergy. *Nutrients* **2022**, *14*, 2137. https://doi.org/10.3390/nu14102137

Received: 29 April 2022
Accepted: 18 May 2022
Published: 20 May 2022

Publisher's Note: MDPI stays neutral with regard to jurisdictional claims in published maps and institutional affiliations.

Food-induced anaphylaxis is an immediate adverse reaction, primarily triggered by the cross-linking of allergen-specific immunoglobulin (Ig) E bound to the high-affinity IgE receptor (FcεRI) on mast cells (MCs) after re-exposure to the same food allergen. Patients with an IgE-mediated food allergy often suffer from a variety of symptoms, e.g., gastrointestinal, skin, lungs, and, in the worst case, anaphylaxis. The site where food antigens are firstly taken up, either the skin or the gut, may cause sensitization against this food antigen [1]. Sensitization in the gut can take place due to increased intestinal permeability, and in the skin, a disrupted skin barrier is often responsible for higher exposure to allergens, which consequently leads to increased sensitization. This is often the case in patients with atopic dermatitis. In these patients, often 20% of the body area is affected, mostly caused by mutations in filaggrin (FLG) null, which encodes for the epidermal protein FLG. The IL-33 levels in these patients are high, mostly caused by scratching. This also increases the degranulation of mast cells and intestinal permeability. Van Splunter et al. described increased interleukin (IL)-33 levels in serum, which activate dendritic cells (DCs) and interleukin 2 (ILC2) cells [2]. Furthermore, cutaneous sensitization induces thymic stromal lymphopoietin (TSLP) activation of basophils, and the production of IL-4, IL-5 and IL-13, leading to a reduced gut barrier for food allergens and an IgE-mediated degranulation of MCs. This illustrates the existence of a skin-to-gut crosstalk, in which damaged skin can promote food-induced anaphylaxis by driving intestinal MC expansion.

Sensitization to food allergens can be measured with the Skin Prick test (SPT) and/or specific serum IgE (sIgE). Unfortunately, standardized commercial food allergen extracts for SPT are less readily available. Furthermore, due to globalization, the number of foods that causes an allergic reaction is increasing. In addition to the SPT, the Prick-to-Prick test (PTP) is also very commonly used to measure sensitization. The PTP test shows high sensitivity and specificity, but it is not very practical, as fresh fruits have to be available at the department. One alternative for commercial extracts and fresh fruits for PTP might be to prepare homemade (HM) extracts through standardized protocols. Recently, S. Terlouw et al. performed a clinical trial in 54 food-allergic patients comparing SPT results with commercial and home-made extracts [3]. Extracts form hazelnut, walnut, apple, peach and almond were compared. The intraclass correlation coefficient between the SPT results of both extract methods was strong for hazelnut, moderate for apple and peanut and weak for the other allergens. Many SPT's with almond were positive without causing symptoms in the patients. In contrast, results with home-made peach extract showed high agreement with the peach-specific allergic symptoms. The homemade extract consists of a few drops of juice that are rapidly produced from the whole peach and stored in small aliquots at −20 °C. This method mimics the PTP method. In that way, many small aliquots from different fruits and vegetables, and even from fresh herbs, can be available every day at the clinic.

The evaluation of patients with a possible food allergy starts with an extensive food specific medical history. The standardized diet history tool published by S. Skypala et al. provides a practical approach to support food allergy diagnosis, ensuring that all relevant information is captured and interpreted in a robust manner [4]. Although the combination of the allergy-focused diet history with positive sensitization to the specific food allergen in SPT and/or sIgE measurements often leads to a clear diagnosis, in many cases, discrepancies occur. The only method (gold standard) to finally confirm a food allergy is to perform a double-blind, placebo-controlled food challenge (DBPCFC). This is very time-consuming, expensive and at risk for anaphylaxis, but in some cases, it is indispensable and therefore commonly used. Masking the food for a DBPCFC is not always possible, and therefore, in some cases, the patient is blinded and uses a nose clip. This method is called single blinded. A good example of this kind of test is the pear study performed by de Jong et al. [5]. In the Netherlands, research has been conducted to measure differences in allergenic properties between various cultivars, e.g., pear cultivars. Also in the Netherlands, in early 2007, the "Santana apple" showed reduced allergenic properties because of its lower Mal d 1 levels. The Santana apple caused significantly fewer allergic symptoms in apple-allergic individuals than the Golden Delicious and Topaz apples [6]. Unfortunately, this lower allergenicity could not be measured for a new "Cepuna" pear cultivar. The results of challenges with the new 'Cepuna' pear were comparable with the more common "Conference" pear. The only (non-significant) difference in favour of the Cepuna pear was that it caused less objective symptoms and less severe symptoms after consumption. The highest dose used in the challenge was 200 g, which comes close to a whole pear. Doses and volumes of foods often lead to discussions among allergists and dieticians. In challenges, the patient reacts to a certain dose, but the question remains how to translate that to normal consumption. When the patient reacts to 100 mg of protein, is that comparable to a spoon, a bite, a sip or a cup? For the management of food allergies, we should be aware that there are no standard definitions. Recently, M. Kok et al. estimated sizes of bit and sip for milk, egg, peanut and hazelnut in selected age groups: 2–3, 4–6 and 19–30 years [7]. The results could be compared with ED10 and ED50 (10% and 50% of the allergic subjects react with objective symptoms) [8]. Only one food contained less estimated allergenic protein per portion when comparing the amount of milk in foods to the ED10 for milk. This was the case for four foods: for egg, peanut and hazelnut none of the foods contained less than the ED10. This means that all the other foods will provoke allergic reactions in allergic patients who belong to the 10% most clinically sensitive individuals. The protein content in a single bit or sip contained a sufficient amount of allergenic protein in all cases to elicit an allergic reaction.

Doses and servings become more important since the early introduction of foods is advised for the prevention of food allergy. S. Filep et al. published doses of specific allergens in "early introduction foods" (EIF) for the prevention of food allergy for 17 major food allergens [9]. Cumulative allergen doses for each EIF were estimated using serving sizes and consumption recommendations provided by the manufacturer. For early introduction of foods, as well as of introduction of foods after a negative food challenge, the doses are of high importance [10]. The starting dose for introduction after a negative food challenge should not exceed the highest dose that was given during the food challenge. In individual cases, an introduction schedule can be provided to the patient for home introduction, and in other cases, the doses should be given at the outpatient clinic. Furthermore, regular telephone calls are important to follow-up the patient. Two studies comparing introduction with and without a structured protocol showed significant differences: vd Valk et al. [11] and JAM Emons et al. [12]. The latter study showed only 8% of failed introductions versus 52% in the earlier study by Valk et al. So, protocols and follow-ups are mandatory to successfully introduce the food into the daily diet of the patient.

Since 2011, when J.S. Kim et al. [13] published a paper proving that dietary "baked milk" accelerates the resolution of cow's milk allergy (CMA) in children, many trials have studied the effects of processing of foods, e.g., baking and drying. Apparently, (dry) heating and glycation of cow's milk protein (Maillard reaction) have been shown to alter its digestibility and immunogenicity, and consequently, CMA children are able to consume this form of cow's milk (CM). Moreover, "baked milk" products (using dry heating) have been shown to accelerate the resolution of cow's milk allergy. The study of Zenker et al. Investigated specific peptide profiles of CM proteins heated at low and high temperatures after simulated infant in vitro digestion and compared this to non-treated CM [14]. This study showed that during simulated infant in vitro digestion of milk that was dry heated in the presence of lactose, different peptide profiles are generated. High-temperature dry heating had the largest effects on peptide generation, resulting in much lower numbers of peptides with lower sequence coverage. Moreover, a much lower number of sIgE-binding epitopes and a larger proportion of glycated sIgE-binding epitopes and T-cell epitopes in heated samples indicated that the immunogenicity and allergenicity of these samples could be affected.

Many studies have investigated the tolerance-inducing effect of baked milk, but the form of the product (e.g., cake, bread, cheese or pizza) and the precise heating process were found to be highly variable. For the introduction of, e.g., milk and egg, so-called milk and egg ladders can be used, but even the latest literature from Venter et al. in 2022 [15] does not give detailed information on the exact baking temperature or baking time of the products. Even protein content of the several doses is unknown. FrieslandCampina (Amersfoort, the Netherlands) developed a standardized dry-heated CM protein powder, with an exact baking temperature and time, and the method is accurately described in the article [16]. To test the new baked milk (HP) powder, challenge-proven CMA children were included (3 months–3 years), and the HP powder was introduced in incremental doses by dissolving it in the child's daily milk formula. Seventy-two percent (18/25) of the children tolerated the HP product, and seven children experienced adverse events. These results are comparable with the baked milk studies. The group that does not tolerate the baked milk product most likely has a more severe or even a persistent CM allergy. Currently, a randomized placebo-controlled study is being carried out in 10 different children hospitals using this HP powder to measure the tolerance-inducing capacities of the product. The results are expected by the end of 2022. The Maillard reaction (MR) can affect the sensitization properties of allergens in patients. The process is widely studied in CM allergy, but studies with other food allergens can hardly be found. Wheat flour is an important component of many baked goods, and during the baking process, wheat protein may also undergo the MR because sugars are usually present. However, reports on the allergenicity change in wheat proteins after glycation are rare. This was also concluded by Gou et al. in a recent review [17]. An important allergen of wheat is gluten, especially in the form of glutenin. Methylglyoxal (MGO) has the highest reactivity as intermediate in the MR. The project of Wang et al. aimed to determine the effect of MGO on the allergenicity of glutenin based on the BALB/c mouse model pre-sensitized to native glutenin, heated glutenin and MGO-glutenin, in order [18]. The digestibility and changes in the structure of glutenin and gut microflora in mice were analysed to elucidate the detailed mechanism by which the potential for allergic reaction is reduced as a result of MGO decoration. The current research results show that glutenin could alleviate the resulting allergic reaction in mice after MGO decoration. This study provides a theoretical basis for alleviating glutenin allergic reactions through processing which should be confirmed in clinical trials in humans.

To further investigate whether food components or processed food components have effects on the adaptive immune system, intervention studies are widely suggested. The debate is still far from consensual, in particular on skewing the immune system towards a more homeostatic situation by means of inducing production of higher numbers of Tregs, which can decrease the number of T-helper 2 (TH2) cells and consequently decrease, e.g., IL-4 and IL-5 production. Lately, the supplementation of brown seaweed is presented in literature as having modulating properties on adaptive immune responses. The article by Kamunde et al. showed a highly significant increased total plasma antioxidant capacity in fish [19]. These reactive oxygen species are known to be important drivers of inflammation. Recently, E.M. Olsthorn published an extensive review on brown seafood supplementation and its effects on allergy and inflammation and their consequences [20]. They conside the seaweed effects by enhanced production of IL-1 and TNF, as well as secondary cy tokines, such as IL-10. IL-10 specifically has a clear immuno-suppressive effect. Allerger immunotherapy induces IL-10-producing type 2 innate lymphoid cells, which are strongl associated with a clinical response by modulating grass pollen allergy [21]. In this ligh better-designed human studies applying individual seaweed constituents, as well as whol seaweed (extracts), will provide more insight into the applicability of brown seaweed as ar immune-modulatory nutritional intervention strategy.

Author Contributions: N.W.d.J. and H.J.W. wrote the manuscript. All authors have read and agreec to the published version of the manuscript.

Funding: This research received no external funding.

Conflicts of Interest: The authors declare no conflict of interest.

References

1. Kulis, M.D.; Smeekens, J.M.; Immormino, R.M.; Moran, T.P. The airway as a route of sensitization to peanut: An update to the dual allergen exposure hypothesis. *J. Allergy Clin. Immunol.* **2021**, *148*, 689–693. [CrossRef] [PubMed]
2. Van Splunter, M.; Liu, L.; van Neerven, R.J.J.; Wichers, H.J.; Hettinga, K.A.; de Jong, N.W. Mechanisms Underlying the Skin-Gut Cross Talk in the Development of IgE-Mediated Food Allergy. *Nutrients* **2020**, *12*, 3830. [CrossRef] [PubMed]
3. Terlouw, S.; van Boven, F.E.; Borsboom-van Zonneveld, M.; de Graaf-In 't Veld, C.; van Splunter, M.E.; van Daele, P.L.A. van Maaren, M.S.; Schreurs, M.W.J.; de Jong, N.W. Homemade Food Allergen Extracts for Use in Skin Prick Tests in the Diagnosis of IgE-Mediated Food Allergy: A Good Alternative in the Absence of Commercially Available Extracts? *Nutrients* **2022**, *14*, 475 [CrossRef] [PubMed]
4. Skypala, I.J.; Venter, C.; Meyer, R.; deJong, N.W.; Fox, A.T.; Groetch, M.; Oude Elberink, J.N.; Sprikkelman, A.; Diamandi, L. Vlieg-Boerstra, B.J.; et al. The development of a standardised diet history tool to support the diagnosis of food allergy. *Clin. Transl. Allergy* **2015**, *5*, 7. [CrossRef]
5. De Jong, N.W.; Terlouw, S.; van Boven, F.E.; van Maaren, M.S.; Schreurs, M.W.J.; van den Berg-Somhorst, D.; Esser, D. Bastiaan-Net, S. Birch Pollen Related Pear Allergy: A Single-Blind Oral Challenge TRIAL with 2 Pear Cultivars. *Nutrients* **2021**, *13*, 1355. [CrossRef]
6. Kootstra, H.S.; Vlieg-Boerstra, B.J.; Dubois, A.E. Assessment of the reduced allergenic properties of the Santana apple. *Ann. Allergy Asthma Immunol.* **2007**, *99*, 522–525. [CrossRef]
7. Kok, M.; Compagner, A.; Panneman, I.; Sprikkelman, A.; Vlieg-Boerstra, B. A Food, a Bite, a Sip: How Much Allergen Is in That? *Nutrients* **2021**, *13*, 587. [CrossRef]
8. Blom, W.M.; Vlieg-Boerstra, B.J.; Kruizinga, A.G.; van der Heide, S.; Houben, G.F.; Dubois, A.E. Threshold dose distributions for 5 major allergenic foods in children. *J. Allergy Clin. Immunol.* **2013**, *131*, 172–179. [CrossRef]
9. Filep, S.; Chapman, M.D. Doses of Specific Allergens in Early Introduction Foods for Prevention of Food Allergy. *J. Allergy Clin. Immunol. Pract.* **2022**, *10*, 150–158.e3. [CrossRef]
10. Fleischer, D.M.; Conover-Walker, M.K.; Christie, L.; Burks, A.W.; Wood, R.A. Peanut allergy: Recurrence and its management. *J. Allergy Clin. Immunol.* **2004**, *114*, 1195–1201. [CrossRef]
11. Van der Valk, J.P.; Gerth van Wijk, R.; Vergouwe, Y.; de Jong, N.W. Failure of introduction of food allergens after negative oral food challenge tests in children. *Eur. J. Pediatr.* **2015**, *174*, 1093–1099. [CrossRef] [PubMed]
12. Emons, J.; van Gunst, M.; Liem, O.; Landzaat, L.; Arends, N. Improving Successful Introduction after a Negative Food Challenge Test: How to Achieve the Best Result? *Nutrients* **2020**, *12*, 2731. [CrossRef] [PubMed]
13. Kim, J.S.; Nowak-Węgrzyn, A.; Sicherer, S.H.; Noone, S.; Moshier, E.L.; Sampson, H.A. Dietary baked milk accelerates the resolution of cow's milk allergy in children. *J. Allergy Clin. Immunol.* **2011**, *128*, 125–131.e2. [CrossRef] [PubMed]
14. Zenker, H.E.; Wichers, H.J.; Tomassen, M.M.M.; Boeren, S.; De Jong, N.W.; Hettinga, K.A. Peptide Release after Simulated Infant In Vitro Digestion of Dry Heated Cow's Milk Protein and Transport of Potentially Immunoreactive Peptides across the Caco-2 Cell Monolayer. *Nutrients* **2020**, *12*, 2483. [CrossRef]
15. Venter, C.; Meyer, R.; Ebisawa, M.; Athanasopoulou, P.; Mack, D.P. Food allergen ladders: A need for standardization. *Pediatr. Allergy Immunol.* **2022**, *33*, e13714. [CrossRef]
16. De Jong, N.W.; van Splunter, M.E.; Emons, J.A.M.; Hettinga, K.A.; Gerth van Wijk, R.; Wichers, H.J.; Savelkoul, H.F.J.; Sprikkelman, A.B.; van Neerven, R.J.J.; Liu, L.; et al. Introduction of Heated Cow's Milk Protein in Challenge-Proven Cow's Milk Allergic Children: The iAGE Study. *Nutrients* **2022**, *14*, 629. [CrossRef]
17. Gou, J.; Liang, R.; Huang, H.; Ma, X. Maillard Reaction Induced Changes in Allergenicity of Food. *Foods* **2022**, *11*, 530. [CrossRef]
18. Wang, Y.; Li, X.; Wu, S.; Dong, L.; Hu, Y.; Wang, J.; Zhang, Y.; Wang, S. Methylglyoxal Decoration of Glutenin during Heat Processing Could Alleviate the Resulting Allergic Reaction in Mice. *Nutrients* **2020**, *12*, 2844. [CrossRef]

9. Kamunde, C.; Sappal, R.; Melegy, T.M. Brown seaweed (AquaArom) supplementation increases food intake and improves growth, antioxidant status and resistance to temperature stress in Atlantic salmon, Salmo salar. *PLoS ONE* **2019**, *14*, e0219792. [CrossRef]
0. Olsthoorn, S.E.M.; Wang, X.; Tillema, B.; Vanmierlo, T.; Kraan, S.; Leenen, P.J.M.; Mulder, M.T. Brown Seaweed Food Supplementation: Effects on Allergy and Inflammation and Its Consequences. *Nutrients* **2021**, *13*, 2613. [CrossRef]
1. Golebski, K.; Layhadi, J.A.; Sahiner, U.; Steveling-Klein, E.H.; Lenormand, M.M.; Li, R.C.Y.; Bal, S.M.; Heesters, B.A.; Vilà-Nadal, G.; Hunewald, O.; et al. Induction of IL-10-producing type 2 innate lymphoid cells by allergen immunotherapy is associated with clinical response. *Immunity* **2021**, *54*, 291–307.e7. [CrossRef] [PubMed]

nutrients

MDPI

Review

Mechanisms Underlying the Skin-Gut Cross Talk in the Development of IgE-Mediated Food Allergy

Marloes van Splunter [1,†], Liu Liu [1,†], R.J. Joost van Neerven [2,3], Harry J. Wichers [4], Kasper A. Hettinga [5] and Nicolette W. de Jong [1,*]

1 Internal Medicine, Allergology & Clinical Immunology, Erasmus Medical Centre, 3000 CA Rotterdam, The Netherlands; m.e.vansplunter@erasmusmc.nl (M.v.S.); l.liu@hagaziekenhuis.nl (L.L.)
2 Cell Biology and Immunology, Wageningen University & Research Centre, 6708 WD Wageningen, The Netherlands; joost.vanneerven@wur.nl
3 FrieslandCampina, 3818 LE Amersfoort, The Netherlands
4 Wageningen Food & Biobased Research, Wageningen University & Research Centre, 6708 WG Wageningen, The Netherlands; harry.wichers@wur.nl
5 Food Quality & Design Group, Wageningen University & Research Centre, 6708 WG Wageningen, The Netherlands; kasper.hettinga@wur.nl
* Correspondence: n.w.dejong@erasmusmc.nl; Tel.: +31-621697954
† These authors contributed equally to this work.

Received: 9 November 2020; Accepted: 12 December 2020; Published: 15 December 2020

Abstract: Immune-globulin E (IgE)-mediated food allergy is characterized by a variety of clinical entities within the gastrointestinal tract, skin and lungs, and systemically as anaphylaxis. The default response to food antigens, which is antigen specific immune tolerance, requires exposure to the antigen and is already initiated during pregnancy. After birth, tolerance is mostly acquired in the gut after oral ingestion of dietary proteins, whilst exposure to these same proteins via the skin, especially when it is inflamed and has a disrupted barrier, can lead to allergic sensitization. The crosstalk between the skin and the gut, which is involved in the induction of food allergy, is still incompletely understood. In this review, we will focus on mechanisms underlying allergic sensitization (to food antigens) via the skin, leading to gastrointestinal inflammation, and the development of IgE-mediated food allergy. Better understanding of these processes will eventually help to develop new preventive and therapeutic strategies in children.

Keywords: skin-gut-axis; cutaneous sensitization; food allergy; atopic dermatitis; microbiota

1. Introduction

Immune-globulin E (IgE)-mediated food allergy is characterized by a variety of clinical entities within the gastrointestinal tract, skin, and lungs, as well as systemically as anaphylaxis. IgE is the hallmark of allergic sensitization and, therefore, the most important antibody isotype in patients with atopic diseases. Sensitization is the process that leads to the presence of food-specific IgE in the serum and the skin, which predisposes to the development of food allergy. Several animal models have shown that epicutaneous allergen exposure, prior to oral challenge with the same antigen, induces allergic responses in the gastrointestinal tract [1–4]. Observations that children with atopic dermatitis (AD) and altered epidermal barrier function are more often sensitized to food antigens led to the hypothesis that exposure to a low dose of food antigen on the skin leads to sensitization, whereas early oral ingestion of food antigens (in a high dose) mediates tolerance. This hypothesis was first described by the group of Gideon Lack as the dual allergen hypothesis [5], and was recently updated and reviewed [6]. Several clinical trials in atopic children have shown evidence supporting the dual allergen hypothesis. AD is often the first manifestation of the atopic march and clear positive correlations have been seen between early-onset eczema (particularly for ages less than three months) and more severe eczema and the risk of developing food allergy [7]. Furthermore, AD can progress into asthma, a process mediated by thymic stromal lymphopoietin (TSLP) as demonstrated in mouse models of experimental asthma [8–10]. Next to this, AD is found to further increase the effect of exposure and sensitization to food allergens [11]. In the prospective LEAP (Learning Early About Allergy to Peanut trial) and LEAP-on studies, early introduction of peanut to 4–11-month-old children, with high risk of developing peanut allergy, significantly reduced the risk of developing peanut allergy by the age of 5 [12]. This high-risk population is identified as infants having AD and/or egg allergy, without having an established peanut allergy [13]. To interfere with the effect of AD, children with AD were extensively treated for their eczema with topical corticosteroids or emollients until remission in the PETIT study. The treatment of eczema was combined with either an early introduction of egg white or placebo (4–5 months of age) in a two-step protocol and resulted in a lower prevalence of egg allergy (9%) compared to children that were given placebo (38% egg allergic) at the age of 1 year [14]. This shows that optimal eczema treatment of AD patients, which resulted in lower SCORAD and POEM scores in both groups, is in itself not enough to prevent sensitization to food proteins. In contrast, early oral introduction of food does prevent sensitization to food proteins, as described in the dual allergen hypothesis [5]. So, optimal eczema treatment of AD patients can contribute in the prevention of sensitization; however, this only works in addition to oral introduction of food allergens.

How prior allergic sensitization via the skin progresses to food-induced anaphylaxis is not fully elucidated. Food-induced anaphylaxis is an immediate, adverse reaction, primarily triggered by cross-linking of allergen-specific IgE bound to the high-affinity IgE receptor (FcεRI) on mast cells (MCs) after re-exposure to the same food allergen [15]. However, only some individuals develop anaphylaxis, while others do not, independently of allergen-specific serum IgE levels. This suggests that other mechanisms than solely allergen-specific IgE are involved in the cascade of symptoms seen in food allergy. In this review, we will focus on the molecular and cellular mechanisms supporting the dual allergen exposure hypothesis and recent advances in understanding the interaction between immune responses in the skin and in the gut in the development of food allergy.

2. Skin Barrier and Skin Sensitization

The function of the skin epithelium is to provide a permeability barrier to maintain water and electrolytes homeostasis and an immune barrier, which facilitates commensal, but not pathogenic bacteria [16]. The skin is composed of epidermis and dermis. The epidermis is subdivided into the stratum corneum on the outside and inwards the stratum corneum is followed by the stratum granulosum, stratum spinosum, and stratum basale [17]. The stratum corneum is composed of keratinocytes differentiated into corneocytes and contains, among others: keratin filaments, filaggrin, and lipids [17]. Tight junctions are located in the stratum granulosum and are sealing the keratinocytes of the stratum corneum providing the permeability barrier [17]. Deeper into the skin through the epidermal barrier, the human dermis contains numerous immune cell types, such as Langerhans cells, mast cells, adaptive resident lymphocytes, and innate lymphoid cells (ILCs), together constituting the immune barrier.

Atopic dermatitis (AD) is a comorbid condition, which often precedes food allergy in patients. Furthermore, AD is a common inflammatory skin disease, which often develops during infancy and proceeds into adulthood. It has a relapsing character with pruritus eczematous flares. The pathophysiology of AD is multifactorial and includes genetic predisposition leading towards a defective skin barrier, dysregulated immune response, and microbial dysbiosis. Furthermore, environmental factors, such as allergens, micro-organisms, and toxins, influence the disease development [16,17]. Patients with AD have a significantly higher risk of developing food allergy [7]. This suggests that the skin is an important site of food allergen sensitization. In allergic sensitization via the skin, food antigens cross the disrupted epithelial barrier and mediate the release of danger signals and inflammatory cytokines

TSLP and interleukin 33 (IL-33) through epithelial cells. These cytokines activate dendritic cells, which induce the differentiation of naive CD4+ T cells into a T helper cell 2 (Th2) phenotype. Clinical studies indicate that children who suffer from IgE-mediated food allergy are most likely sensitized through the gastro-intestinal tract and/or the skin in early infancy. Hill et al. has demonstrated in a multicenter large cohort study that early AD onset and severity are associated with high levels of IgE to food allergens, such as milk, egg, and peanut [18].

The most important genetic risk factors for AD are filaggrin (FLG) null mutations, which encodes for the epidermal protein FLG [19]. However, this FLG mutation alone is neither sufficient nor necessary to drive the development of AD. Patients with AD who carry the FLG mutation tend to have early onset, severe, and persistent skin disease and are more likely to be sensitized to multiple (food) allergens and to develop asthma [19]. Although a decreased barrier function is associated with increased intradermal allergen exposure, the mechanism by which this leads to allergic sensitization is not fully understood. Transepidermal water loss (TEWL) is a marker of epidermal dysfunction and highly correlated with altered epidermal lipid composition and structure in AD, independently of FLG mutation [20]. Increased TEWL at the age of 2 days was found to be correlated with AD and with being allergic to food later in life, at the age of one and two years, respectively. These results further emphasize the likelihood of the skin as an important site for sensitization at an early age [21,22].

An intact epithelial barrier which prevents the entry of antigens, pathogens, and irritants, and thereby the production of inflammatory cytokines, is important in the maintenance of homeostasis. The importance of an intact epithelial barrier is emphasized by the finding in human subjects that mutations in genes, encoding proteins that are involved in skin barrier integrity, such as FLG and SPINK5, are independent risk factors for peanut allergy [22–25]. Interestingly, the odds ratio for FLG mutations and peanut allergy is even stronger than for AD (5.3 vs. 3.1) [19]. Therefore, disrupted barrier function by FLG mutations alone or by AD in general leads to enhanced sensitization. However, allergen-specific IgE levels induced by sensitization do not correlate with the prevalence of food allergy and type 2 inflammatory reactions

Key messages:

- Skin is an important permeability and immune barrier.
- Disrupted skin barrier leads to increased sensitization to food allergens in the skin.

3. Environmental Factors Induce Sensitization to Food Allergens via the Skin

Other factors may also play an important role in the process of sensitization. Walker et al. have demonstrated that skin barrier mutations, together with exposure to environmental allergens, such as *Alternaria alternata* or house dust mite (HDM) extract, were required to drive the development of food allergen sensitization and anaphylaxis [26]. The exposure to environmental allergens was done after the skin of mice was wiped with 4% sodium dodecyl sulfate (SDS) as detergent, to resemble the use of cleaning wipes on infants, and this turned out to be essential for the absorbance of the topical applied environmental allergens. SDS is a key ingredient of soap, which can degrade corneodesmosomes and thereby reduce integrity of the stratum corneum resulting in type 2 immune responses [27]. Next to this, Cayrol et al. showed proteases from a whole range of allergens including *A. alternata* and HDM can process IL-33 full-length into a more biological active form of IL-33 inducing type responses [28]. In this paper, it was even suggested that the cleavage of full-length IL-33 by allergen proteases is used as an allergen sensing system. Exposure to environmental allergens and detergents may happen prior to the development of atopic dermatitis, as well as decreases the development of tolerance during oral consumption of the food allergen [26]. In patients with AD, epicutaneous application of HDM was shown to induce TSLP expression in both lesioned and unaffected skin [29].

The importance of oral tolerance to food allergens is emphasized by the study of Han et al., where the development of food allergy could be blocked when the allergen was ingested by mice prior to skin exposure [1]. Strid et al. found that epicutaneous exposure to peanut protein 20 days prior to ingestion of a tolerogenic dose of peanut protein completely abolished oral tolerance induction in mice, whereas epicutaneous exposure 6 days prior to ingestion only partly disrupted oral tolerance induction [30]. Even in already oral tolerant mice, epicutaneous exposure of peanut protein resulted in increased IL-4 levels and increased peanut-specific IgE levels, thus demonstrating an increase in sensitization to peanut [30].

In humans, Leung et al. showed through RNA sequencing that non-lesioned skin of 62 children with AD and food allergy had unique properties associated with an immature skin barrier and type 2 immune activation [31]. Patients with AD and food allergy exhibited a high dendritic cell activation in their non-lesioned skin, which is comparable to that of the lesioned skin of all AD participants. Furthermore, FLG was found to be downregulated in both lesioned and non-lesioned skin of patients with AD [31,32]. Taken together, a decreased skin barrier function (possibly induced by detergents and intrinsic genetic defects), in combination with exposure of the skin to food allergens with meals and dust containing HDM, *A alternata*, or *Staphylococcus aureus*, likely synergize to promote sensitization to food allergens and the subsequent development of food allergy.

Key messages:

- Detergents and environmental allergens, like house dust mite or *Alternaria alternata* allergens, can disrupt skin barrier.
- Cutaneous exposure of allergens prior to ingestion leads to increased sensitization.
- Tolerance is induced if allergens are ingested prior to cutaneous exposure.

4. TSLP-Mediated Type 2 Inflammation in the Skin

Thymic stromal lymphopoietin (TSLP) is an epithelial cytokine, expressed mainly by epithelial cells of the skin, lungs, and intestine [33,34]. TSLP, in mice, was shown to be induced by cutaneous exposure to food antigens and upon skin barrier disruption [2,3,35]. In a Korean birth cohort, skin epithelial expression of TSLP at two months of age has been linked to the development of AD at 24 months of age [36]. TSLP is found to regulate naive T cell differentiation towards an inflammatory phenotype by conditioning dendritic cell (DC) maturation as antigen presenting cells [37]. These TSLP-DCs induce a unique type of Th2 cells through the OX-40 ligand that produces the classical type 2 pro-inflammatory cytokines (IL-4, IL-5, and IL-13) together with tumor necrosis factor (TNF)-α and no production of IL-10 [34,38]. Furthermore, in lesioned human AD skin samples, it was shown that high TSLP production leads to activation and migration of Langerhans cells from the epidermis towards the dermis and an increase of activated DCs in the dermis [34].

In mice, TSLP-activated DCs express OX40L, as well, and it was shown that OX40L-OX40 interaction between DC and T cells induced IL-3 production by naive T cells, resulting both in recruitment of basophils in the skin-draining lymph nodes, as well as IL-4 expression of T cells [39]. In mice with an atopic dermatitis-like skin, cutaneous food allergen sensitization induces an expansion of TSLP-elucidated basophils in the skin, which is sufficient to promote the development of IgE-mast cell mediated food allergy after oral antigen exposure [3,4]. Moreover, clinical signs of food allergy are significantly reduced after epicutaneous sensitization in mice whose basophils cannot produce IL-4. In addition, IL-4 depletion in epicutaneous sensitized mice results in a diminished IgE-mediated anaphylaxis response upon an oral challenge in mice [40]. Taken together, this suggests a critical role for IL-4 derived from TSLP-induced basophils in the sensitization to food allergens in the skin, and the development of food allergy. The importance of TSLP is also noted in eosinophilic esophagitis, a food allergy-associated inflammatory disease, where skin-derived TSLP results in basophil-mediated disease activation in humans, which was IgE-independent (based on mice experiments) [2].

Key message:

- Disrupted skin barrier leads to increased sensitization to food allergens in the skin. This process is mediated by TSLP-induced DC and basophils, producing IL-4 and resulting in enhanced type 2 responses.

5. Major Role for Type 2 Innate Lymphoid Cells (ILC2) and Epithelial Cytokines in the Development of Food Allergy

In recent years, it has become clear that particularly ILC type 2 cells (ILC2s) play an important role in food allergy and these cells are considered as the innate counterparts of adaptive T helper 2 cells. Barrier epithelial cells, such as skin keratinocytes, lung cells, and intestinal epithelial cells, are found to be crucial in recruiting these immune cells by producing chemokines. AD-like disorders can even be induced by overexpression of the chemokine CCL17 by keratinocytes [41]. Furthermore, barrier cells can determine type 2 immunity by controlling the activation of DCs and ILC2s through the secretion of the epithelium-derived cytokines TSLP, IL-25, and IL-33 [41].

One of these barrier cells are the so-called tuft cells. Tuft cells (or brush cells) produce IL-25, a distinct IL-17 cytokine member (IL-17E), upon inflammation and these epithelial cells are located in the intestine and trachea [36,42]. Murine strains that lack the IL-25 receptor are found to be more resistant to developing IgE-mediated food allergy after oral intake [1,43]. IL-25 stimulation, together with CD4+Th2 cells, that are induced after allergic sensitization, cause ILC2s to produce large amounts of IL-5 and IL-13, resulting in the development of food allergy in mice [43]. Furthermore, IL-13 produced by ILC2s and/or Th2 cells can promote the differentiation and expansion of tuft cells, resulting in a positive feedback loop [42].

ILC2 can activate dendritic cells and promote a Th2 cell-mediated immune response, and expand in an antigen-independent manner in the presence of TSLP, IL-25, and IL-33 [41]. TSLP is hereby the most important factor for ILC2 survival, whereas IL-33 mainly results in ILC2 activation, although the combination of IL-25, IL-33, and TSLP results in the highest cytokine production [44]. As a result, ILC2s produce large quantities of Th2 cytokines, such as IL-5, IL-9, IL-13, and, to a lesser extent, IL-4, as reviewed by Reference [36,45]. In addition, in particular the cytokines IL-4 and IL-13, can disrupt allergen-specific regulatory T cell (Treg) induction and proliferation, resulting in fewer Tregs and a decrease in their suppressive functions [46]. In addition to this, allergen-specific Tregs were found to have a more Th2-skewed profile, with the production of IL-4 in both mice and human [46]. Furthermore, these ILC2-derived cytokines can enhance mucosal mast cell activation and ILC2s can be activated by mast cells in an IgE-dependent way, creating a positive feedback loop, thereby further promoting the induction of food allergies in mice [47,48].

IL-33 is another epithelial cytokine and is constitutively expressed in high levels in epithelial cells. IL-33 is released whenever cells are activated via adenosine triphosphate (ATP) or when cells are damaged or become necrotic [49]. Not only keratinocytes produce IL-33; fibroblasts, endothelial cells and epithelial cells produce IL-33, as well. Whether immune cells are bona fide producers of IL-33 is debated as often only IL-33 mRNA expression is reported [50]. During inflammation, as is the case in AD, IL-33 levels are elevated in skin lesions [51,52] and serum [53]. These serum IL-33 levels correlate with AD severity [36,53]. IL-33 and IL-4 can both downregulate FLG in keratinocytes and thereby further affect the skin barrier and possible entrance of allergens [54]. In an AD-like mouse model, it was shown that IL-33 could induce the atopic march and gastrointestinal allergy, independently from TSLP [55].

The IL-33 receptor IL33R/ST2 (suppression of tumorigenicity 2) is found to be increased in skin lesions of patients with AD [51]. Galand et al. found that IL-33 is released after mechanical skin injury in mice and induces IgE-mediated mast cell degranulation, although IL-33 had no direct effect on specific IgE levels in serum or on Th2 responses [56]. In humans, IL-33 mRNA expression is also increased after tape stripping of the skin [56]. Tape stripping of the skin is used as a model for scratching. Besides, IL-33 activates mast cell degranulation in humans in vitro [57]. In patients with AD, more ILC2 cells are found in skin biopsies from lesions compared to skin biopsies of healthy donors [52,58]. In a mouse model, ILC2s proved to be necessary for the development of an AD-like phenotype, even independently of the adaptive immune system [58]. In AD patients, ILC2 cells have a higher expression of receptors for IL-25, IL-33, and TSLP [52]. When stimulating skin-derived ILC2 cells from healthy donors ex vivo, only IL-33 or the combination IL-33, together with IL-25 and TSLP, induced type 2 cytokines IL-5 and IL-13, but no IL-4 [52]. Besides, IL-33R expression was upregulated after IL-33 stimulation and IL-33 was more potent than TSLP to induce migration of ILC2s.

Next to this, it was shown in an in vivo experimental model that HDM allergic patients have a higher infiltration of lymphocytes and (ST2-positive) ILC2 cells and a higher IL-4, IL-5, and IL-13 level in blister fluid upon intra-epidermal injection of HDM compared to healthy subjects [52]. Furthermore, it was confirmed in a mouse model that ST2-positive skin DCs drive the development of Th2 responses to peanut, resulting in peanut allergy upon epicutaneous peanut exposure [59]. So, IL-33 is involved in acute reactions to consumed food by acting directly on mast cells and enhancing IgE-mediated activation, as well as inducing ILC2 cells and activation of DCs that drive Th2 cell responses [52,56,59]. On the other hand, in mice, it is shown that IL-33 can induce epithelial tissue repair by activating the production of amphiregulin by ILC2s [60,61] or Tregs [62].

In short, activation of ILC2s by local epithelial cytokines IL-33 and TSLP has been shown to play a major role in the development of food allergy. While Th2 cells, by producing IL-4, IL-5, and IL-13, were initially believed to be the only major players driving the type 2 immune response, our current knowledge indicates that the type 2 immune response is mediated by the cooperative actions of Th2 cells and ILC2s and can be induced by scratching. Neutralizing these type 2 cells or their secreted cytokines via, e.g., monoclonal antibodies, used in anti-IL-4 or anti-IL-5 therapy can be a useful approach for patients with an already disrupted skin barrier who have an increased chance of developing food-induced allergy.

Key messages:

- Skin damage results in the release of IL-33, TSLP.
- Specifically, IL-33 cells activate DCs and ILC2 cells.
- Through activating ILC2 cells and DCs, epithelial cytokines, e.g., TSLP and IL-25, can mediate a type 2 inflammation reaction in an antigen independent manner.

6. How Can Pruritus Lead to Food-Induced Anaphylaxis?

Pruritus or itch is the unpleasant sensation that causes an urge to scratch [63]. Pruritus can have multiple causes, such as local nerve fiber compression or degeneration of nerve fiber in the peripheral or central nerve system or, in the case of dermatological pruritus, due to type 2 immune responses in the skin [63,64].

IL-33 is very important in the crosstalk between the skin and gut. In the study of Savinko et al., 'scratching' of the skin by tape stripping affected 10% of the mouse total body surface area, which resulted in a significant 2-fold increase in circulating levels of IL-33 [51]. In AD patients, around 20% of the total body surface area is affected, and the median IL-33 level in serum from patients with AD is more than 10-fold that of healthy control subjects [53,65]. Interestingly, in mice, this IL-33 increases the number of mucosal MCs in the small intestine via ILC2 activation [56,66]. Furthermore IL-33 enhances IgE-mediated degranulation of MCs in the gut, which leads to the development of an anaphylactic response to ingested food allergens. MCs and basophils are essential in anaphylactic responses by releasing mediators into the circulation [15]. However, not all sensitized individuals who have food allergen-specific IgE antibodies develop food allergy, and serum concentration is not a predictive marker for allergy severity [67]. Altogether, these results indicate that IL-33 released on mechanical skin injury as a replacement for scratching can potentially target ST2-expressing cells, including MCs at distant sites.

Intestinal MC expansion is associated with susceptibility to food-induced anaphylaxis and increased intestinal MC load correlates with an increased severity of food-induced anaphylaxis [68]. Furthermore, tape stripped epicutaneously sensitized mice and not orally immunized mice show expansion of intestinal MCs and IgE-mediated anaphylaxis after a single oral antigen challenge [69]. Tape stripping in mice induces intestinal tuft cells to produce IL-25 at the same time. This IL-25 activates and expands ILC2s in skin and small intestine and mediates the release of IL-4 and IL-13, which in turn activates tuft cells to produce IL-25 in a positive feedback loop [42,66]. ILC2-produced

IL-4 and IL-13 was essential, and not IL-5 and IL-9, to increase the intestinal MC load in this mouse model, which was all independent of T cells [66]. Intestinal MCs control intestinal permeability and, therefore, systemic absorption of food antigen and food anaphylaxis; see Figures 1 and 2 [68,70]. Furthermore, it is known that sensitization towards food antigens can also occur throughout the gastro-intestinal tract [71]. Especially the increased intestinal permeability could lead to enhanced sensitization to food antigens in the intestinal tract, potentially followed by an allergic response or even an anaphylactic response to these food antigens. These effects of IL-25 and IL-33 in combination with TSLP-induced IL-4 production by basophils in the skin all result in enhanced IgE-mediated mast cell degranulation in the intestines; hence, IL-25, IL-33, and TSLP are key players in the skin-to-gut axis. That all three epithelial cytokines, IL-25, IL-33, and TSLP, play a role in the induction of food allergy was proven by Khodoun et al. In mice, they showed that only treatment with a cocktail of the three monoclonal antibodies against IL-25, IL-33, and TSLP, and not a single treatment, was sufficient to inhibit development of murine food allergy [72].

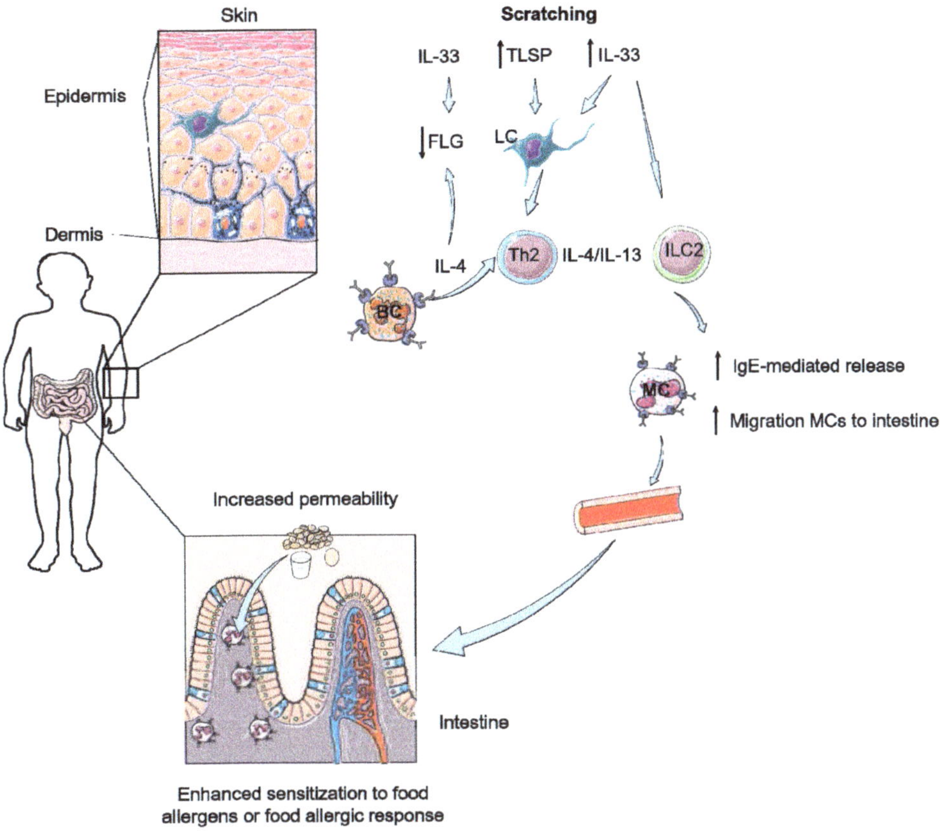

Figure 1. Scratching can further result in a decreased barrier function of the skin and the intestine. Scratching elicits thymic stromal lymphopoietin (TSLP) and IL-33 responses in skin activating Langerhans' cells (LC), innate lymphoid type 2 cells (ILC2) and T helper 2 cells (Th2). Furthermore, IL-4 production of basophils (BC) enhances the type 2 responses and leads do a decrease of filaggrin (FLG) expression in combination with IL-33. Due to type 2 responses both IgE-mediated release of mast cells (MC) and migration of MCs to the intestine is increased. This results in an increased permeability of the intestine and therefore of an influx of food allergens, potentially leading to enhanced sensitization or allergic responses to these food antigens.

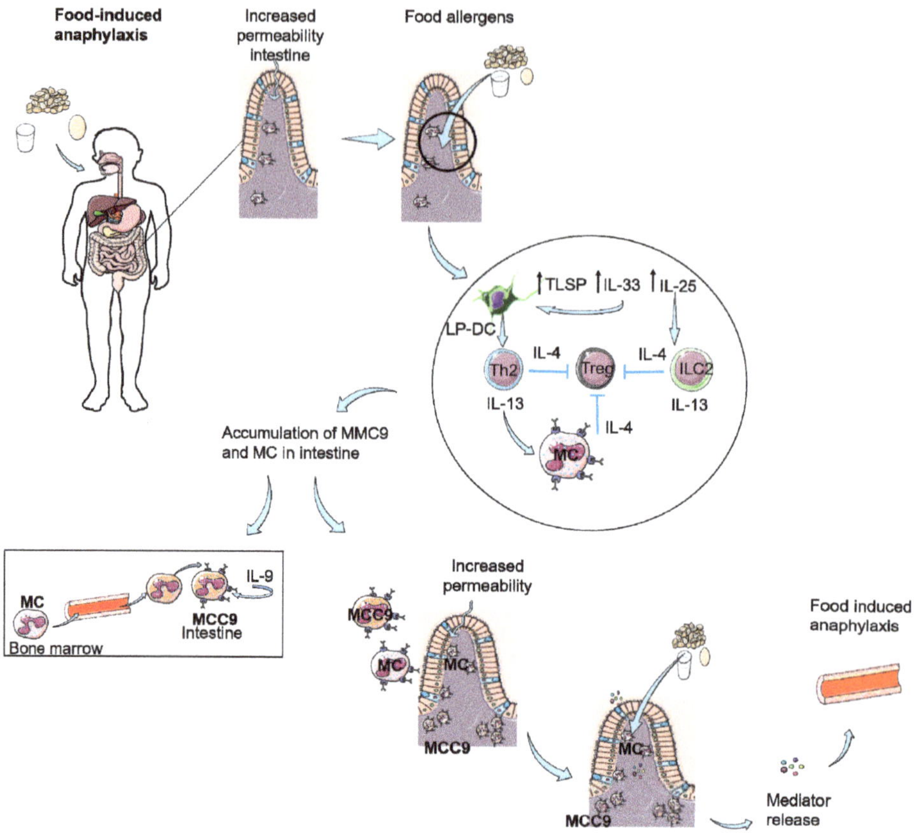

Figure 2. Food-induced anaphylaxis as a result of scratching or decreased skin barrier function. Intestinal permeability is increased due to influx of mast cells (MC) as a result of scratching or damaged skin. This results an increased entrance of food antigens in the intestine eliciting the production of TSLP, IL-33, and IL-25. IL-25 can activate ILC2 cells and IL-33 and TSLP activate dendritic cells in the lamina propia of the intestine (LP-DC), which activate Th2 cells. Th2 cells and ILC2 cells produce IL-4 and IL-13, resulting in inhibition of Tregs and stimulation of MCs. This leads to an accumulation of (sensitized) MC and IL-9 producing mucosal mast cells (MMC9) in the intestine, which causes an increased permeability of the intestine. Food allergens can passage the epithelial barrier, resulting in IgE-mediated degranulation of the MCs and as a result of mediator release, food-induced anaphylaxis.

Additionally, the number of intestinal MCs was higher in duodenal biopsies of patients with AD compared to non-AD patients [66]. The authors conclude that increased intestinal MCs and permeability can be elicited by scratching, which play an important role in promoting food anaphylaxis in patients with AD. Therefore, interventions that inhibit scratching may be useful in dampening the severity of food allergy in these patients by decreasing their intestinal MC load [66].

A specific group of intestinal mucosal mast cells are found to be associated with IgE-mediated food allergy [73]. These cells have been identified as IL-9-producing mucosal MCs (MMC9s) and are mainly located in the lamina propria of the small intestine in mice [74]. So far, in human subjects, an increase in IL-9 producing cells are identified in duodenal biopsies of food allergic patients compared to healthy controls, based on qPCR [74]. Furthermore, in humans, expression levels of IL-9, IL-13, and MC-specific transcripts are associated with food allergic patients who develop comorbid allergic diseases, such as eczema and urticaria [74]. However, no flow cytometric analysis or immunohistochemistry analysis has been performed of these IL-9 producing cells, hence we do not know if these cells are real MMC9 cells. Therefore, all data further discussed in this section is based on murine models. MMC9s function as type-2-promoting innate myeloid cells by producing IL-9 and IL-13 cytokines in response to IL-33, but not IL-25, and secrete histamine and other MC mediators upon antigen-induced IgE-complex crosslinking [74]. In mucosal tissues, MC expansion is dependent on the Th2 and Th9 cytokines IL-3 and IL-9 [73]. Allergic symptoms are reduced in IL-9-deficient mice, whereas intestinal mastocytosis, intestinal permeability, and intravascular leakage are observed in mice overexpressing IL-9, leading to a predisposition to oral antigen sensitization [75]. Furthermore, cross-linking on the surface of an antigen-specific IgE/FcεR-complex promotes the proliferation of MMC9s and MCs [46,74]. Of note, MMC9s seem to derive from mast cell progenitor cells from the bone marrow and are able to mature into mucosal mast cells with a reduced IL-9 production [74]. Levels of both MMC9s and MCs are increased after repeated intragastric ovalbumin (OVA) challenges from a basal level of 0.5% to 9% of total mononuclear cells in the small intestine in sensitized mice, resulting in the development of experimentally induced food allergy [74]. In this experimental food allergy model, mice are sensitized twice with OVA (day 0 and day 12) and are intragastrically challenged six times with OVA between day 25 and day 36, resulting in a food allergic reaction towards OVA. In addition, Th2 cells are increasing at the same time from 0.5 to 4%, and Chen et al. show that IL-4 and Th2 cells are required for the induction of MMC9s resulting in experimentally induced food allergy [74,76]. However, ILC2 and basophil levels remained constant. Similar results were obtained in skin-sensitized mice. In addition, intestinal MCs can provide an IL-4 signal to induce regulatory T cell reprogramming toward a Th2-cell-like lineage, resulting in the impairment of regulatory T cell function and the loss of tolerance [46]. In addition to Th2 cell activation, naive T cells are shown to differentiate to Th9. Th9 cells also secrete IL-9 cytokines and further promote the accumulation of tissue residing mast cells in mice [77].

In summary, scratching affects the barrier function of the skin and leads to the release of IL-33 in the skin and increased IL-33 levels systemically; see Figure 1. A reduced barrier function of the skin due to FLG mutation or detergents, sometimes in combination with adjuvant activities of microbial ligands, ultimately result in the induction of IL-33 and TSLP. TSLP and/or IL- 33 released by keratinocytes synergize with IL-25 released by intestinal tuft cells to expand ILC2s and increase their expression/production of IL-4 and/or IL-13, as depicted in Figures 1 and 2. ILC2-derived IL-4 and IL-13 target MCs to cause their expansion in the gastrointestinal mucosa, increase IgE-dependent degranulation of MCs and stimulate DCs to reduce allergen-specific Tregs. Furthermore, cutaneous sensitization induces TSLP activation of basophils that produce IL-4, production of type 2 cytokines IL-5 and IL-13, leading, as well, to an accumulation of mast cells in the intestine [3,30,40]. The accumulation of MCs in the gut results in a reduced gut barrier and hence a higher permeability of the gut barrier for food allergens, which leads to an enhanced risk for IgE-mediated degranulation of MCs and for the development of anaphylactic responses after exposure to food allergens. Taken together, these observations illustrate the existence of a skin-to-gut crosstalk in which mechanical skin injury can promote food-induced anaphylaxis by driving intestinal MC expansion, in addition to facilitating sensitization to food allergens.

Key messages:

- Scratching induces enhanced IL-33 levels in the skin and in serum.
- IL-33 together with IL-4 and Th2 cells are able to induce accumulation of mast cells and IL-9 producing mucosal mast cells (MMC9) in the intestine.
- IL-33 results in more IgE-mediated degranulation of these MCs and MMC9 cells, leading to food allergy.
- Scratching increases numbers of intestinal mast cells and increased permeability of the intestines resulting in the development of food allergy.
- A skin-to-gut axis is inevitable as food allergy symptoms in the intestine apparently can be induced by increased IL-33 levels in serum, which is induced by a damaged skin barrier due to scratching or AD.

7. The Role of Skin Microbiota in the Development of Food Allergy

In the development of AD, two hypotheses have been popular throughout the past decades. First, the 'inside to outside-hypothesis' was developed, in which the gut microbiota and the immune system were responsible for the decreased skin barrier function, leading to AD and allergy [78,79]. Later on, the 'outside to inside-hypothesis' became more dominant, in which skin barrier dysfunction was seen as a driver of AD development, which in turn leads to the activation of the immune system, resulting in a further reduced and affected skin barrier [80]. This review is mostly based on the 'outside to inside-hypothesis' and therefore we started from the skin barrier and its relation to allergy. The skin epithelial barrier is colonized with microbiota, and this microbiota diversifies throughout life [81]. Atopic dermatitis is in general associated with a lower diversity in the skin microbiota [82]. Staphylococcus aureus (*S. aureus*) colonization is highly associated with atopic dermatitis and eczema severity [83–85] and is found to cause a (partial) reduction of microbial diversity [82]. In the LEAP and LEAP-on studies, *S. aureus* colonization was found in children at 4 to 11 months of age (~22% on skin and ~18% in nose), which decreased to 8% on skin and 18% in nose at an age of 60 months [85]. In contrast, a recent birth cohort study revealed that in 1-year-old infants with AD, there was no dysbiosis in microbial communities and these infants' microbiome were not (yet) colonized by *S aureus*. However, AD-affected children had less commensal Staphylococci compared to healthy children [81]. *S. aureus* is found to release δ-toxin, which triggers degranulation of mouse-derived mast cells in vitro and promotes both innate and adaptive type 2 responses in vivo in mice [86]. Pre-incubation of allergen-specific IgE on mouse-derived mast cells even resulted in a synergistic degranulation effect of *S. aureus* derived δ-toxin in the absence of antigen [86]. Exposure of mouse skin to Staphylococcus enterotoxin B (SEB), together with food allergens (soy, ovalbumin or peanut), can induce the Th2 phenotype via IL-33 stimulation of skin-draining DCs and induce food allergy [59]. Interestingly, not all food allergens needed an exogenous adjuvant (SEB or cholera toxin): cow's milk allergen α-lactalbumin, green bean, and soy did need adjuvants to be able to induce sensitization, whereas cashew nut and peanut had intrinsic 'adjuvant activities' themselves [59]. Furthermore, only the combination of SEB and ovalbumin and not the single treatments resulted in Th2 responses in mice, and local mast cell activation and degranulation in the jejunum of these mice [87]. Next to SEB, other pathogenic factors, such as staphylococcal peptidoglycan or pertussis toxin, induced a Th2 polarization but not lipopolysaccharide (LPS) [87]. In line with these results, infiltrating T cells specific for SEB have been found in skin of AD patients [84]. These results underline that exposure of food allergens on skin, sometimes in combination with adjuvants in the form of microbial ligands or non-microbial ligands (e.g., detergents), is needed to develop food allergy.

A possible role for *S. aureus* colonization in the development of food allergy has also been proposed in human studies. An increase in the relative abundance of *S. aureus* in non-lesioned skin of AD patients with and without a food allergy compared to non-atopic controls was found in the study Leung et al. Next to this, a trend was observed of increased relative abundance of *S. aureus* in lesioned skin of AD patients with a food allergy compared to AD patients without a food allergy [31]. In the LEAP and LEAP-on studies, *S. aureus* colonization was related to more persistent egg white allergy and higher chances of having a peanut allergy at 60 months of age [85]. Furthermore, higher levels of specific-IgE levels to egg white, cow's milk, and peanut were found [85]. Interestingly, these associations were independent of eczema severity.

Key messages:

- *Staphylococcus aureus* colonization is related to reduce microbial diversity in the skin and increased prevalence of atopic dermatitis and food allergy.

8. The Role of Intestinal Microbiota on the Development of Atopy and Atopic Dermatitis

Next to the skin-to-gut axis, there has also recently been more interest in the role of the gut microbiota in skin diseases, such as acne, psoriasis, and atopic dermatitis [88–90]. Another link between the skin and gut is the use of epicutaneous immunotherapy in the treatment of food allergy, where patches with food allergens are placed on intact skin for 8 to 48 h [91]. As we focus in this review on food allergens, we will only discuss the role of microbes in the gut and its importance for the development of atopy. Especially, the role of gut microbiota in AD is well investigated and could contribute to the understanding of the development of food allergies, as well.

In newborns, microbial colonization is dependent on maternal diet during pregnancy, type of delivery, drinking breastmilk or not, antibiotic use (pre- and postnatal) and environmental exposure, as reviewed by Perdijk and Marsland [92]. In children, colonization by E. coli in the gut at the age of 1 month was related to higher odds for the prevalence of eczema, but not for developing atopic dermatitis at 2 years. Infants colonized with Clostridium difficile had a higher risk of developing eczema, atopic dermatitis, recurrent wheeze, and atopic sensitization at 2 years of age [93]. No effect of Bifidobacteria or Lactobacilli colonization was found on the development of eczema, atopic dermatitis, wheeze, or sensitization [93]. Other studies showed that allergic children had lower prevalence of fecal Bifidobacteria [94–96], Lactobacilli [94,97], and a higher prevalence of *S. aureus* [94,98] and Clostridium [97,99] compared to non-allergic children. For atopic dermatitis, some studies found a decrease in Bifidobacteria [100,101], but no difference in the microbiota was found between AD patients with and without food-specific IgE [102] or in AD patients with matched controls [103–105]. In contrast, some studies link a reduced microbial biodiversity to the development of AD [106,107] or to atopy in general [104], although no significant effect was found for AD in this study. To induce changes in the microbiota, intervention trials have been performed with different strains of probiotics that reduced the development of atopic dermatitis [108–115] or reduced sensitization towards egg white [116], although some studies found no effect [117].

Commensal bacteria are important regulators for mucosal immunity by influencing epithelial barrier function, decreasing TSLP-production in skin via induction of Tregs in the skin and maintain homeostasis between effector and regulatory T cells in the skin, as reviewed by Salem et al. [88]. Lactobacillus casei administration was found to affect differentiation from CD8+ T cells into skin effector cells, decreased homing of these T cells to skin upon stimulation in mice, and increased the number of Tregs in the skin [118]. Oyoshi et al. found that an allergic reaction of the skin was caused by CD4+ T cells of orally sensitized mice that expressed a gut-homing profile ($\alpha4\beta7+$) and in the draining lymph nodes switched to a skin homing profile (CCR4+) upon cutaneous exposure by OVA [119]. In addition, in children with a peanut allergy, peanut-specific T cells with a skin homing capacity showed higher proliferation compared to gut-homing peanut-specific T cells, indicating that sensitization had taken place in the skin [120]. Reducing migration capabilities of effector T cells to the skin by microbiota, while increasing Tregs in the skin, is important in preventing the development of allergic reactions in the skin.

Another way in which commensal bacteria can have an effect on allergy is by the production of short chain fatty acids (SCFAs). Acetate, propionate and butyrate are SCFAs produced by bacteria in the colon upon fermentation of non-digestible fibers. These SCFA regulate mucosal barrier function and can regulate immune responses both in the gut, as well as in the lung and skin [121–125]. Mice fed a high fiber diet have an increase in circulating SCFAs and showed reduced allergic inflammation in the airways [126]. In a birth cohort study, children with the highest levels of butyrate and propionate at one year of age had lower sensitization to allergens at six years of age [127]. In a recent study, human peripheral blood mononuclear cell –derived mast cells were incubated with different SCFAs in vitro [128]. Propionate and butyrate, but not acetate, were able to inhibit both IgE-mediated and non-IgE-mediated mast cell degranulation in a concentration dependent manner [128]. Furthermore, AD patients had lower SCFA production compared to control patients and several studies found that SCFA have antimicrobial effect and in particular propionate has an antimicrobial effect on *S. aureus* in vitro, as reviewed by Salem et al. [88].

So, there is no conclusive evidence that specific microbial species are responsible for the development of allergy or atopic dermatitis; nevertheless, there seems to be a crosstalk between the gut microbiota, its metabolites and the skin.

Key messages:

- There is no conclusive evidence that specific microbial species are responsible for the development of allergy or atopic dermatitis.
- Short chain fatty acids produced by intestinal microbiota are linked to reduced allergic inflammation.

9. Future Human Research Priorities

In this review a few important mechanisms are described that can play a role in the sensitization to food and food allergy, which are proven in murine models but not yet in humans. For the MMC9 cells there is circumstantial evidence that they are present in humans [74]. However, no flow cytometric analysis or immunohistochemistry has been performed on duodenal biopsies of food allergic patients to confirm the existence of MMC9 cells in humans.

In mice it was proven that only a cocktail of the three monoclonal antibodies against IL-25, IL-33, and TSLP can inhibit the development of food allergy in mice [72]. To our knowledge, this approach has not been tested in humans yet, although it could be very beneficial for AD patients in general, as well.

The role of environmental allergens, such as exposure to detergents as SDS, followed by HDM allergens or *Alternaria alternate*, is investigated in mice. In AD patients, exposure to HDM increased TSLP release in the skin [29], but no combination was made with detergents or other environmental allergens. Detergents are tested in vitro in human epidermal keratinocytes, and these resulted in decreased tight junction formation and barrier function of epidermal keratinocytes [129]. Therefore, it would be very interesting to investigate the effect of exposure to a combination of environmental factors, such as detergents and HDM in healthy volunteers and AD patients in a double-blind placebo-controlled study. In this set-up, skin barrier function can be addressed and atopy development could be followed as outcome. Next to this, large cohort-studies should be carried out to investigate which of the factors: skin barrier function, carriage of *S. aureus*, and having AD, is responsible for the increased sensitivity to food allergens [21]. Consequently, treatment of the skin can be optimized and sensitization to food allergens via the skin can be prevented. If these studies are performed in very young children the effect of oral ingestion prior to skin exposure of allergens should be included.

10. Conclusions

In this review, we highlighted the role of the skin in the development of IgE-mediated food allergy. Furthermore, we summarized the cellular and molecular mechanisms in the skin-to-gut crosstalk in the development of IgE-mediated food allergy. The site where food antigens are firstly taken up, either the skin or the gut, may cause sensitization (skin) or tolerance (gut) against this food antigen. However, sensitization towards food antigens can potentially also take place in the intestine as the result of an increased intestinal permeability. Prevention of scratching the skin is an important therapeutic target to prevent impaired skin barrier. Evidence in mouse models and clinical studies suggest that, if the skin barrier can be improved and/or the inflammation of AD can be proactively prevented, in combination with early introduction of food antigens, then the incidence of food allergy and possibly other forms of allergic diseases might be decreased.

Author Contributions: Conceptualization, M.v.S., L.L.; methodology, M.v.S., L.L. and N.W.d.J.; writing—original draft preparation, M.v.S., L.L., R.J.J.v.N. and N.W.d.J.; writing—review and editing, M.v.S., L.L., R.J.J.v.N., H.J.W., K.A.H. and N.W.d.J. All authors have read and agreed to the published version of the manuscript.

Funding: This research received no external funding.

Conflicts of Interest: The authors declare no conflict of interest. RJ Joost van Neerven is in addition to his academic affiliation also employed by FrieslandCampina.

Glossary of Terms

ATP	Adenosine triphosphate
AD	atopic dermatitis
DC	dendritic cell
FLG	Filaggrin
HDM	House dust mite
Ig	Immune globuline
IL	interleukin
ILC	innate lymphoid cells
LPS	lipopolysaccharide
MC	mast cells
MMC9	IL-9-producing mucosal mast cell
OVA	ovalbumin
SCFA	short chain fatty acids
SDS	sodium dodecyl sulfate
S. aureus	*Staphylococcus aureus*
TEWL	transepidermal water loss
Th2	T helper cell 2
TSLP	thymic stromal lymphopoietin
Treg	regulatory T cell

References

1. Han, H.; Thelen, T.D.; Comeau, M.R.; Ziegler, S.F. Thymic stromal lymphopoietin-mediated epicutaneous inflammation promotes acute diarrhea and anaphylaxis. *J. Clin. Investig.* **2014**, *124*, 5442–5452. [CrossRef]
2. Noti, M.; Wojno, E.D.T.; Kim, B.S.; Siracusa, M.C.; Giacomin, P.R.; Nair, M.G.; Benitez, A.J.; Ruymann, K.R.; Muir, A.B.; Hill, D.A.; et al. Thymic stromal lymphopoietin-elicited basophil responses promote eosinophilic esophagitis. *Nat. Med.* **2013**, *19*, 1005–1013. [CrossRef]
3. Noti, M.; Kim, B.S.; Siracusa, M.C.; Rak, G.D.; Kubo, M.; Moghaddam, A.E.; Sattentau, Q.A.; Comeau, M.R.; Spergel, J.M.; Artis, D. Exposure to food allergens through inflamed skin promotes intestinal food allergy through the thymic stromal lymphopoietin-basophil axis. *J. Allergy Clin. Immunol.* **2014**, *133*, 1390–1399.e6. [CrossRef]
4. Hussain, M.; Borcard, L.; Walsh, K.P.; Pena Rodriguez, M.; Mueller, C.; Kim, B.S.; Kubo, M.; Artis, D.; Noti, M. Basophil-derived IL-4 promotes epicutaneous antigen sensitization concomitant with the development of food allergy. *J. Allergy Clin. Immunol.* **2018**, *141*, 223–234.e5. [CrossRef]
5. Lack, G. Early exposure hypothesis: Where are we now? *Clin. Transl. Allergy* **2011**, *1*, S71. [CrossRef]
6. Brough, H.A.; Nadeau, K.C.; Sindher, S.B.; Alkotob, S.S.; Chan, S.; Bahnson, H.; Leung, D.Y.M.; Lack, G. Epicutaneous sensitization in the development of food allergy: What is the evidence and how can this be prevented? *Allergy* **2020**, *75*, 2185–2205. [CrossRef]
7. Martin, P.E.; Eckert, J.K.; Koplin, J.J.; Lowe, A.J.; Gurrin, L.C.; Dharmage, S.C.; Vuillermin, P.; Tang, M.L.K.K.; Ponsonby, A.-L.L.; Matheson, M.; et al. Which infants with eczema are at risk of food allergy? Results from a population-based cohort. *Clin. Exp. Allergy* **2015**, *45*, 255–264. [CrossRef] [PubMed]
8. Han, H.; Xu, W.; Headley, M.B.; Jessup, H.K.; Lee, K.S.; Omori, M.; Comeau, M.R.; Marshak-Rothstein, A.; Ziegler, S.F. Thymic stromal lymphopoietin (TSLP)-mediated dermal inflammation aggravates experimental asthma. *Mucosal Immunol.* **2012**, *5*, 342–351. [CrossRef]
9. Leyva-Castillo, J.M.; Hener, P.; Jiang, H.; Li, M. TSLP Produced by Keratinocytes Promotes Allergen Sensitization through Skin and Thereby Triggers Atopic March in Mice. *J. Investig. Dermatol.* **2013**, *133*, 154–163. [CrossRef]
10. Zhang, Z.; Hener, P.; Frossard, N.; Kato, S.; Metzger, D.; Li, M.; Chambon, P. Thymic stromal lymphopoietin overproduced by keratinocytes in mouse skin aggravates experimental asthma. *Proc. Natl. Acad. Sci. USA* **2009**, *106*, 1536–1541. [CrossRef]

11. Brough, H.A.; Liu, A.H.; Sicherer, S.; Makinson, K.; Douiri, A.; Brown, S.J.; Stephens, A.C.; Irwin McLean, W.H.; Turcanu, V.; Wood, R.A.; et al. Atopic dermatitis increases the effect of exposure to peanut antigen in dust on peanut sensitization and likely peanut allergy. *J. Allergy Clin. Immunol.* **2015**, *135*, 164–170.e4. [CrossRef] [PubMed]
12. Du Toit, G.; Roberts, G.; Sayre, P.H.; Bahnson, H.T.; Radulovic, S.; Santos, A.F.; Brough, H.A.; Phippard, D.; Basting, M.; Feeney, M.; et al. Randomized trial of peanut consumption in infants at risk for peanut allergy. *N. Engl. J. Med.* **2015**, *372*, 803–813. [CrossRef]
13. Du Toit, G.; Roberts, G.; Sayre, P.H.; Plaut, M.; Bahnson, H.T.; Mitchell, H.; Radulovic, S.; Chan, S.; Fox, A.; Turcanu, V.; et al. Identifying infants at high risk of peanut allergy: The Learning Early about Peanut Allergy (LEAP) screening study. *J. Allergy Clin. Immunol.* **2013**, *131*, 135–143.e12. [CrossRef]
14. Natsume, O.; Kabashima, S.; Nakazato, J.; Yamamoto-Hanada, K.; Narita, M.; Kondo, M.; Saito, M.; Kishino, A.; Takimoto, T.; Inoue, E.; et al. Two-step egg introduction for prevention of egg allergy in high-risk infants with eczema (PETIT): A randomised, double-blind, placebo-controlled trial. *Lancet* **2017**, *389*, 276–286. [CrossRef]
15. Peavy, R.D.; Metcalfe, D.D. Understanding the mechanisms of anaphylaxis. *Curr. Opin. Allergy Clin. Immunol.* **2008**, *8*, 310–315. [CrossRef]
16. Schleimer, R.P.; Berdnikovs, S. Etiology of epithelial barrier dysfunction in type 2 inflammatory disease. *J. Allergy Clin. Immunol.* **2017**, *139*, 1752–1761. [CrossRef] [PubMed]
17. Egawa, G.; Kabashima, K. Multifactorial skin barrier deficiency and atopic dermatitis: Essential topics to prevent the atopic march. *J. Allergy Clin. Immunol.* **2016**, *138*, 350–358.e1. [CrossRef] [PubMed]
18. Hill, D.J.; Hosking, C.S.; De Benedictis, F.M.; Oranje, A.P.; Diepgen, T.L.; Bauchau, V.; Warner, J.O.; Naspitz, C.K.; Simons, F.E.R.; Diepgen, T.L.; et al. Confirmation of the association between high levels of immunoglobulin E food sensitization and eczema in infancy: An international study. *Clin. Exp. Allergy* **2008**, *38*, 161–168. [CrossRef]
19. Irvine, A.D.; McLean, W.H.I.; Leung, D.Y.M. Filaggrin mutations associated with skin and allergic diseases. *N. Engl. J. Med.* **2011**, *365*, 1315–1327. [CrossRef]
20. Van Smeden, J.; Janssens, M.; Kaye, E.C.J.; Caspers, P.J.; Lavrijsen, A.P.; Vreeken, R.J.; Bouwstra, J.A. The importance of free fatty acid chain length for the skin barrier function in atopic eczema patients. *Exp. Dermatol.* **2014**, *23*, 45–52. [CrossRef]
21. Kelleher, M.M.; Dunn-Galvin, A.; Hourihane, J.O.O.B.; Murray, D.; Campbell, L.E.; McLean, W.H.I.I.; Irvine, A.D. Skin barrier dysfunction measured by transepidermal water loss at 2 days and 2 months predates and predicts atopic dermatitis at 1 year. *J. Allergy Clin. Immunol.* **2015**, *135*, 930–935.e1. [CrossRef] [PubMed]
22. Kelleher, M.M.; Dunn-Galvin, A.; Gray, C.; Murray, D.M.; Kiely, M.; Kenny, L.; McLean, W.H.I.; Irvine, A.D.; Hourihane, J.O. Skin barrier impairment at birth predicts food allergy at 2 years of age. *J. Allergy Clin. Immunol.* **2016**, *137*, 1111–1116.e8. [CrossRef] [PubMed]
23. Brown, S.J.; Asai, Y.; Cordell, H.J.; Campbell, L.E.; Zhao, Y.; Liao, H.; Northstone, K.; Henderson, J.; Alizadehfar, R.; Ben-Shoshan, M.; et al. Loss-of-function variants in the filaggrin gene are a significant risk factor for peanut allergy. *J. Allergy Clin. Immunol.* **2011**, *127*, 661–667. [CrossRef] [PubMed]
24. Kusunoki, T.; Okafuji, I.; Yoshioka, T.; Saito, M.; Nishikomori, R.; Heike, T.; Sugai, M.; Shimizu, A.; Nakahata, T. SPINK5 polymorphism is associated with disease severity and food allergy in children with atopic dermatitis. *J. Allergy Clin. Immunol.* **2005**, *115*, 636–638. [CrossRef]
25. Brough, H.A.; Simpson, A.; Makinson, K.; Hankinson, J.; Brown, S.; Douiri, A.; Belgrave, D.C.M.; Penagos, M.; Stephens, A.C.; McLean, W.H.I.; et al. Peanut allergy: Effect of environmental peanut exposure in children with filaggrin loss-of-function mutations. *J. Allergy Clin. Immunol.* **2014**, *134*, 867–875.e1. [CrossRef]
26. Walker, M.T.; Green, J.E.; Ferrie, R.P.; Queener, A.M.; Kaplan, M.H.; Cook-Mills, J.M. Mechanism for initiation of food allergy: Dependence on skin barrier mutations and environmental allergen costimulation. *J. Allergy Clin. Immunol.* **2018**, *141*, 1711–1725.e9. [CrossRef]
27. Akdis, C.A.; Arkwright, P.D.; Brüggen, M.C.; Busse, W.; Gadina, M.; Guttman-Yassky, E.; Kabashima, K.; Mitamura, Y.; Vian, L.; Wu, J.; et al. Type 2 immunity in the skin and lungs. *Allergy* **2020**, *74*, 1582–1605. [CrossRef]
28. Cayrol, C.; Duval, A.; Schmitt, P.; Roga, S.; Camus, M.; Stella, A.; Burlet-Schiltz, O.; Gonzalez-De-Peredo, A.; Girard, J.P. Environmental allergens induce allergic inflammation through proteolytic maturation of IL-33. *Nat. Immunol.* **2018**, *19*, 375–385. [CrossRef]

29. Landheer, J.; Giovannone, B.; Mattson, J.D.; Tjabringa, S.; Bruijnzeel-Koomen, C.A.F.M.; McClanahan, T.; De Waal Malefyt, R.; Knol, E.; Hijnen, D. Epicutaneous application of house dust mite induces thymic stromal lymphopoietin in nonlesional skin of patients with atopic dermatitis. *J. Allergy Clin. Immunol.* **2013**, *132*, 1252–1254. [CrossRef]
30. Strid, J.; Hourihane, J.; Kimber, I.; Callard, R.; Strobel, S. Epicutaneous exposure to peanut protein prevents oral tolerance and enhances allergic sensitization. *Clin. Exp. Allergy* **2005**, *35*, 757–766. [CrossRef]
31. Leung, D.Y.M.; Calatroni, A.; Zaramela, L.S.; LeBeau, P.K.; Dyjack, N.; Brar, K.; David, G.; Johnson, K.; Leung, S.; Ramirez-Gama, M.; et al. The nonlesional skin surface distinguishes atopic dermatitis with food allergy as a unique endotype. *Sci. Transl. Med.* **2019**, *11*, eaav2685. [CrossRef]
32. Pellerin, L.; Henry, J.; Hsu, C.-Y.; Balica, S.; Jean-Decoster, C.; Méchin, M.-C.; Hansmann, B.; Rodriguez, E.; Weindinger, S.; Schmitt, A.-M.; et al. Defects of filaggrin-like proteins in both lesional and nonlesional atopic skin. *J. Allergy Clin. Immunol.* **2013**, *131*, 1094–1102. [CrossRef] [PubMed]
33. Reche, P.A.; Soumelis, V.; Gorman, D.M.; Clifford, T.; Liu, M.; Travis, M.; Zurawski, S.M.; Johnston, J.; Liu, Y.-J.; Spits, H.; et al. Human Thymic Stromal Lymphopoietin Preferentially Stimulates Myeloid Cells. *J. Immunol.* **2001**, *167*, 336–343. [CrossRef] [PubMed]
34. Soumelis, V.; Reche, P.A.; Kanzler, H.; Yuan, W.; Edward, G.; Homey, B.; Gilliet, M.; Ho, S.; Antonenko, S.; Lauerma, A.; et al. Human epithelial cells trigger dendritic cell–mediated allergic inflammation by producing TSLP. *Nat. Immunol.* **2002**, *3*, 673–680. [CrossRef]
35. Oyoshi, M.K.; Larson, R.P.; Ziegler, S.F.; Geha, R.S. Mechanical injury polarizes skin dendritic cells to elicit a T(H)2 response by inducing cutaneous thymic stromal lymphopoietin expression. *J. Allergy Clin. Immunol.* **2010**, *126*, 976–984.e5. [CrossRef] [PubMed]
36. Kim, J.; Kim, B.E.; Lee, J.; Han, Y.; Jun, H.Y.; Kim, H.; Choi, J.; Leung, D.Y.M.; Ahn, K. Epidermal thymic stromal lymphopoietin predicts the development of atopic dermatitis during infancy. *J. Allergy Clin. Immunol.* **2016**, *137*, 1282–1285.e4. [CrossRef]
37. Liu, Y.-J.; Soumelis, V.; Watanabe, N.; Ito, T.; Wang, Y.-H.; de Waal Malefyt, R.; Omori, M.; Zhou, B.; Ziegler, S.F. TSLP: An Epithelial Cell Cytokine that Regulates T Cell Differentiation by Conditioning Dendritic Cell Maturation. *Annu. Rev. Immunol.* **2007**, *25*, 193–219. [CrossRef]
38. Ito, T.; Wang, Y.-H.; Duramad, O.; Hori, T.; Delespesse, G.J.; Watanabe, N.; Qin, F.X.-F.; Yao, Z.; Cao, W.; Liu, Y.-J. TSLP-activated dendritic cells induce an inflammatory T helper type 2 cell response through OX40 ligand. *J. Exp. Med.* **2005**, *202*, 1213–1223. [CrossRef]
39. Leyva-Castillo, J.M.; Hener, P.; Michea, P.; Karasuyama, H.; Chan, S.; Soumelis, V.; Li, M. Skin thymic stromal lymphopoietin initiates Th2 responses through an orchestrated immune cascade. *Nat. Commun.* **2013**, *4*, 2847. [CrossRef]
40. Hsieh, K.-Y.; Tsai, C.-C.; Wu, C.H.H.; Lin, R.-H. Epicutaneous exposure to protein antigen and food allergy. *Clin. Exp. Allergy* **2003**, *33*, 1067–1075. [CrossRef]
41. Hammad, H.; Lambrecht, B.N. Barrier Epithelial Cells and the Control of Type 2 Immunity. *Immunity* **2015**, *43*, 29–40. [CrossRef] [PubMed]
42. von Moltke, J.; Liang, H.-E.; Locksley, R.M. Tuft-cell-derived IL-25 regulates an intestinal ILC2–epithelial response circuit. *Nature* **2016**, *529*, 221–225. [CrossRef] [PubMed]
43. Lee, J.-B.; Chen, C.-Y.; Liu, B.; Mugge, L.; Angkasekwinai, P.; Facchinetti, V.; Dong, C.; Liu, Y.-J.; Rothenberg, M.E.; Hogan, S.P.; et al. IL-25 and CD4+Th2 cells enhance ILC2-derived IL-13 production that promotes IgE-mediated experimental food allergy. *J. Allergy Clin. Immunol.* **2016**, *137*, 1216–1225.e5. [CrossRef] [PubMed]
44. Camelo, A.; Rosignoli, G.; Ohne, Y.; Stewart, R.A.; Overed-sayer, C.; Sleeman, M.A.; May, R.D. IL-33, IL-25, and TSLP induce a distinct phenotypic and activation profile in human type 2 innate lymphoid cells. *Blood Adv.* **2017**, *1*, 577–589. [CrossRef] [PubMed]
45. Artis, D.; Spits, H. The biology of innate lymphoid cells. *Nature* **2015**, *517*, 293–301. [CrossRef]
46. Noval Rivas, M.; Burton, O.T.; Wise, P.; Charbonnier, L.M.; Georgiev, P.; Oettgen, H.C.; Rachid, R.; Chatila, T.A. Regulatory T cell reprogramming toward a Th2-Cell-like lineage impairs oral tolerance and promotes food allergy. *Immunity* **2015**, *42*, 512–523. [CrossRef]
47. Burton, O.T.; Medina Tamayo, J.; Stranks, A.J.; Miller, S.; Koleoglou, K.J.; Weinberg, E.O.; Oettgen, H.C. IgE promotes type 2 innate lymphoid cells in murine food allergy. *Clin. Exp. Allergy* **2018**, *48*, 288–296. [CrossRef]

48. Noval Rivas, M.; Burton, O.T.; Oettgen, H.C.; Chatila, T. IL-4 production by group 2 innate lymphoid cells promotes food allergy by blocking regulatory T-cell function. *J. Allergy Clin. Immunol.* **2016**, *138*, 801–811.e9. [CrossRef]
49. Liew, F.Y.; Girard, J.P.; Turnquist, H.R. Interleukin-33 in health and disease. *Nat. Rev. Immunol.* **2016**, *16*, 676–689. [CrossRef]
50. Cayrol, C.; Girard, J.P. Interleukin-33 (IL-33): A nuclear cytokine from the IL-1 family. *Immunol. Rev.* **2018**, *281*, 154–168. [CrossRef]
51. Savinko, T.; Matikainen, S.; Saarialho-Kere, U.; Lehto, M.; Wang, G.; Lehtimäki, S.; Karisola, P.; Reunala, T.; Wolff, H.; Lauerma, A.; et al. IL-33 and ST2 in Atopic Dermatitis: Expression Profiles and Modulation by Triggering Factors. *J. Investig. Dermatol.* **2012**, *132*, 1392–1400. [CrossRef] [PubMed]
52. Salimi, M.; Barlow, J.L.; Saunders, S.P.; Xue, L.; Gutowska-Owsiak, D.; Wang, X.; Huang, L.C.; Johnson, D.; Scanlon, S.T.; McKenzie, A.N.J.; et al. A role for IL-25 and IL-33-driven type-2 innate lymphoid cells in atopic dermatitis. *J. Exp. Med.* **2013**, *210*, 2939–2950. [CrossRef] [PubMed]
53. Tamagawa-Mineoka, R.; Okuzawa, Y.; Masuda, K.; Katoh, N. Increased serum levels of interleukin 33 in patients with atopic dermatitis. *J. Am. Acad. Dermatol.* **2014**, *70*, 882–888. [CrossRef] [PubMed]
54. Seltmann, J.; Roesner, L.M.; Von Hesler, F.W.; Wittmann, M.; Werfel, T. IL-33 impacts on the skin barrier by downregulating the expression of filaggrin. *J. Allergy Clin. Immunol.* **2015**, *135*, 1659–1661.e4. [CrossRef]
55. Han, H.; Roan, F.; Johnston, L.K.; Smith, D.E.; Bryce, P.J.; Ziegler, S.F. IL-33 promotes gastrointestinal allergy in a TSLP-independent manner. *Mucosal Immunol.* **2018**, *11*, 394–403. [CrossRef]
56. Galand, C.; Leyva-Castillo, J.M.; Yoon, J.; Han, A.; Lee, M.S.; McKenzie, A.N.J.; Stassen, M.; Oyoshi, M.K.; Finkelman, F.D.; Geha, R.S. IL-33 promotes food anaphylaxis in epicutaneously sensitized mice by targeting mast cells. *J. Allergy Clin. Immunol.* **2016**, *138*, 1356–1366. [CrossRef]
57. Parveen, S.; Saravanan, D.B.; Saluja, R.; Elden, B.T. IL-33 mediated amplification of allergic response in human mast cells. *J. Recept. Signal Transduct.* **2019**, *39*, 359–367. [CrossRef]
58. Kim, B.S.; Siracusa, M.C.; Saenz, S.A.; Noti, M.; Monticelli, L.A.; Sonnenberg, G.F.; Hepworth, M.R.; Van Voorhees, A.S.; Comeau, M.R.; Artis, D. TSLP elicits IL-33-independent innate lymphoid cell responses to promote skin inflammation. *Sci. Transl. Med.* **2013**, *5*, 170ra16. [CrossRef]
59. Tordesillas, L.; Goswami, R.; Benedé, S.; Grishina, G.; Dunkin, D.; Järvinen, K.M.; Maleki, S.J.; Sampson, H.A.; Berin, M.C. Skin exposure promotes a Th2-dependent sensitization to peanut allergens. *J. Clin. Investig.* **2014**, *124*, 4965–4975. [CrossRef]
60. Monticelli, L.A.; Osborne, L.C.; Noti, M.; Tran, S.V.; Zaiss, D.M.W.; Artis, D. IL-33 promotes an innate immune pathway of intestinal tissue protection dependent on amphiregulin-EGFR interactions. *Proc. Natl. Acad. Sci. USA* **2015**, *112*, 10762–10767. [CrossRef]
61. Monticelli, L.A.; Sonnenberg, G.F.; Abt, M.C.; Alenghat, T.; Ziegle, C.G.K.; Doering, T.A.; Angelosanto, J.M.; Laidlaw, B.J.; Yang, C.Y.; Sathaliyawala, T.; et al. Innate lymphoid cells promote lung tissue homeostasis following acute influenza virus infection. *Nat. Immunol.* **2012**, *12*, 1045–1054. [CrossRef]
62. Arpaia, N.; Green, J.A.; Moltedo, B.; Arvey, A.; Hemmers, S.; Yuan, S.; Treuting, P.M.; Rudensky, A.Y. A Distinct Function of Regulatory T Cells in Tissue Protection. *Cell* **2015**, *162*, 1078–1089. [CrossRef] [PubMed]
63. Misery, L.; Brenaut, E.; Le Garrec, R.; Abasq, C.; Genestet, S.; Marcorelles, P.; Zagnoli, F. Neuropathic pruritus. *Nat. Rev. Neurol.* **2014**, *10*, 408–416. [CrossRef] [PubMed]
64. Seegräber, M.; Srour, J.; Walter, A.; Knop, M.; Wollenberg, A. Dupilumab for treatment of atopic dermatitis. *Expert Rev. Clin. Pharmacol.* **2018**, *11*, 467–474. [CrossRef] [PubMed]
65. Sugarman, J.L.; Fluhr, J.W.; Fowler, A.J.; Bruckner, T.; Diepgen, T.L.; Williams, M.L. The Objective Severity Assessment of Atopic Dermatitis Score. *Arch. Dermatol.* **2003**, *139*, 1417–1422. [CrossRef]
66. Leyva-Castillo, J.M.; Galand, C.; Kam, C.; Burton, O.; Gurish, M.; Musser, M.A.; Goldsmith, J.D.; Hait, E.; Nurko, S.; Brombacher, F.; et al. Mechanical Skin Injury Promotes Food Anaphylaxis by Driving Intestinal Mast Cell Expansion. *Immunity* **2019**, *50*, 1262–1275.e4. [CrossRef]
67. Fleischer, D.M.; Bock, S.A.; Spears, G.C.; Wilson, C.G.; Miyazawa, N.K.; Gleason, M.C.; Gyorkos, E.A.; Murphy, J.R.; Atkins, D.; Leung, D.Y.M. Oral Food Challenges in Children with a Diagnosis of Food Allergy. *J. Pediatr.* **2011**, *158*, 578–583.e1. [CrossRef]

68. Ahrens, R.; Osterfeld, H.; Wu, D.; Chen, C.-Y.; Arumugam, M.; Groschwitz, K.; Strait, R.; Wang, Y.-H.; Finkelman, F.D.; Hogan, S.P. Intestinal mast cell levels control severity of oral antigen-induced anaphylaxis in mice. *Am. J. Pathol.* **2012**, *180*, 1535–1546. [CrossRef]
69. Bartnikas, L.M.; Gurish, M.F.; Burton, O.T.; Leisten, S.; Janssen, E.; Oettgen, H.C.; Beaupré, J.; Lewis, C.N.; Austen, K.F.; Schulte, S.; et al. Epicutaneous sensitization results in IgE-dependent intestinal mast cell expansion and food-induced anaphylaxis. *J. Allergy Clin. Immunol.* **2013**, *131*, 451–460.e6. [CrossRef]
70. Lee, H.; Park, J.H.; Park, D.I.; Kim, H.J.; Cho, Y.K.; Sohn, C.I.; Jeon, W.K.; Kim, B.I.; Chae, S.W. Mucosal mast cell count is associated with intestinal permeability in patients with diarrhea predominant irritable bowel syndrome. *J. Neurogastroenterol. Motil.* **2013**, *19*, 244–250. [CrossRef]
71. Sampson, H.A.; O'Mahony, L.; Burks, A.W.; Plaut, M.; Lack, G.; Akdis, C.A. Mechanisms of food allergy. *J. Allergy Clin. Immunol.* **2018**, *141*, 11–19. [CrossRef] [PubMed]
72. Khodoun, M.V.; Tomar, S.; Tocker, J.E.; Wang, Y.H.; Finkelman, F.D. Prevention of food allergy development and suppression of established food allergy by neutralization of thymic stromal lymphopoietin, IL-25, and IL-33. *J. Allergy Clin. Immunol.* **2018**, *141*, 171–179.e1. [CrossRef] [PubMed]
73. Kurashima, Y.; Kiyono, H. New era for mucosal mast cells: Their roles in inflammation, allergic immune responses and adjuvant development. *Exp. Mol. Med.* **2014**, *46*, e83. [CrossRef] [PubMed]
74. Chen, C.Y.; Lee, J.B.; Liu, B.; Ohta, S.; Wang, P.Y.; Kartashov, A.V.; Mugge, L.; Abonia, J.P.; Barski, A.; Izuhara, K.; et al. Induction of Interleukin-9-Producing Mucosal Mast Cells Promotes Susceptibility to IgE-Mediated Experimental Food Allergy. *Immunity* **2015**, *43*, 788–802. [CrossRef] [PubMed]
75. Forbes, E.E.; Groschwitz, K.; Abonia, J.P.; Brandt, E.B.; Cohen, E.; Blanchard, C.; Ahrens, R.; Seidu, L.; McKenzie, A.; Strait, R.; et al. IL-9- and mast cell-mediated intestinal permeability predisposes to oral antigen hypersensitivity. *J. Exp. Med.* **2008**, *205*, 897–913. [CrossRef]
76. Tomar, S.; Ganesan, V.; Sharma, A.; Zeng, C.; Waggoner, L.; Smith, A.; Kim, C.H.; Licona-Limón, P.; Reinhardt, R.L.; Flavell, R.A.; et al. IL-4-BATF signaling directly modulates IL-9 producing mucosal mast cell (MMC9) function in experimental food allergy. *J. Allergy Clin. Immunol.* **2020**. [CrossRef]
77. Sehra, S.; Yao, W.; Nguyen, E.T.; Glosson-Byers, N.L.; Akhtar, N.; Zhou, B.; Kaplan, M.H. TH9 cells are required for tissue mast cell accumulation during allergic inflammation. *J. Allergy Clin. Immunol.* **2015**, *136*, 433–440.e1. [CrossRef]
78. Chamlin, S.L.; Kao, J.; Frieden, I.J.; Sheu, M.Y.; Fowler, A.J.; Fluhr, J.W.; Williams, M.L.; Elias, P.M. Ceramide-dominant barrier repair lipids alleviate childhood atopic dermatitis: Changes in barrier function provide a sensitive indicator of disease activity. *J. Am. Acad. Dermatol.* **2002**, *47*, 198–208. [CrossRef]
79. Leung, D.Y.M. Atopic dermatitis: New insights and opportunities for therapeutic intervention. *J. Allergy Clin. Immunol.* **2000**, *105*, 860–876. [CrossRef]
80. Elias, P.M.; Feingold, K.R. Does the tail wag the dog? Role of the barrier in the pathogenesis of inflammatory dermatoses and therapeutic implications. *Arch. Dermatol.* **2001**, *137*, 1079–1081.
81. Kennedy, E.A.; Connolly, J.; Hourihane, J.O.; Fallon, P.G.; McLean, W.H.I.; Murray, D.; Jo, J.-H.; Segre, J.A.; Kong, H.H.; Irvine, A.D. Skin microbiome before development of atopic dermatitis: Early colonization with commensal staphylococci at 2 months is associated with a lower risk of atopic dermatitis at 1 year. *J. Allergy Clin. Immunol.* **2017**, *139*, 166–172. [CrossRef]
82. Kong, H.H.; Oh, J.; Deming, C.; Conlan, S.; Grice, E.A.; Beatson, M.A.; Nomicos, E.; Polley, E.C.; Komarow, H.D.; Mullikin, J.; et al. Temporal shifts in the skin microbiome associated with disease flares and treatment in children with atopic dermatitis. *Genome Res.* **2012**, *22*, 850–859. [CrossRef] [PubMed]
83. Rudikoff, D.; Lebwohl, M. Atopic dermatitis. *Lancet* **1998**, *351*, 1715–1721. [CrossRef]
84. Bunikowski, R.; Mielke, M.E.A.; Skarabis, H.; Worm, M.; Anagnostopoulos, I.; Kolde, G.; Wahn, U.; Renz, H. Evidence for a disease-promoting effect of Staphylococcus aureus-derived exotoxins in atopic dermatitis. *J. Allergy Clin. Immunol.* **2000**, *105*, 814–819. [CrossRef] [PubMed]
85. Tsilochristou, O.; du Toit, G.; Sayre, P.H.; Roberts, G.; Lawson, K.; Sever, M.L.; Bahnson, H.T.; Radulovic, S.; Basting, M.; Plaut, M.; et al. Association of Staphylococcus aureus colonization with food allergy occurs independently of eczema severity. *J. Allergy Clin. Immunol.* **2019**, *144*, 494–503. [CrossRef] [PubMed]
86. Nakamura, Y.; Oscherwitz, J.; Cease, K.B.; Chan, S.M.; Muñoz-Planillo, R.; Hasegawa, M.; Villaruz, A.E.; Cheung, G.Y.C.; McGavin, M.J.; Travers, J.B.; et al. Staphylococcus δ-toxin induces allergic skin disease by activating mast cells. *Nature* **2013**, *503*, 397–401. [CrossRef] [PubMed]

87. Yang, P.-C.; Xing, Z.; Berin, C.M.; Soderholm, J.D.; Feng, B.-S.; Wu, L.; Yeh, C. TIM-4 expressed by mucosal dendritic cells plays a critical role in food antigen-specific Th2 differentiation and intestinal allergy. *Gastroenterology* **2007**, *133*, 1522–1533. [CrossRef] [PubMed]
88. Salem, I.; Ramser, A.; Isham, N.; Ghannoum, M.A. The gut microbiome as a major regulator of the gut-skin axis. *Front. Microbiol.* **2018**, *9*, 1459. [CrossRef]
89. Lee, S.-Y.; Lee, E.; Park, Y.M.; Hong, S.J. Microbiome in the gut-skin axis in atopic dermatitis. *Allergy Asthma Immunol. Res.* **2018**, *10*, 354–362. [CrossRef]
90. O'Neill, C.A.; Monteleone, G.; McLaughlin, J.T.; Paus, R. The gut-skin axis in health and disease: A paradigm with therapeutic implications. *BioEssays* **2016**, *38*, 1167–1176. [CrossRef]
91. Esposito, S.; Isidori, C.; Pacitto, A.; Salvatori, C.; Sensi, L.; Frati, F.; Di Cara, G.; Marcucci, F. Epicutaneous immunotherapy in rhino-conjunctivitis and food allergies: A review of the literature. *J. Transl. Med.* **2018**, *16*, 329. [CrossRef] [PubMed]
92. Perdijk, O.; Marsland, B.J. The microbiome: Toward preventing allergies and asthma by nutritional intervention. *Curr. Opin. Immunol.* **2019**, *60*, 10–18. [CrossRef] [PubMed]
93. Penders, J.; Thijs, C.; van den Brandt, P.A.; Kummeling, I.; Snijders, B.; Stelma, F.; Adams, H.; van Ree, R.; Stobberingh, E.E. Gut microbiota composition and development of atopic manifestations in infancy: The KOALA Birth Cohort Study. *Gut* **2007**, *56*, 661–667. [CrossRef]
94. Björkstén, B.; Naaber, P.; Sepp, E.; Mikelsaar, M. The intestinal microflora in allergic Estonian and Swedish 2-year-old children. *Clin. Exp. Allergy* **1999**, *29*, 342–346. [CrossRef] [PubMed]
95. Björkstén, B.; Sepp, E.; Julge, K.; Voor, T.; Mikelsaar, M. Allergy development and the intestinal microflora during the first year of life. *J. Allergy Clin. Immunol.* **2001**, *108*, 516–520. [CrossRef] [PubMed]
96. Yap, G.C.; Loo, E.X.L.; Aw, M.; Lu, Q.; Shek, L.P.C.; Lee, B.W. Molecular analysis of infant fecal microbiota in an Asian at-risk cohort-correlates with infant and childhood eczema. *BMC Res. Notes* **2014**, *7*, 166. [CrossRef] [PubMed]
97. Sepp, E.; Julge, K.; Mikelsaar, M.; Björkstén, B. Intestinal microbiota and immunoglobulin E responses in 5-year-old Estonian children. *Clin. Exp. Allergy* **2005**, *35*, 1141–1146. [CrossRef] [PubMed]
98. Adlerberth, I.; Strachan, D.P.; Matricardi, P.M.; Ahrné, S.; Orfei, L.; Åberg, N.; Perkin, M.R.; Tripodi, S.; Hesselmar, B.; Saalman, R.; et al. Gut microbiota and development of atopic eczema in 3 European birth cohorts. *J. Allergy Clin. Immunol.* **2007**, *120*, 343–350. [CrossRef]
99. Kalliomäki, M.; Kirjavainen, P.; Eerola, E.; Kero, P.; Salminen, S.; Isolauri, E. Distinct patterns of neonatal gut microflora in infants in whom atopy was and was not developing. *J. Allergy Clin. Immunol.* **2001**, *107*, 129–134. [CrossRef]
100. Watanabe, S.; Narisawa, Y.; Arase, S.; Okamatsu, H.; Ikenaga, T.; Tajiri, Y.; Kumemura, M. Differences in fecal microflora between patients with atopic dermatitis and healthy control subjects. *J. Allergy Clin. Immunol.* **2003**, *111*, 587–591. [CrossRef]
101. Johansson, M.A.; Sjögren, Y.M.; Persson, J.O.; Nilsson, C.; Sverremark-Ekström, E. Early colonization with a group of Lactobacilli decreases the risk for allergy at five years of age despite allergic heredity. *PLoS ONE* **2011**, *6*, e23031. [CrossRef] [PubMed]
102. Kendler, M.; Uter, W.; Rueffer, A.; Shimshoni, R.; Jecht, E. Comparison of fecal microflora in children with atopic eczema/dermatitis syndrome according to IgE sensitization to food. *Pediatr. Allergy Immunol.* **2006**, *17*, 141–147. [CrossRef] [PubMed]
103. Gore, C.; Munro, K.; Lay, C.; Bibiloni, R.; Morris, J.; Woodcock, A.; Custovic, A.; Tannock, G.W. Bifidobacterium pseudocatenulatum is associated with atopic eczema: A nested case-control study investigating the fecal microbiota of infants. *J. Allergy Clin. Immunol.* **2008**, *121*, 135–140. [CrossRef] [PubMed]
104. Bisgaard, H.; Li, N.; Bonnelykke, K.; Chawes, B.L.K.; Skov, T.; Paludan-Müller, G.; Stokholm, J.; Smith, B.; Krogfelt, K.A. Reduced diversity of the intestinal microbiota during infancy is associated with increased risk of allergic disease at school age. *J. Allergy Clin. Immunol.* **2011**, *128*, 646–652. [CrossRef]
105. Storrø, O.; Øien, T.; Langsrud, O.; Rudi, K.; Dotterud, C.; Johnsen, R. Temporal variations in early gut microbial colonization are associated with allergen-specific immunoglobulin E but not atopic eczema at 2 years of age. *Clin. Exp. Allergy* **2011**, *41*, 1545–1554. [CrossRef]
106. Wang, M.; Karlsson, C.; Olsson, C.; Adlerberth, I.; Wold, A.E.; Strachan, D.P.; Martricardi, P.M.; Åberg, N.; Perkin, M.R.; Tripodi, S.; et al. Reduced diversity in the early fecal microbiota of infants with atopic eczema. *J. Allergy Clin. Immunol.* **2008**, *121*, 129–134. [CrossRef]

107. Ismail, I.H.; Oppedisano, F.; Joseph, S.J.; Boyle, R.J.; Licciardi, P.V.; Robins-Browne, R.M.; Tang, M.L.K. Reduced gut microbial diversity in early life is associated with later development of eczema but not atopy in high-risk infants. *Pediatr. Allergy Immunol.* **2012**, *23*, 674–681. [CrossRef]
108. Kalliomäki, M.; Salminen, S.; Arvilommi, H.; Kero, P.; Koskinen, P.; Isolauri, E. Probiotics in primary prevention of atopic disease: A randomised placebo-controlled trial. *Lancet* **2001**, *357*, 1076–1079. [CrossRef]
109. Kalliomäki, M.; Salminen, S.; Poussa, T.; Arvilommi, H.; Isolauri, E. Probiotics and prevention of atopic disease: 4-year follow-up of a randomised placebo-controlled trial. *Lancet* **2003**, *361*, 1869–1871. [CrossRef]
110. Rautava, S.; Kainonen, E.; Salminen, S.; Isolauri, E. Maternal probiotic supplementation during pregnancy and breast-feeding reduces the risk of eczema in the infant. *J. Allergy Clin. Immunol.* **2012**, *130*, 1355–1360. [CrossRef]
111. Enomoto, T.; Sowa, M.; Nishimori, K.; Shimazu, S.; Yoshida, A.; Yamada, K.; Furukawa, F.; Nakagawa, T.; Yanagisawa, N.; Iwabuchi, N.; et al. Effects of bifidobacterial supplementation to pregnant women and infants in the prevention of allergy development in infants and on fecal microbiota. *Allergol. Int.* **2014**, *63*, 575–585. [CrossRef] [PubMed]
112. Lau, S.; Gerhold, K.; Zimmermann, K.; Ockeloen, C.W.; Rossberg, S.; Wagner, P.; Sulser, C.; Bunikowski, R.; Witt, I.; Wauer, J.; et al. Oral application of bacterial lysate in infancy decreases the risk of atopic dermatitis in children with 1 atopic parent in a randomized, placebo-controlled trial. *J. Allergy Clin. Immunol.* **2012**, *129*, 1040–1047. [CrossRef] [PubMed]
113. Kim, J.Y.; Kwon, J.H.; Ahn, S.H.; Lee, S.I.; Han, Y.S.; Choi, Y.O.; Lee, S.Y.; Ahn, K.M.; Ji, G.E. Effect of probiotic mix (*Bifidobacterium bifidum, Bifidobacterium lactis, Lactobacillus acidophilus*) in the primary prevention of eczema: A double-blind, randomized, placebo-controlled trial. *Pediatr. Allergy Immunol.* **2010**, *21*, e386–e393. [CrossRef] [PubMed]
114. Bertelsen, R.J.; Brantsæter, A.L.; Magnus, M.C.; Haugen, M.; Myhre, R.; Jacobsson, B.; Longnecker, M.P.; Meltzer, H.M.; London, S.J. Probiotic milk consumption in pregnancy and infancy and subsequent childhood allergic diseases. *J. Allergy Clin. Immunol.* **2014**, *133*, 165–171.e8. [CrossRef]
115. Jeong, K.; Kim, M.; Jeon, S.A.; Kim, Y.H.; Lee, S. A randomized trial of Lactobacillus rhamnosus IDCC 3201 tyndallizate (RHT3201) for treating atopic dermatitis. *Pediatr. Allergy Immunol.* **2020**, *31*, 783–792. [CrossRef]
116. Abrahamsson, T.R.; Jakobsson, T.; Böttcher, M.F.; Fredrikson, M.; Jenmalm, M.C.; Björkstén, B.; Oldaeus, G. Probiotics in prevention of IgE-associated eczema: A double-blind, randomized, placebo-controlled trial. *J. Allergy Clin. Immunol.* **2007**, *119*, 1174–1180. [CrossRef]
117. Taylor, A.L.; Dunstan, J.A.; Prescott, S.L. Probiotic supplementation for the first 6 months of life fails to reduce the risk of atopic dermatitis and increases the risk of allergen sensitization in high-risk children: A randomized controlled trial. *J. Allergy Clin. Immunol.* **2007**, *119*, 184–191. [CrossRef]
118. Chapat, L.; Chemin, K.; Dubois, B.; Bourdet-Sicard, R.; Kaiserlian, D. Lactobacillus casei reduces CD8+ T cell-mediated skin inflammation. *Eur. J. Immunol.* **2004**, *34*, 2520–2528. [CrossRef]
119. Oyoshi, M.K.; Elkhal, A.; Scott, J.E.; Wurbel, M.; Hornick, J.L.; Campbell, J.J.; Geha, R.S. Epicutaneous challenge of orally immunized mice redirects antigen-specific gut-homing T cells to the skin. *J. Clin. Investig.* **2011**, *121*, 2210–2220. [CrossRef]
120. Chan, S.M.H.; Turcanu, V.; Stephens, A.C.; Fox, A.T.; Grieve, A.P.; Lack, G. Cutaneous lymphocyte antigen and α4β7 T-lymphocyte responses are associated with peanut allergy and tolerance in children. *Allergy Eur. J. Allergy Clin. Immunol.* **2012**, *67*, 336–342. [CrossRef]
121. Koh, A.; De Vadder, F.; Kovatcheva-Datchary, P.; Bäckhed, F. From dietary fiber to host physiology: Short-chain fatty acids as key bacterial metabolites. *Cell* **2016**, *165*, 1332–1345. [CrossRef] [PubMed]
122. Thorburn, A.N.; McKenzie, C.I.; Shen, S.; Stanley, D.; Macla, L.; Mason, L.J.; Roberts, L.K.; Wong, C.H.Y.; Shim, R.; Robert, R.; et al. Evidence that asthma is a developmental origin disease influenced by maternal diet and bacterial metabolites. *Nat. Commun.* **2015**, *6*, 7320. [CrossRef] [PubMed]
123. Thio, C.L.P.; Chi, P.Y.; Lai, A.C.Y.; Chang, Y.J. Regulation of type 2 innate lymphoid cell–dependent airway hyperreactivity by butyrate. *J. Allergy Clin. Immunol.* **2018**, *142*, 1867–1883.e12. [CrossRef] [PubMed]
124. Sanford, J.A.; Zhang, L.J.; Williams, M.R.; Gangoiti, J.A.; Huang, C.M.; Gallo, R.L. Inhibition of HDAC8 and HDAC9 by microbial short-chain fatty acids breaks immune tolerance of the epidermis to TLR ligands. *Sci. Immunol.* **2016**, *1*, eaah4609. [CrossRef]
125. Schwarz, A.; Bruhs, A.; Schwarz, T. The Short-Chain Fatty Acid Sodium Butyrate Functions as a Regulator of the Skin Immune System. *J. Investig. Dermatol.* **2017**, *137*, 855–864. [CrossRef]

126. Trompette, A.; Gollwitzer, E.S.; Yadava, K.; Sichelstiel, A.K.; Sprenger, N.; Ngom-Bru, C.; Blanchard, C.; Junt, T.; Nicod, L.P.; Harris, N.L.; et al. Gut microbiota metabolism of dietary fiber influences allergic airway disease and hematopoiesis. *Nat. Med.* **2014**, *20*, 159–166. [CrossRef]
127. Roduit, C.; Frei, R.; Ferstl, R.; Loeliger, S.; Westermann, P.; Rhyner, C.; Schiavi, E.; Barcik, W.; Rodriguez-Perez, N.; Wawrzyniak, M.; et al. High levels of butyrate and propionate in early life are associated with protection against atopy. *Allergy Eur. J. Allergy Clin. Immunol.* **2019**, *74*, 799–809. [CrossRef]
128. Folkerts, J.; Redegeld, F.; Folkerts, G.; Blokhuis, B.; van den Berg, M.P.M.; de Bruijn, M.J.W.; van IJcken, W.F.J.; Junt, T.; Tam, S.; Galli, S.J.; et al. Butyrate inhibits human mast cell activation via epigenetic regulation of FcεRI-mediated signaling. *Allergy* **2020**, *75*, 1966–1978. [CrossRef]
129. Xian, M.; Wawrzyniak, P.; Rückert, B.; Duan, S.; Meng, Y.; Sokolowska, M.; Globinska, A.; Zhang, L.; Akdis, M.; Akdis, C.A. Anionic surfactants and commercial detergents decrease tight junction barrier integrity in human keratinocytes. *J. Allergy Clin. Immunol.* **2016**, *138*, 890–893.e9. [CrossRef]

Publisher's Note: MDPI stays neutral with regard to jurisdictional claims in published maps and institutional affiliations.

nutrients MDPI

Article

A Food, a Bite, a Sip: How Much Allergen Is in That?

Melanie Kok [1], Astrid Compagner [1], Ina Panneman [2], Aline Sprikkelman [3] and Berber Vlieg-Boerstra [1,4,*]

1 Department of Nutrition and Dietetics, Hanze University of Applied Sciences, 9713 GZ Groningen, The Netherlands; melaniekok02@hotmail.com (M.K.); astrid_compagner@hotmail.com (A.C.)
2 The Food College, Practice for Diet, Nutrition, Sports and Food Technology, 8651 AE IJlst, The Netherlands; hetvoedingscollege@online.nl
3 Department of Pediatric Pulmonology and Allergy, University Medical Center Groningen, University of Groningen, 9700 RB Groningen, The Netherlands; a.b.sprikkelman@umcg.nl
4 Department of Paediatrics, OLVG Hospital, 1090 HM Amsterdam, The Netherlands
* Correspondence: b.vlieg-boerstra@olvg.nl; Tel.: +31-62-096-5612

Abstract: Detailed information about the amount of allergenic protein ingested by the patient prior to an allergic reaction yields valuable information for the diagnosis, guidance and management of food allergy. However, the exact amount of ingredients is often not declared on the label. In this study the feasibility was studied for estimating the amount of allergenic protein from milk, eggs, peanuts and hazelnuts in frequently consumed composite and non-composite foods and per bite or sip size in different age groups in the Netherlands. Foods containing milk, egg, peanut or hazelnut most frequently consumed were selected for the age groups 2–3, 4–6 and 19–30 years. If the label did not yield clear information, the amount of allergenic protein was estimated based on food labels. Bite or sip sizes were determined in these age groups in 30 different foods. The amount of allergenic protein could be estimated in 47/70 (67%) of composite foods, which was complex. Estimated protein content of milk, egg, peanut and hazelnut was 2–3 g for most foods but varied greatly from 3 to 8610 mg and may be below threshold levels of the patient. In contrast, a single bite or sip can contain a sufficient amount of allergenic protein to elicit an allergic reaction. Bite and sip sizes increased with age. In every day practice it is hard to obtain detailed and reliable information about the amount of allergenic protein incorporated in composite foods. We encourage companies to disclose the amount of common allergenic foods on their labels.

Keywords: diet history; food allergy; allergenic protein; thresholds; eliciting dose; bite size; cow's milk; hen's egg; peanut; hazelnut

Citation: Kok, M.; Compagner, A.; Panneman, I.; Sprikkelman, A.; Vlieg-Boerstra, B. A Food, a Bite, a Sip: How Much Allergen Is in That? *Nutrients* **2021**, *13*, 587. https://doi.org/10.3390/nu13020587

Academic Editor: Linda Monaci
Received: 13 January 2021
Accepted: 2 February 2021
Published: 10 February 2021

1. Introduction

For health care professionals who are involved in food allergies, detailed information about the amount of allergenic protein ingested by the patient prior to an allergic reaction yields valuable information for the diagnosis, guidance and management of the food allergy. This information can be obtained by a detailed allergy-focused diet history. One of the aims of the diet history in allergies is to identify suspected foods by linking symptoms to foods [1–4]. Information about the type and amount ingested which elicited allergic reactions, as well as the severity of the reaction, helps to estimate the clinical sensitivity of the patient and the risk for severe reactions. It is generally accepted that the higher the amount ingested, the more severe the expected allergenic reaction [1,5]. A low eliciting dose is assumed to reflect a higher clinical sensitivity [6] and may therefore be an indication for prescription of an epinephrine auto-injector [7]. This is important information for the design of the oral food challenge test for diagnosis and may lead to more stringent dietary advice. In contrast, in certain patients a high eliciting dose may lead to less stringent dietary advice [8].

The amount of allergenic protein ingested should be estimated by the health care professional from both the portion size ingested and from the amount of allergenic protein present in the food [2,3].

Often the patient has not consumed a full portion of a food but may only have taken one or a few bites or sips from the food until the reaction developed. Thus, in that case the health care professional should estimate the amount of allergenic protein ingested from the size of the bite or sip taken from the food. To our knowledge, no studies have been performed on the bite or sip sizes of foods containing allergenic ingredients.

Secondly, the health care professional should estimate the amount of allergenic protein in the food ingested [2,3]. However, in the majority of composite foods in which the protein content is delivered by several allergenic and non-allergenic ingredients, the exact amount of ingredients is not declared on the label.

The presence of fourteen major food allergens should be fully disclosed on the label in clear wordings according to European regulations. These are milk (including lactose) egg, soy, peanut, tree nuts, gluten, fish, shellfish, mollusks, celery, mustard, lupin, sesame and sulphite [9]. Risk-based approaches to managing allergens in foods are currently being developed by the food industry and regulatory authorities to support food-allergic consumers to avoid ingesting their problem food [10,11]. In non-composite foods or foods having only one protein source, the amount of allergenic ingredients can be derived from the label, e.g., milk contains 3.5% protein from cow's milk. However, in composite foods most labels do not yield information on the amount of allergenic ingredients unless explicitly stated (e.g., Nutella contains 13% hazelnuts). Thus, most foods lack these data which does not allow the physician or dietitian to accurately estimate the amount of allergenic protein ingested prior to an allergic reaction.

Oral food challenges are the preferred test to establish the diagnosis of food allergy [1,2,4,12]. During oral food challenges, the suspected food is administered to the patient in incrementing amounts with 15–20 minutes time intervals in an open, single-blind or double-blind fashion. Inter-individual thresholds to food allergens widely differ between patients, for reasons not yet fully understood. Patients may react to tiny amounts, such as crumbs of peanut or egg, or to higher doses up to full portions of the allergenic food. Therefore, 6 to 8 dose incremental scales in oral food challenges range from 1 mg protein to more than 4 g protein of the allergenic food, reflecting a full portion size [1,2,12]. Information about the clinical sensitivity of the patient is important for the design of the oral food challenge. Reactions to small amounts in history require increased safety measures during oral food challenges, such as selection of the challenge setting and a lower starting dose [12].

The oral food challenge yields information about the threshold, i.e., the amount of allergenic food eliciting symptoms, as well as the severity of symptoms, although it is recognized that threshold levels in oral food challenges in a clinical setting may be different from threshold levels in everyday life and may not be reproducible [13]. It was recently shown that co-factors such as lack of sleep and physical exercise significantly decrease threshold levels [14].

Once the threshold dose in an oral food challenge is established, insight in the amount of allergenic protein in foods could allow patients with mild symptoms and a high threshold level to expand their diets with foods containing small amounts well below their threshold levels in the absence of known co-factors. However, lack of this information does not allow the dietitian or patient to select foods with allergenic protein below their thresholds to expand the diet of the patient. So far, a more practical approach has been chosen, for example in patients who have passed a baked milk or baked egg challenge. These patients are advised to introduce foods with milk or egg listed as the third ingredient on the label or further down the list [15]. Alternatively, recipes are provided by dietitians to cook or bake their own products with the tolerated amounts of protein incorporated in the recipe [15].

The aims of this study were (1) to study if it is feasible to estimate the amount of allergenic protein from milk, egg, peanut and hazelnut in frequently consumed composite and non-composite foods per portion, per100 g food, and per bite or sip size in different age groups in the Netherlands, and (2) to discuss why it is important to have detailed information of the amount of allergenic protein in foods in the diagnosis and management of food allergies.

2. Materials and Methods

2.1. The Amount of Allergenic Protein in Foods

Based on the Dutch National Food Consumption Survey 2011, the most frequently consumed foods were selected for the assessment of the amount of allergenic protein [16,17]. Foods containing milk, egg, peanut or hazelnut consumed by 1% or more of consumers in the age groups 2–3 years, 4–6 years and 19–30 years were selected. Subsequently, the amount of allergenic protein in the selected foods was estimated according to an algorithm (Figure 1), including different methods.

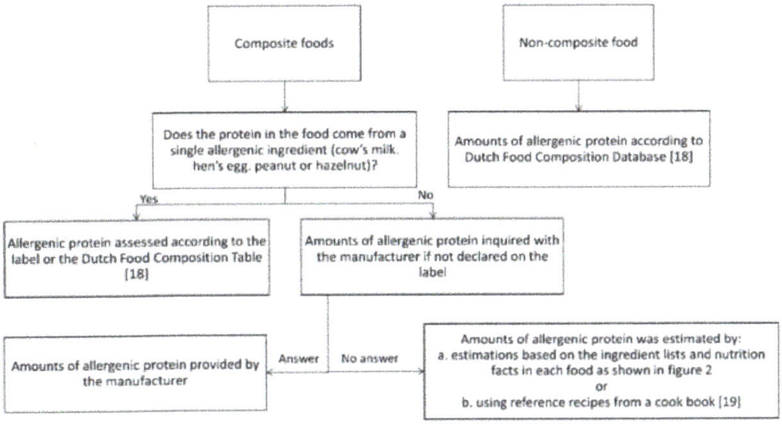

Figure 1. Assessment of the amount of allergenic protein [18,19].

In this study we defined non-composite foods as foods with only one ingredient or having only one ingredient yielding protein. We defined composite foods as foods with multiple ingredients yielding protein.

For non-composite foods, data were derived from the Dutch Food Composition Database 2011 (NEVO) [18] or food labels. For composite foods, the labels were checked for declaration of the amount of allergenic ingredients. If this was not declared, the manufacturer was contacted. When the required information was not provided by the manufacturer, the amount of allergenic protein was, as a non-validated method, estimated by calculations based on the ingredients lists and nutrition facts as follows (Figure 2):

First, it is a given fact that the ingredients on the label are listed in descending order according to their predominance by weight. Second, the nutrition facts (protein, fat, carbohydrates, energy) for each ingredient were relisted per 100 g. Third, the nutrition facts per 100 g were complete when the amount of ingredients was specified on the label (e.g., 13 g of hazelnut, indicated in yellow in Figure 2). Fourth, the amount of the other ingredients was estimated by trial and error until, fifth, the sum of the macronutrients of the ingredients approximated the nutrition facts on the label as closely as possible. Finally, if this method was not feasible, the amount of estimated allergenic protein was based on reference recipes from a Dutch cookbook [19].

Nutella							
Nutritional value Nutella / 100 g		Nutritional value according to label				Result calculation	
		Protein (g) / 100 g	Carbohydrates (g) / 100 g	Fat (g) / 100 g	Kcal / 100 g	Cow's milk (g) / 100 gram	Hazelnut (g) / 100 gram
		6.0	57.6	31.6	546	1.8	3.2 (2.3 + 0.9)
Nutrional value of ingredients / 100 g	Ingredients	Protein (g) / 100 g	Carbohydrates (g) / 100 g	Fat (g) / 100 g	Kcal / 100 g	Reference	
	Sugar	0.0	100.0	0.0	400	[18]	
	Vegetable oil	0.0	0.0	100.0	900		
	Hazelnuts	14.0	6.0	69.0	717	[18]	
	Low fat cacao	19.8	10.8	24.5	236		
	Skimmed milk powder	35.0	50.0	1.0	349		
	Whey powder	85.0	10.0	5.0	425		
	Lecithins	10.0	35.0	50.0	620		
	Vanilla						
Nutritional value per ingredient / 100 g Nutella	Ingredients	Protein (g) / 100 g	Carbohydrates (g) / 100 g	Fat (g) / 100 g	Kcal / 100 g	Ingredients (g)	
	Sugar	0.0	46.0	0.0	184	46.0	
	Vegetable oil	0.0	0.0	20.0	180	20.0	
	Hazelnuts	1.8	0.8	9.0	93	13.0	
	Low fat cacao	1.5	7.4	1.8	18	7.4	
	Skimmed milk powder	2.3	3.3	0.1	23	6.6	
	Whey powder	0.9	0.1	0.1	4	1.0	
	Lecithins	0.1	0.2	0.3	3	0.5	
	Vanilla					0.5	
	Total	6.6	57.8	31.3	505	95.0	

Procedure
1. Listed in descending order
2. Nutritional value of ingredients / 100 gram
3. Amount specified on label
4. Estimation of remaining ingredients by trial and error until best of totals
5. Total nutritional value according to label
6. Result: Estimated amount of allergenic protein

Figure 2. Estimation of the amount of allergen from the label [18].

Following these assessments, manufacturers were contacted to verify the results of the assessment of the amount of allergenic protein and were requested to comment on our findings.

Results were compared with ED10 and ED50 values for milk, egg, peanut and hazelnut as established in Dutch children and adults by Blom et al. and Klemans et al. [20,21]. ED10 and ED50 is the amount of allergenic protein to which, respectively, 10% and 50% of the allergic subjects react with objective symptoms.

2.2. *Assessment of Bite and Sip Sizes in Different Age Groups*

2.2.1. Selection of Foods

For the three age groups, the top 1% of the most frequently consumed foods containing milk, egg, peanut or hazelnut, as established by the National Food Consumption Survey [16,17], were selected and were allocated into food groups. The four foods most frequently used from each of the food groups were selected for the assessment of bite and sip sizes.

2.2.2. Study Population and Measurements of Bites and Sips

Healthy 2 to 3-year-old children from a preschool, 4 to 6-year-old children from two primary schools, and 19 to 30-year-old students in a nutrition and dietetics faculty were included. Study participants with a food allergy or other conditions that could affect the food intake were excluded.

In the 2–3-year-old children, foods were administered to the children for a bite or sip one by one while playing games. The 4–6-year-old children were asked to take a bite or sip without any instruction and without emphasis on this task to mimic regular bite and sip sizes as closely as possible. Each food was tested in 2–19 individuals in each age group. Each child received a maximum of eight foods. The adults were informed about the purpose of the study and were asked to take a single bite or sip of the food. The adults were asked to test all foods. The food was weighed before and after every bite or sip.

2.2.3. Pilot Study

A pilot study was performed prior to the study at the preschool and in one of the primary schools to test the feasibility of children taking bites or sips. The following essential findings were included in our methods: (1) to keep the attention of the children, all the foods were displayed on site to speed up the process; (2) to ensure a good appetite, the study was performed just before lunch or dinner time; and (3) to imitate the natural meal setting. The study was performed in subgroups of 4–6-year-old children.

2.2.4. Statistics

The results of the study were processed in SPSS. For each food the median intake was calculated in the different age groups and compared using the Mann-Whitney test, as well as the differences in intake between men and women in each age category.

3. Results

3.1. *The Amount of Allergenic Protein in Foods*

Ninety-seven foods were selected: 27 non-composite foods for which the amount of allergenic protein was determined using the Dutch NEVO Database [18] or the label (Table 1), and 70 composite foods (Table 2).

Table 1. Amount of estimated allergenic protein in most frequently consumed non-composite foods in mg or ml per portion, mg or ml per 100 g and mg or ml per median bite or sip size in different age groups.

Food	Composite or Non-Composite Food	Amount of Protein (mg or ml/Portion)	Amount of Protein (mg or ml)/100g)	Amount of Protein (mg or ml)/Median Bite or Sip Size 2–3 Years	Amount of Protein (mg or ml)/Median Bite or Sip Size 4–6 years	Amount of Protein (mg or ml)/Median Bite or Sip Size 19–30 years)
			COW'S MILK			
			Cheese			
Cottage cheese [18]	Non-composite food	1680	11,200	100	100	290
Goat cheese fresh [18]	Non-composite food	2010	13,400	120	120	350
Cheese spread 20+ [18]	Non-composite food	2550	17,000	150	150	440
Brie 60+ [18]	Non-composite food	3400	17,000	Nd	Nd	Nd
Goat cheese hard [18]	Non-composite food	4480	22,400	250	250	760
Gouda cheese 48+ [18]	Non-composite food	4560	22,800	250	250	780
Gouda cheese 20+ (low-fat) [18]	Non-composite food	6840	34,200	380	380	1160
			Milk, Milk Products, Milk Replacers and Ice Cream			
Coffee creamer, powder, low-fat [18]	Non-composite food	50	2000	Nd	Nd	Nd
Coffee creamer full fat [18]	Non-composite food	50	8100	Nd	Nd	Nd
Whipping cream [18]	Non-composite food	230	2300	Nd	Nd	Nd
Crème fraiche [18]	Non-composite food	330	2200	Nd	Nd	Nd
Crème fraiche demi [18]	Non-composite food	450	3000	Nd	Nd	Nd
Sour cream [18]	Non-composite food	470	3100	Nd	Nd	Nd
Fromage frais full fat 8.2% [18]	Non-composite food	1420	7100	Nd	Nd	Nd
Fromage frais low-fat 0.5% [18]	Non-composite food	2020	10,100	610	610	1370
Fromage frais half fat 4.6% [18]	Non-composite food	2300	11,500	690	690	1560
Nutrilon 2 Infant Formula (Nutricia)	Non-composite food	2800	1400	Nd	Nd	Nd

Table 1. *Cont.*

Food	Composite or Non-Composite Food	Amount of Protein (mg or ml/Portion)	Amount of Protein (mg or ml)/100g)	Amount of Protein (mg or ml)/Median Bite or Sip Size 2–3 Years	Amount of Protein (mg or ml)/Median Bite or Sip Size 4–6 years	Amount of Protein (mg or ml)/Median Bite or Sip Size 19–30 years)
3.5% Full fat yoghurt [18]	Non-composite food	5550	3700	220	220	500
Yoghurt, low-fat 0.3% [18]	Non-composite food	6150	4100	250	250	560
Yoghurt, low-fat 1.5% [18]	Non-composite food	6750	4500	270	270	610
Buttermilk [18]	Non-composite food	7500	3000	Nd	Nd	Nd
Whole Milk 3.5% [18]	Non-composite food	8250	3300	83	264	1056
Semi skimmed milk 1.5% [18]	Non-composite food	8500	3400	85	274	1088
Skimmed milk 0.1% [18]	Non-composite food	9250	3700	93	296	1184
			Fat, Oil and Sauce			
Butter, salted [18]	Non-composite food	40	700	<10	<10	<10
			HEN'S EGG			
			Egg			
Boiled egg [18]	Non-composite food	6200	12,300	308	615	1476
			PEANUT			
			Spread			
Peanut butter (Calvé)	Non-composite food	3200	21,420	210	190	560

mg, milligram; ml, milliliter; g, gram; Nd, no data.

For these 70 composite foods, 37 different food manufacturers and two supermarket chains were contacted by telephone and email. Only four different manufacturers provided the required data for four foods.

The amount of allergenic protein of the remaining 66 foods was estimated by the method depicted in Figure 2. The amount of allergenic protein could be estimated in 47/70 (67%) of the composite foods and are listed in Table 2. In 15/47 (32%) of the included composite foods, at least one allergenic ingredient was quantified on the label (e.g., Nutella, 13% hazelnut). For 19/70 (27%) of the composite foods, it was unfeasible to assess the amount of allergenic protein because the nutritional value of the main ingredients could not be estimated. These foods were excluded from further analyses.

Five of the 35 manufacturers responded when verifying these results: three confirmed that the estimated amounts were correct for margarine, filled milk chocolate bar with hazelnuts and hazelnut chocolate bar. Two confirmed that the estimated amounts were incorrect, namely for beef salad and tortellini. According to the manufacturer, beef salad contained 1 g of egg protein per portion instead of 0.5 g according to our estimation. For tortellini, the content of egg protein was 1.19 g per portion instead of 0.91 g per portion. The remaining 30 manufacturers either did not respond or responded but did not confirm or reject the amounts estimated and indicated that they were not willing to share the amount of allergenic protein of their products.

It was found that the actual or estimated amounts of allergenic protein varied widely in foods, and as expected, was highest in non-composite foods (Table 1). Of the non-composite foods with milk, the highest amounts of milk protein per portion were found in skimmed milk 0.1%, semi-skimmed milk 1.5%, whole milk 3.5%, buttermilk, and low-fat Gouda cheese: 9250 mg, 8500 mg, 8250 mg, 7500 mg, and 6840 mg, respectively. The lowest amounts of milk protein per portion were observed for whipping cream, coffee creamer and butter: 230 mg, 50 mg, and 40 mg, respectively. Peanut butter yielded 3200 mg peanut protein per portion.

Table 2. Amount of estimated allergenic protein in most frequently consumed composite foods in mg or ml per portion, mg or ml per 100 g and mg or ml per median bite or sip size in different age groups.

Food	Composite or Non-Composite Food	Amount of Protein (mg or ml/Portion)	Amount of Protein (mg or ml)/100g)	Amount of Protein (mg or ml)/Median Bite or Sip Size 2–3 Years	Amount of Protein (mg or ml)/Median Bite or Sip Size 4–6 years	Amount of Protein (mg or ml)/Median Bite or Sip Size 19–30 years)
COW'S MILK						
Bread and Crackers						
Currant bread **(Jumbo)	Composite food	590	1700	70	70	100
White bread (Jumbo *)	Composite food	770	1700	30	30	100
Spread						
Chocolate hazelnut spread (Nutella)	Composite food	470	3160	30	30	80
Cake and Biscuits						
Filled Biscuit (Biscuit fourré **) (Jumbo *)	Composite food	8.75	175	<10	<10	<10
Syrup waffle (Jumbo *)	Composite food	70	180	<10	<10	<10
Waffle (Jumbo *)	Composite food	90	180	<10	<10	20
Penny waffle (Jumbo *)	Composite food	117	780	<10	<10	20
Eclair with whipped cream filling (Roomsoesje **) (Jumbo *)	Composite food	190	1550	Nd	Nd	Nd
Apple flan and crumble topping (Jumbo *)	Composite food	430	430	Nd	Nd	Nd
Cake [19]	Composite food	550	1830	70	70	160
Cheesecake with fromage frais (Dr. Oetker *)	Composite food	2000–6000	2000–6000	Nd	Nd	Nd
Vegetables						
Creamed spinach frozen (Iglo)	Composite food	600	1250	50	250	130
Milk, Milk Products, Milk Replacers and Ice Cream						
Ice cream dairy, Cornetto Classic (Ola)	Composite food	882	1470	Nd	Nd	Nd

Table 2. *Cont.*

Food	Composite or Non-Composite Food	Amount of Protein (mg or ml/Portion)	Amount of Protein (mg or ml)/100g)	Amount of Protein (mg or ml)/Median Bite or Sip Size 2–3 Years	Amount of Protein (mg or ml)/Median Bite or Sip Size 4–6 years	Amount of Protein (mg or ml)/Median Bite or Sip Size 19–30 years)
Vanilla custard full fat (Friesland Campina *)	Composite food	3600	2400	140	140	460
Ice cream dairy, cream based (Hertog)	Composite food	3690	2460	Nd	Nd	Nd
Baby Porridge vanilla (Pyjama–papje **) (Nestlé)	Composite food	3840	1920	Nd	Nd	Nd
Composite Meals						
Infant jarred food: Lasagna with vegetables (Olvarit)	Composite food	1100	550	Nd	Nd	Nd
Pancakes) [19]	Composite food	1491	2130	90	60	Nd
Soup						
Chinese Tomato soup, canned (Unox)	Composite food	110	40	Nd	Nd	<10
Sweets and Chocolate						
Foam sweets banana flavor (Bananen schuimpjes **) (Haribo)	Composite food	6	120*	Nd	Nd	Nd
Fudge Caramel Vanilla (Lonka)	Composite food	80	1575	Nd	Nd	Nd
Chocolate bar with hazelnuts (Verkade)	Composite food	180	3510	100	70	210
Filled milk chocolate bar with hazelnuts (BonBon Bloc Praliné milk **) (Cote d'Or)	Composite food	650	4310	130	090	260
Belgium chocolate (Zeevruchten bonbon **) (Isaura)	Composite food	650	4320	130	90	260

Table 2. *Cont.*

Food	Composite or Non-Composite Food	Amount of Protein (mg or ml/Portion)	Amount of Protein (mg or ml)/100g)	Amount of Protein (mg or ml)/Median Bite or Sip Size 2–3 Years	Amount of Protein (mg or ml)/Median Bite or Sip Size 4–6 years	Amount of Protein (mg or ml)/Median Bite or Sip Size 19–30 years)
Fat, Oil and Sauce						
Low-fat margarine (Gouda's Glorie *)	Composite food	4	80*	<10	<10	<10
Tzatziki (Remia)	Composite food	110	740	Nd	Nd	30
Gravy, powdered (Knorr)	Composite food	110	700	Nd	Nd	Nd
Bechamel sauce [19]	Composite food	1300	8680	Nd	Nd	Nd
Meat and Poultry						
Hamburger (Mora)	Composite food	780	1060	20	20	100
Ragout, beef, canned (Unox)	Composite food	175	350	Nd	Nd	Nd
HEN'S EGG						
Bread and Crackers						
Round toast (Bolletje *)	Composite food	5	50	Nd	Nd	Nd
Round toast, whole wheat (Bolletje *)	Composite food	30	300	Nd	Nd	Nd
Cake and Biscuits						
Syrup waffle (Jumbo *)	Composite food	3	6	<10	<10	<10
Penny waffle (Jumbo *)	Composite food	3	33	<10	<10	Nd
Marzipan and chocolate cake (Mergpijpje **) (Jumbo *)	Composite food	100	980	Nd	Nd	Nd
Chocolate coated marsh mellow (Schuimzoenen **) (Buys)	Composite food	140	1400	Nd	Nd	Nd
Eclair with whipped cream filling (Roomsoesje **) (Jumbo *)	Composite food	220	1720	Nd	Nd	Nd
Cake [19]	Composite food	390	1300	50	50	120

Table 2. *Cont.*

Food	Composite or Non-Composite Food	Amount of Protein (mg or ml/Portion)	Amount of Protein (mg or ml)/100g)	Amount of Protein (mg or ml)/Median Bite or Sip Size 2–3 Years	Amount of Protein (mg or ml)/Median Bite or Sip Size 4–6 years	Amount of Protein (mg or ml)/Median Bite or Sip Size 19–30 years)
Dutch sponge cake (Eierkoek **) (AH)	Composite food	660	2210	90	90	200
Waffle (Jumbo *)	Composite food	1970	3940	118	158	355
Pasta						
Ravioli (Grand Italia)	Composite food	1190	2380	Nd	Nd	Nd
Tortellini (Grand Italia)	Composite food	1190	2380	Nd	Nd	Nd
Milk, Milk Products, Milk Replacers and Ice Cream						
Ice cream dairy, Cornetto Classic (Ola)	Composite food	3	5	Nd	Nd	Nd
Snacks, Meals						
Beef salad (Johma)	Composite food	50	30	Nd	Nd	Nd
Composite Dishes						
Egg roll, chicken and ham (Mora)	Composite food	1100	630	Nd	Nd	Nd
Pancakes [19]	Composite food	8610	12,300	492	369	Nd
Fat, Oil and Sauce						
Salad cream 25% oil (Slasaus **) (Remia)	Composite food	20	150	Nd	Nd	Nd
Sauce for chips 35% oil (Fritessaus **) (Remia)	Composite food	50	334	<10	20	10
Mayonnaise (Remia)	Composite food	130	840	<10	500	30
PEANUT						
Cake and Biscuits						
Peanut cookie (Jumbo *)	Composite food	630	6300	63	63	Nd
Snack Food						

Table 2. *Cont.*

Food	Composite or Non-Composite Food	Amount of Protein (mg or ml/Portion)	Amount of Protein (mg or ml)/100g)	Amount of Protein (mg or ml)/Median Bite or Sip Size 2–3 Years	Amount of Protein (mg or ml)/Median Bite or Sip Size 4–6 years	Amount of Protein (mg or ml)/Median Bite or Sip Size 19–30 years)
Japanese rice cracker mix with peanuts (Davis)	Composite food	600	3020	Nd	Nd	Nd
Coated peanuts (Duyvis)	Composite food	2720	13,610	Nd	Nd	Nd
Sweets and Chocolate						
Candy bar, Snickers	Composite food	1210	6050	180	180	Nd
M&M's, chocolate with peanut	Composite food	1160	5800	Nd	Nd	Nd
Peanuts coated with milk chocolate (Chocopinda's **) (Jumbo *)	Composite food	1260	6300	Nd	Nd	Nd
Fats, Oils and Savory Sauces						
Peanut sauce (Wijko)	Composite food	1820	12,100	120	610	420
HAZELNUT						
Spread						
Chocolate hazelnut spread (Nutella)	Composite food	270	1820	20	20	50
Cake and Biscuits						
Penny waffle (Jumbo *)	Composite food	7	70	<10	<10	Nd
Cereals						
Muesli (Jumbo *)	Composite food	60	140	Nd	Nd	<10
Milk, Milk Products, Milk Replacers and Ice Cream						
Ice cream dairy, Cornetto Classic (Ola)	Composite food	168	280	Nd	Nd	Nd
Sugar, Sweets, Chocolate and Sweet Sauces						
Belgium chocolate (Zeevruchten bonbon **) (Isaura)	Composite food	380	2520	80	580	Nd

* Amounts of protein (Dutch Food Composition Database 2011 (NEVO, 2011)) [18] per bite are derived from the bite sizes of wheat bread and reference portion sizes for spread, such as Nutella and peanut butter; ** Amounts of protein (NEVO, 2011) [18] are derived from a comparable food in the NEVO table; mg, milligram; ml, milliliter; g, gram; Nd, no data.

Of the composite foods with milk (Table 2), the highest amounts of milk protein per portion were found in cheesecake, baby porridge, ice cream, vanilla custard and pancakes up to 6000 mg, 3840 mg, 3690 mg, 3600 mg, and 1491 mg, respectively. Relatively low amounts of milk protein were found in low-fat margarine, foam sweets banana flavor and filled biscuit, and creamer: 4 mg, 6 mg, 8 mg, and 75 mg of milk protein per portion, respectively.

Of the composite foods with egg (Table 2), the highest amounts of eggprotein per portion were found in pancakes, waffles, ravioli and tortellini: 8610 mg, 1970 mg, 1190 mg and 1190 mg, respectively. Low amounts of egg protein were found in round toast, syrup waffles, penny waffles and Cornetto ice cream: 5 mg, 3 mg, 3 mg, and 3 mg of egg protein per portion, respectively.

Of the composite foods with peanut (Table 2), the amounts of peanut protein per portion varied between 630 mg (peanut cookie) and 2720 mg (coated peanuts).

Of the composite foods with hazelnut (Table 2), the amounts of hazelnut protein per portion varied between 380 mg (Belgium bonbon) and 7 mg (penny waffle).

3.2. Comparison of the Estimated Amount of Allergenic Protein to ED10 and ED50

The estimated amount of allergenic milk-, egg-, peanut- and hazelnut-protein per portion were compared to the ED10 and ED50 in children for objective symptoms as established in a Dutch population by Blom et al. [20]. Additionally, the estimated amount of peanut protein per portion was compared to the ED10 and ED50 in children and adults for objective and subjective symptoms by Klemans et al. [21].

- Milk

None of the selected composite or non-composite foods contained less estimated milk protein per portion than the ED10 (4.24 mg), except low-fat margarine. Nine foods contained less estimated milk protein per portion than the ED 50 (156 mg). The other foods contained higher estimated amounts.

- Egg

Four foods contained less estimated egg protein per portion than the ED10 (5.82 mg), while ten foods contained less estimated egg protein per portion than the ED50 (199 mg). All the other foods contained more estimated egg protein per portion.

- Peanut

None of the foods contained less estimated peanut protein per portion than the ED10 (4.42 mg) by Blom [20], the ED10 in children (18.6 mg) and in adults (13.7 mg) by Klemans [21] or the ED50 in children (67,3 mg) by Blom [20]. Only one food contained less peanut protein than the ED50 in adults (821 mg) by Klemans [21].

- Hazelnut

None of the foods contained less estimated hazelnut protein per portion than the ED10 (1.38 mg) by Blom [20]. Two foods contained less estimated hazelnut protein per portion than the ED50 (80.6 mg) by Blom [20].

3.3. Assessment of Bite and Sip Sizes in Different Age Groups

Thirty foods were selected: 17 foods for the children 2–3 years of age, 17 foods for the children 4–6 years of age and 19 foods for the adults 19–30 years of age. Several foods were selected for more than one age group. In total, 71 participants were included (41 male (57.7%); 30 females (42.3%)).

In the 2–3-year-old age group, 18 toddlers participated (8 males, 10 females; median 3 years of age). A maximum of eight foods were tested in each child. The sip and bite sizes were close for all foods, except for soft drinks in which the largest median sip size was observed (11 mL) in contrast to milk, in which the smallest median intake was measured (2.5 mL). A large range in bite sizes was measured for pancakes (3.00–9.00 g). There were no significant differences between boys and girls in bite or sip sizes of the selected foods.

In the age group of 4–6-year-old children, 39 children were included (28 males, 11 females; median 4 years). For each child a maximum of eight foods were tested (Table 3). The sip and bite sizes were similar for all foods. Between boys and girls, there was only a significant difference in bite size for pancakes ($p = 0.008$) [19].

Table 3. Median (IQR) bite and sip sizes in different age groups in grams or milliliters.

Type of Food	Median Weight or Volume in Gram or ml * (IQR) *	Number of Participants
	2–3 years of age	
Wheat bread	2.00 (2.00–3.00)	9
Chocolate hazelnut spread, Nutella *	0.86	n.d.
Peanut butter *	0.86	n.d.
Low-fat margarine *	0.29	n.d.
Cheese *	1.14	n.d.
Milk **	2.50 (2.00–4.25)	6
Boiled egg	2.50 (2.00–4.50)	4
Biscuit (Maria biscuit)	1.00 (1.00–1.00)	8
Crisps (Hamka's)	0.50 (0.25–0.60)	5
Currant bread	3.00 (1.25–5.50)	4
Pancake	4.00 (3.00–9.00)	5
Snickers	2.50 (1.88–3.25)	6
Soft drink (Taksi) **	11.00 (8.00–12.25)	6
Vanilla custard	6.00 (−)	3
Cake batter	2.00 (1.00–3.50)	5
Fried egg	3.00 (1.00–3.00)	6
Creamed spinach	4.00 (−)	2
Chicken nuggets	2.00 (1.75–3.50)	6
Mayonnaise	1.00 (1.00–2.00)	6
Milk chocolate	2.50 (1.00–4.00)	10
Muffin	3.00 (1.00–4.50)	9
	4–6 years of age	
Wheat bread	2.00 (1.00–2.00)	17
Hazelnut spread *	0.86	n.d.
Peanut butter *	0.86	n.d.
Low-fat margarine *	0.29	n.d.
Cheese *	1.14	n.d.
Milk **	8.00 (4.00–18.00)	11
Boiled egg	5.00 (3.00–7.75)	16
Biscuit	1.00 (1.00–2.00)	19
Crisps (Hamka's chips)	0.25 (0.1875–0.5425)	18
Currant bread	3.00 (2.00–4.00)	15
Pancake	3.00 (2.00–5.25)	18
Snickers	3.00 (2.00–4.00)	16
Soft drink (Taksi)	8.00 (4.00–12.00)	17

Table 3. *Cont.*

Type of Food	Median Weight or Volume in Gram or ml * (IQR) *	Number of Participants
Vanilla custard	6.00 (3.00–8.50)	13
Cake batter	1.00 (0.50–1.00)	7
Fried egg	3.00 (2.00–4.00)	11
Cream spinach	5.00 (4.00–7.00)	7
Chicken nuggets	2.00 (1.50–4.00)	9
Mayonnaise	0.50 (0.50–1.00)	9
Milk chocolate	2.00 (2.00–3.00)	17
Muffin	4.00 (2.00–6.25)	14
	19–30 years	
Milk **	32.00 (24.00–58.75)	14
Hardboiled egg	12.00 (7.75–16.50)	14
Crisps (Hamka's chips)	2.00 (1.00–2.00)	11
Soft drink (Taksi)	36.00 (30.70–43.75)	14
Fried egg	6.50 (5.00–9.25)	14
Cream spinach	10.00 (8.50–13.00)	14
Muffin	9.00 (6.00–11.25)	14
Soft drink (Rivella)	31.50 (27.25–50.75)	14
Canned Soup	9.00 (8.00–10.00)	12
Cappuccino	23.50 (14.50–37.50)	14
White bread	6.00 (3.75–8.25)	14
Hazelnut spread *	2.57	n.d.
Peanut butter *	2.57	n.d.
Low-fat margarine *	0.86	n.d.
Cheese *	3.43	n.d.
Yogurt with muesli (Cruesli)	19.00 (14.00–21.25)	14
Potato croquette	5.50 (4.75–8.25)	14
Schnitzel	9.00 (8.50–10.00)	14
Sate sauce	3.50 (2.00–6.00)	14
Nougat	5.50 (4.75–7.25)	14
Belgium chocolate	6.00 (3.75–10.00)	14
Spiced biscuit	3.00 (3.00–5.00)	14

n.d.: not done. * Amounts of protein (NEVO, 2011) (12) are derived from the bite sizes of wheat bread and reference portion sizes for spreads, such as Nutella and peanut butter (17) ** 1 g is considered equivalent to 1 milliliter.

In the age group 19–30 years, fourteen adults were included (5 males, 9 females; median age 22 years). There were large differences in the bite and sip sizes for the foods within this age group (Table 3). The largest interquartile range (IQR) was observed for milk (24.00–58.75 mL). Between men and women, significant differences in bite and sip sizes were found for eleven other foods and drinks (p values 0.001 to 0.042).

3.3.1. Differences between the Different Age Groups

In the 2–3-year-old children, the bite and sip sizes for wheat bread and mayonnaise were significantly larger than those in 4–6-year-old children (p = 0.029 and p = 0.012), whereas 4–6-year-old children had significantly larger sip sizes for milk (p = 0.010).

For the foods tested in all age groups, the bite and sip sizes of the 19–30-year-old adults were significantly larger compared to the 2–3-year-old children and 4–6-year-old children for all foods.

3.3.2. Amount of Protein per Bite or Sip

In Tables 1 and 2 it is shown that a single bite or sip of many foods contains sufficient amounts of allergenic protein to elicit an allergic reaction.

4. Discussion

This study aimed to test the feasibility of estimating the amount of allergenic protein in frequently consumed foods, as estimated per 100 g, per portion and per bite and sip sizes in different age groups for improved diagnosis and management of food allergies. For non-composite foods, the amount of allergenic protein could easily be derived from the label or food composition tables, as all the protein was delivered by one allergenic ingredient. For composite foods we showed that it is very hard to obtain detailed information about the amount of allergenic protein. Through a lot of effort, the allergenic protein content of many composite foods may at best be estimated, however, true amounts of allergenic protein values may be somehow different.

For most composite foods depicted in Table 2, the amounts of allergenic protein are estimates rather than established amounts of protein. Based on our estimations, most composite foods contain less than 2–3 g of allergenic protein, except a few products that contain higher amounts such as cheesecake, baby porridge, ice cream, vanilla custard and pancakes.

For 19/70 (27%) of the composite foods, the amount of allergenic protein could not be estimated due to lack of detailed information on the label or lack of information from the manufacturer.

Four manufacturers provided us the required protein amounts of four foods (6%). For the other 47/70 (67%) composite foods, the amount of allergenic protein could be estimated using a non-validated method. Only 2/70 (3%) of the composite foods fully disclosed the amount of all allergenic ingredients (Nutella and peanut butter).

If detailed data on the amount of allergenic ingredients were provided by the manufacturer, these data could increase the quality of the diagnosis and management of patients with food allergies. First, full disclosure of not only the presence of allergens but also the amount of allergenic protein on the labels would allow for quantitative risk assessment in diet history and diagnosis. The health care professional could better assess how much allergenic protein is ingested prior to the allergic reaction. These data would help to establish the sensitivity of the patient for the allergenic food in question and, if necessary, sustain decision-making on extra safety measures during oral food challenges in highly sensitive patients. In addition, patients having reacted (severely) to small amounts in history will receive stringent dietary avoidance advice.

Second, detailed data on the amount of allergenic ingredients would support the decision-making for epinephrine auto-injector prescription in clinically sensitive patients.

Third, detailed data on the amount of allergenic ingredients would enable individually tailored dietary advice in food allergic patients. It would allow patients who had a mild reaction to try higher doses in oral food challenge tests to safely introduce foods with small amounts into their diet well below their thresholds. This could include the use of foods containing precautionary labeling, such as "may contain traces of ... " [8] or foods containing small amounts of an allergen listed in the ingredient list. This information would allow the dietitian or patient to select foods with allergenic protein below their thresholds to expand the diet. Patients tolerating baked milk and baked egg could introduce products with baked egg and milk into their diets. Lastly, milk and egg ladders, practical tools developed by dietitians to introduce foods at home [22], could be adapted based on the amount of allergen listed on the label.

However, due to a lack of detailed information on the label, the estimations in this study on the amount of allergenic protein do not allow for detailed advice in everyday clinical practice. We therefore encourage companies to disclose the amount of common allergenic foods on their labels.

We do not expect that improved quantitative risk assessment in dietary history will precisely predict the threshold dose during an oral challenge, as exposure in daily life occurs in uncontrolled conditions. Previous studies have shown a lack of correlation between the severity of reactions at home and thresholds or severity during oral food challenges [6,13,14]. This may be due to an incomplete diet history with a lack of data on the exact amount of ingested allergenic food, because thresholds in oral food challenges vary over time and because of co-factors such as sleep deprivation and physical exercise [13,14].

For clinical relevance we compared the amount of estimated allergenic protein per portion with the ED10 and ED50 for allergens as established by several authors [20,21] in the Dutch population. For milk, only one food contained less estimated allergenic protein per portion when comparing the amount of milk in foods to the ED10 for milk; for egg this was found for four foods, while for peanut and hazelnut none of the foods contained less than the ED10 [20,21]. This means that, theoretically, all the other foods will provoke allergic reactions in allergic patients who belong to the 10% most clinically sensitive individuals.

When comparing allergenic protein contents with the ED50, nine foods containing milk, ten foods containing egg, no foods containing peanut and two foods containing hazelnut had allergenic amounts per portion below the ED50 in children. Thus, when taking a diet history, inconsistent reactions may be explained by low amounts of allergenic protein in food, except for peanut. This is even more true when only one of a few bites or sips are taken from the food instead of a full portion.

This study showed a clear difference in bite or sip sizes between the different age groups. As expected, the median bite size increases with age. This difference was significant when comparing the adults with the two younger age groups. We also observed 19–30-year-old men having a larger bite and sip sizes for all types of food than women. We showed that a single bite or sip of many foods contain sufficient amounts of allergenic protein to elicit an allergic reaction.

In the literature, there are some data available about bite and sip sizes, however most studies are performed in adults and in obese versus lean study participants to study the effects of portion size and hunger or satiety on bite or sip sizes [23–26]. Bite sizes increase with increasing portion size [23,25] and body mass index [24,26]. In our study, regular portions were administered and none of the study participants were extremely obese. Bite and sip size in men were larger than in women [23–25], as was found in our study. Our data on bite and sip sizes in both children and adults may further enhance the assessment of the intake of allergenic protein consumed.

Our study has several limitations. We used a non-validated method to assess the amounts of allergenic protein in composite foods. We are not aware of a validated approach, and quantitative measurement of allergenic protein in foods was beyond the scope of this study. We also did not use a power analysis to determine the number of study participants for bite and sip sizes. Therefore, the study participants we used to study bite and sip sizes may not be representative for the different age groups. Bite and sip sizes should therefore be interpreted with caution.

5. Conclusions

In conclusion, in everyday practice it is hard to obtain detailed and reliable information about the amount of allergenic protein incorporated in composite foods. Yet, this study provides some insight into the estimated amount of allergenic protein in a large number of commonly consumed foods per portion, per 100 g and also per bite or sip size in the Netherlands, as established using a non-validated method. Diet history may be inconsistent in less sensitive patients as they may not react to foods containing low amounts of allergenic protein. In contrast, a single bite or sip can contain sufficient amount of an allergenic protein to elicit an allergic reaction. Bite and sip sizes increased with age. Disclosure of the amount of allergenic protein on labels would improve quantitative risk assessment in diet history in clinical practice, as well as dietary management of food allergies by allowing patients to introduce foods into their diet that they tolerate.

Author Contributions: Conceptualization, B.V.-B.; methodology, B.V.-B., A.S. and I.P.; formal analysis, M.K., A.C. and I.P.; writing, M.K. and A.C.; B.V.-B., A.S. and I.P., supervision, B.V.-B. and I.P. All authors have read and agreed to the published version of the manuscript.

Funding: This research received no external funding.

Institutional Review Board Statement: The study was conducted according to the guidelines of the Declaration of Helsinki, and ethical review and approval were waived for this study, according to the Dutch Law.

Informed Consent Statement: Informed consent was obtained from the parents of the children and by the adult study participants. Additionally, children and adults had to be willing to take a bite or sip from the selected foods voluntarily. The schools and preschool gave their consent for this study, too.

Data Availability Statement: The data presented in this study are available on request from the corresponding author. A data sharing agreement will be requested.

Acknowledgments: The authors would like to thank the study participants, the schools and the preschool for their willingness to participate in the involved studies. We thank Isabel Skypala for her critical comments on the manuscript.

Conflicts of Interest: The authors declare no conflict of interest.

References

1. Boyce, J.A.; Assa'ad, A.; Burks, W.A.; Jones, S.M.; Sampson, H.A.; Wood, R.A.; Plaut, M.; Cooper, S.F.; Fenton, M.J.; Arshad, S.H.; et al. Guidelines for the diagnosis and management of food allergy in the United States: Report of the NIAID-sponsored expert panel. *J. Allergy Clin. Immunol.* **2010**, *126*, S1–S58. [CrossRef] [PubMed]
2. Muraro, A.; Werfel, T.; Hoffmann-Sommergruber, K.; Roberts, G.A.; Beyer, K.; Bindslev-Jensen, C.; Cardona, V.; Dubois, A.; Dutoit, G.; Eigenmann, P.; et al. EAACI food allergy and anaphylaxis guidelines: Diagnosis and management of food allergy. *Allergy* **2014**, *69*, 1008–1025. [CrossRef] [PubMed]
3. Skypala, I.J.; Venter, C.; Meyer, R.; de Jong, N.W.; Fox, A.T.; Groetch, M.; Elberink, J.N.O.; Sprikkelman, A.B.; Diamandi, L.; Diamandi, L.; et al. The development of a standardised diet history tool to support the diagnosis of food allergy. *Clin. Transl. Allergy* **2015**, *5*, 7. [CrossRef] [PubMed]
4. Eigenmann, P.; Atanaskovic-Markovic, M.; Hourihane, J.O.; Lack, G.; Lau, S.; Matricardi, P.M.; Muraro, A.; Baranova, L.N.; Nieto, A.; Papadopoulos, N.G.; et al. Testing children for allergies: why, how, who and when. *Pediatr. Allergy Immunol.* **2013**, *24*, 195–209. [CrossRef]
5. Turner, P.J.; Baumert, J.L.; Beyer, K.; Boyle, R.J.; Chan, C.-H.; Clark, A.; Crevel, R.W.R.; DunnGalvin, A.; Fernández-Rivas, M.; Gowland, M.H.; et al. Can we identify patients at risk of life-threatening allergic reactions to food? *Allergy* **2016**, *71*, 1241–1255. [CrossRef] [PubMed]
6. Van der Zee, T.; Dubois, A.; Kerkhof, M.; van der Heide, S.; Vlieg-Boerstra, B.J. The eliciting dose of peanut in double-blind, placebo-controlled food challenges decreases with increasing age and specific IgE in children and young adults. *J. Allergy Clin. Immunol.* **2011**, *128*, 1031–1036. [CrossRef] [PubMed]
7. Dubois, A.E.J.; Turner, P.J.; Hourihane, J.; Ballmer-Weber, B.; Beyer, K.; Chan, C.H.; Gowland, M.H.; O'Hagan, S.; Regent, L.; Remington, B.; et al. How does dose impact on the severity of food-induced allergic reactions, and can this improve risk assessment for allergenic foods? Report from an ILSI Europe Food Allergy Task Force Expert Group and Workshop. *Allergy* **2018**, *73*, 1383–1392. [CrossRef] [PubMed]
8. Brough, H.A.; Turner, P.J.; Wright, T.; Fox, A.T.; Taylor, S.L.; Warner, J.O.; Lack, G. Dietary management of peanut and tree nut allergy: What exactly should patients avoid? *Clin. Exp. Allergy* **2015**, *45*, 859–871. [CrossRef] [PubMed]
9. The European Parliament and the Council of the European Union. REGULATION (EU) No 1169/2011 OF THE EUROPEAN PARLIAMENT AND OF THE COUNCIL of 25 October 2011. Available online: http://eur-lex.europa.eu/LexUriServ/LexUriServ.do?uri=OJ:L:2011:304:0018:0063:EN:PDF (accessed on 11 October 2014).

10. Mills, E.N.C.; Adel-Patient, K.; Bernard, H.; De Loose, M.; Gillard, N.; Huet, A.-C.; Larré, C.; Nitride, C.; Pilolli, R Tranquet, O.; et al. Detection and quantification of allergens in foods and minimum eliciting doses in food-allergic individuals (ThRAll). *J. AOAC Int.* **2019**, *102*, 1346–1353. [CrossRef] [PubMed]
11. Raposo, A.; Pérez, E.; Tinoco de Faria, C.; Carrascosa, C. Allergen management as a key issue in food safety. In *Food Safety and Protection*; Ravishankar Rai, V., Bai, J.A., Eds.; CRC Press: Boca Raton, FL, USA, 2017; pp. 195–242.
12. Sampson, H.A.; Van Wijk, R.G.; Bindslev-Jensen, C.; Sicherer, S.H.; Teuber, S.S.; Burks, A.W.; Dubois, A.E.; Beyer, K.; Eigenmann P.A.; Spergel, J.M.; et al. Standardizing double-blind, placebo-controlled oral food challenges: American Academy of Allergy Asthma & Immunology—European Academy of Allergy and Clinical Immunology PRACTALL consensus report. *J. Allergy Clin Immunol.* **2012**, *130*, 1260–1274. [CrossRef] [PubMed]
13. Glaumann, S.; Nopp, A.; Johansson, S.G.O.; Borres, M.P.; Nilsson, C. Oral peanut challenge identifies an allergy but the peanut allergen threshold sensitivity is not reproducible. *PLoS ONE* **2013**, *8*, e53465. [CrossRef]
14. Dua, S.; Ruiz-Garcia, M.; Bond, S.; Durham, S.R.; Kimber, I.; Mills, C.; Roberts, G.; Skypala, I.; Wason, J.; Ewan, P.; et al. Effect of sleep deprivation and exercise on reaction threshold in adults with peanut allergy: A randomized controlled study. *J. Allergy Clin Immunol.* **2019**, *144*, 1584–1594.e2. [CrossRef] [PubMed]
15. Leonard, S.A.; Caubet, J.C.; Kim, J.S.; Groetch, M.; Nowak-Węgrzyn, A. Baked milk- and egg-containing diet in the management of milk and egg allergy. *J. Allergy Clin. Immunol. Pract.* **2015**, *3*, 13–23. [CrossRef] [PubMed]
16. Van Rossum, C.T.M.; Fransen, H.P.; Verkaik-Kloosterman, J.; Buurma-Rethans, E.J.M.; Ocke, M.C. RIVM Rapport Dutch National Food Consumption Survey 2007—2010: Diet of Children and Adults Aged 7 to 69 Years. 2011. Available online https://www.rivm.nl/bibliotheek/rapporten/350050006.pdf (accessed on 15 September 2013).
17. National Institute for Public Health and the Environment. Food Consumption Survey of Dutch Children (Users and Total Population) Age 2–6 Years. Available online: https://www.rivm.nl/en/dutch-national-food-consumption-survey/overview-surveys/young-children-2-6-years-2005-2006 (accessed on 15 September 2013).
18. National Institute for Public Health and the Environment. NEVO-online. Available online: http://nevo-online.rivm.nl/ (accessed on 30 January 2013).
19. Blommestein, I.; van Eijndhoven, A.; van Mill, J. *Kook Ook*, 8th ed.; VBK Media: Utrecht, The Netherlands, 2008.
20. Blom, W.M.; Vlieg-Boerstra, B.J.; Kruizinga, A.G.; Van Der Heide, S.; Houben, G.F.; Dubois, A.E.J. Threshold dose distributions for 5 major allergenic foods in children. *J. Allergy Clin. Immunol.* **2013**, *131*, 172–179. [CrossRef] [PubMed]
21. Klemans, R.J.B.; Blom, W.M.; van Erp, F.C.; Masthoff, L.J.N.; Rubingh, C.M.; van der Ent, C.K.; Bruijnzeel-Koomen, C.A.F.M.; Houben, G.F.; A Pasmans, S.G.M.; Meijer, Y.; et al. Objective eliciting doses of peanut-allergic adults and children can be combined for risk assessment purposes. *Clin. Exp. Allergy* **2015**, *45*, 1237–1244. [CrossRef] [PubMed]
22. Athanasopoulou, P.; Deligianni, E.; Dean, T.; Dewey, A.; Venter, C. Use of baked milk challenges and milk ladders in clinical practice: a worldwide survey of healthcare professionals. *Clin. Exp. Allergy* **2017**, *47*, 430–434. [CrossRef] [PubMed]
23. Burger, K.S.; Fisher, J.O.; Johnson, S.L. Mechanisms behind the portion size effect: Visibility and bite size. *Obesity* **2011**, *19*, 546–551. [CrossRef] [PubMed]
24. Hill, S.W.; McCutcheon, M.B. Contributions of obesity, gender, hunger, food preference, and body size to bite size, bite speed, and rate of eating. *Appetite* **1984**, *5*, 73–83. [CrossRef]
25. Forde, C.; Van Kuijk, N.; Thaler, T.; De Graaf, C.; Martin, N. Oral processing characteristics of solid savoury meal components, and relationship with food composition, sensory attributes and expected satiation. *Appetite* **2013**, *60*, 208–219. [CrossRef] [PubMed]
26. Hübel, R.; Laessle, R.; Lehrke, S.; Jass, J. Laboratory measurement of cumulative food intake in humans: Results on reliability. *Appetite* **2006**, *46*, 57–62. [CrossRef] [PubMed]

nutrients

MDPI

Article

Homemade Food Allergen Extracts for Use in Skin Prick Tests in the Diagnosis of IgE-Mediated Food Allergy: A Good Alternative in the Absence of Commercially Available Extracts?

Severina Terlouw [1,2,*], Frank E. van Boven [2], Monika Borsboom-van Zonneveld [1], Catharina de Graaf-in 't Veld [1], Marloes E. van Splunter [2], Paul L. A. van Daele [2], Maurits S. van Maaren [2], Marco W. J. Schreurs [3] and Nicolette W. de Jong [2,4]

1 Department of Allergology, Albert Schweitzer Hospital, 3331 LZ Zwijndrecht, The Netherlands; m.vanzonneveld@asz.nl (M.B.-v.Z.); c.degraaf@asz.nl (C.d.G.-i.'t.V.)
2 Internal Medicine, Allergology & Clinical Immunology, Erasmus MC, University Medical Centre, 3015 GD Rotterdam, The Netherlands; f.boven@erasmusmc.nl (F.E.v.B.); m.e.vansplunter@erasmusmc.nl (M.E.v.S.); p.l.a.vandaele@erasmusmc.nl (P.L.A.v.D.); m.vanmaaren@erasmusmc.nl (M.S.v.M.); n.w.dejong@erasmusmc.nl (N.W.d.J.)
3 Laboratory Medical Immunology, Department of Immunology, Erasmus Medical Centre, 3015 GD Rotterdam, The Netherlands; m.schreurs@erasmusmc.nl
4 Department of Pediatric Allergology, Sophia Children Hospital, Erasmus Medical Centre, 3015 GD Rotterdam, The Netherlands
* Correspondence: s.terlouw@erasmusmc.nl; Tel.: +31-640149222

Citation: Terlouw, S.; van Boven, F.E.; Borsboom-van Zonneveld, M.; de Graaf-in 't Veld, C.; van Splunter, M.E.; van Daele, P.L.A.; van Maaren, M.S.; Schreurs, M.W.J.; de Jong, N.W. Homemade Food Allergen Extracts for Use in Skin Prick Tests in the Diagnosis of IgE-Mediated Food Allergy: A Good Alternative in the Absence of Commercially Available Extracts? *Nutrients* **2022**, *14*, 475. https://doi.org/10.3390/nu14030475

Academic Editor: Margarida Castell Escuer

Received: 17 December 2021
Accepted: 19 January 2022
Published: 21 January 2022

Abstract: Introduction: The skin prick test (SPT) is the first step in the diagnosis of an immunoglobulin E (IgE)-mediated food allergy. The availability of commercial food allergen extracts is very limited, resulting in a need for alternative extraction methods of food allergens. The objective of this study was to compare the SPT results of homemade food allergen extracts with commercially available extracts. Methods: Adult patients with a suspected food allergy were included. Food allergen-specific symptoms were scored using a questionnaire. SPTs were performed with homemade and commercially available extracts (ALK-Abelló, Kopenhagen, Denmark) from almond, apple, hazelnut, peach, peanut, and walnut. Serum-specific IgE was measured with ISAC or ImmunoCAP™. Intra-class correlation coefficients (ICC) between the SPT results of both extract methods were calculated. The proportion of agreement with food allergen-specific symptoms was analyzed. Results: Fifty-four patients (mean age 36; range 19–69 years; female/male: 42/12) were included. The intra-class correlation coefficient (ICC) between the SPT results of both extract methods were strong for hazelnut 0.79 (n = 44) and walnut 0.78 (n = 31), moderate for apple 0.74 (n = 21) and peanut 0.66 (n = 28), and weak for almond 0.36 (n = 27) and peach 0.17 (n = 23). The proportion of agreement between SPT results and food allergen-specific symptoms was comparable for homemade and commercially available extracts, except for peach; 0.77 versus 0.36, respectively. Conclusion: In the diagnostic procedures to identify an IgE-mediated food allergy, homemade extracts from hazelnut and walnut appear to be a good alternative in the absence of commercially available food allergen extracts.

Keywords: diagnosis; extracts; food allergy; oral allergy syndrome; skin prick test; specific IgE

1. Introduction

The prevalence of food allergies in Europe is increasing rapidly. Currently, in adults, self-reported symptoms after ingesting different varieties of food are reported by 5.7–61.6% of people, and physician-diagnosed hypersensitivities are reported by 0.2–4.2% of people [1]. Diagnosing a suspected food allergy accurately is of great importance, both to prevent severe allergic reactions and to avoid unnecessary dietary restrictions caused by inaccurate diagnosis. The diagnosis of food allergy involves the use of skin prick tests (SPTs), allergen specific immunoglobulin E (sIgE), and oral food challenges (OFC) as a gold standard [2]. It is generally agreed that the core diagnostic step for type I mediated allergy, the SPT, should be further standardized, and further studies are necessary to define worldwide standards for allergen extracts [3]. In a recent EAACI position paper about in vivo diagnostic test allergens, the importance of reliable allergens was also stressed [4].

The evaluation of patients with a possible food allergy starts with an extensive food specific medical history and a physical examination. The focus should be on possible dietary triggers, the quantity and quality of the ingested food, possible facilitating co-factors around the time of the reaction (exercise, illness, use of medication), and the specific symptoms that led to the allergic reaction [5]. Knowledge of cross-reactivity within protein families would help to decide the ensuing pathway.

The next step in the diagnosis of a food allergy is measuring sensitization to the suspected food allergen by either performing an SPT with the suspected food allergen and/or measuring serum sIgE. SPTs are a quick, reliable, and cheap method to measure sensitization. Although the negative predictive value (NPV) of SPTs often reaches 90% or more [5], false negative SPTs may occur if the used extracts are not standardized or have insufficient quantities of the allergen. In commercially available extracts of fruits and vegetables, the proteins might be destroyed during the manufacturing process, e.g., heating, giving less reliable results [6]. Generally, SPTs with food allergens have high sensitivity but low specificity, and must be interpreted with caution [6], and neither SPT nor sIgE are sufficient to diagnose food allergies on their own [7]. Soares-Weiser et al. (2014) also concluded that SPT and sIgE appear to be sensitive but not specific enough for diagnosing IgE-mediated food allergy, although this may differ between foods [8]. The availability of commercial food allergen extracts is limited, which leads to a need for alternative methods for the extraction of food allergens. One of the alternatives for commercial extracts might be to prepare homemade (HM) extracts through standardized protocols, but the quality of these extracts is unknown. Thus, the objective of this study was to compare SPT results of HM food allergen extracts and commercially available extracts.

2. Materials and Methods

2.1. Study Population

Adult patients with a suspected food allergy for at least one food allergen, visiting the outpatient clinic of the Allergology Department of the Albert Schweitzer hospital, were asked to participate in this study. The suspicion of food allergy was based on the patient's clinical history and a physical examination by an allergist. All participants stopped their anti-histamines for at least 72 h before the SPT. Medical ethical approval was obtained for this study on 1 February 2018, trial number MEC-2017-486, NL61899.078.17. After patients signed an informed consent form, the inclusion took place from September 2018 until December 2020. A food -specific case history was conducted using an extensive food specific standardized questionnaire, which was filled out by the physician during the visit of the patient to the outpatient clinic. Symptoms were defined as the occurrence of oral itching, with or without angioedema of the lips and/or tongue (oral allergy syndrome [OAS]), respiratory symptoms, gastrointestinal symptoms (GI), and/or urticaria (skin symptoms). Inhalant allergies and concomitant medications used were reported. The PRACTALL score list was used to score symptoms and severity [2].

2.2. Skin Prick Tests

Based on the patient's clinical history and the extensive food specific standardized questionnaire, the allergist chose a maximum of 4 food allergens, including the ones that were suspected as causing the food allergy. The SPTs with the chosen food allergens were performed with HM extracts as well as commercial extracts, both containing the same food allergen, in the same patient, at the same time. The food allergen extracts that were used in this study, available for both HM and commercial uses, were: almond, apple, hazelnut, peach, peanut and walnut. The SPT was conducted on the volar surface of the forearm by application of one drop of the allergenic extract to the skin. Subsequently, the dermis was punctured with a disposable standardized skin test needle (ALK-Abelló, Kopenhagen, Denmark), as recommended in the established EAACI guidelines [4]. Dilution buffer (ALK-Abelló, Kopenhagen, Denmark (nr. 002)) was used as a negative control. Mean values of two histamine dihydrochloride 10mg/mL (ALK-Abelló, Kopenhagen, Denmark (nr. 001))-induced wheal sizes were used as a positive control. To avoid puncture technical bias, the same nurse performed all SPTs. SPT results were obtained after 15 min; the contours of the allergen-induced wheal were encircled with a fine-tip pen and transferred to a record sheet by means of translucent tape (ALK-Abelló, Kopenhagen, Denmark). We compared the results of the Histamine Equivalent Prick result (HEP/PAAMOST) [9,10] of SPTs with HM extracts and ALK extracts. In addition to HEP measurement, the allergen-induced mean wheal diameter was measured to decide on positive and negative results (negative <3 mm Ø) according to the EAACI international guidelines [11].

2.3. HM Food Allergen Extracts

The raw material for each tested HM food allergen extract was carefully screened to select the material that best represented the allergen. Nuts and peanuts were fresh, not roasted and not salted, and were bought separately. The raw material was homogenized mechanically, ground with a mortar, and defatted with ether in a Soxhlet, air-dried, and stored at −20 °C until further use. Fruit and vegetables were bought fresh, and after being homogenized in a food processor, pulp was immediately stored in small portions for single use at −20 °C [10]. Pre-treated material of nuts and peanuts was defrosted and tested in a 5% or 10% extract in PBS (negative control; ALK-Abelló, Kopenhagen, Denmark (nr. 1036472)) as described by de Jong et al. [12]. In all cases, the allergens that were tested using HM extracts were compared with commercially available extracts from ALK-Abelló, Kopenhagen, Denmark; almond *Prunus dulcis* 1:20 m/V (nr. 764), apple *Pyrus Malus* spp. 1:20 W/V (nr. 658), hazelnut *Corylus avellana* 1:100 g/V (nr. 761), peach *Prunus persica* 1:20 G/V (nr. 613), peanut *Arachis hypogaea* 1:20 G/V (nr. 762), and walnut *Juglans regia* 1:20 W/V (nr. 766)).

2.4. Serum-Specific IgE

Serum-specific IgE levels were evaluated with the ImmunoCAP™ ISAC multiplex test, when available, or with the regular ImmunoCAP™ monoplex test, according to the manufacturer's instructions (Thermo Fisher Scientific, Uppsala, Sweden). The result of ISAC multiplex sIgE was considered positive when ≥0.30 ISU. ImmunoCAP™ monoplex sIgE results were considered positive when ≥0.35 kU/L.

2.5. Statistical Analysis

Comparison of the SPT-HEP results of the HM and ALK extracts was performed by calculating the intra-class correlation coefficients (ICC) between the HEPs. These coefficients were considered very strong for ICC ≥ 0.9; strong for 0.75 ≤ ICC < 0.9; moderate for 0.5 ≤ ICC < 0.75; and weak for ICC < 0.5 [13]. SPT results were compared for almond, apple, hazelnut, peach, peanut, and walnut. We also compared the numbers of positive (≥3mm) SPT of HM and ALK extracts for the 6 food allergens. We compared the proportion of patients with a positive SPT of HM and of ALK using an exact binomial test. The agreement between qualitative SPT results (positive/negative) with symptoms per food allergen was assessed by calculating the proportion of patients with a positive SPT as well as any specific food allergen-related positive symptom, and a negative SPT with the absence of specific food allergen-related symptoms. Confidence intervals (CI) were calculated for these proportions, with a 0.05 level of significance. All calculations were performed by R (version 4.0.4 https://www.r-project.org, 16 December 2021).

Comparison of qualitative SPT results and sIgE results with food allergen-specific symptoms after consuming the specific food allergen was performed, and sensitivity, specificity, positive predictive value (PPV), negative predictive value (NPV), and the likelihood ratio were also reported.

3. Results

3.1. Study Population

Fifty-four adult patients (mean age 36; range 19-69 years) with a suspected food allergy were included. All participants reported one or more inhalant allergies: 40 (74%) to grass pollen, 51 (94%) to birch pollen, and 31 (57%) to house dust mites. Fifty participants (93%) reported OAS with or without GI symptoms, fourteen participants (26%) reported a skin reaction, and eighteen participants (33%) reported respiratory symptoms after ingestion of the suspected food allergen. Forty-six participants (85%) use any kind of anti-allergic medication. Of this group, forty-five (83%) use anti-histamines, seventeen (31%) use a nose spray, four (7%) use eye drops, thirteen (24%) use lung medication, and six (11%) of the participants need rescue medication (adrenalin). In Table 1, all patient characteristics are summarized.

Table 1. Patient characteristics.

	Patient Characteristics	
Numbers included	54	
Female/male	42	12
Mean age/range	36	19–69
	n	%
Inhalant allergy	54	100
Grass pollen	40	74
Birch pollen	51	94
House dust mite	31	57
Pets	33	61
Medication used	46	85
Anti-histamines	45	83
Corticosteroid nose spray	17	31
Eye drops	4	7
Lung medication	13	24
Adrenaline	6	11
Food allergy symptoms		
GI + OAS	50	93
Skin	14	26
Lung	18	33

n = number; GI = gastro-intestinal symptoms; OAS = oral allergy symptoms; Lung = respiratory symptoms; Skin = skin symptoms.

The total numbers of patients that ever experienced symptoms after ingestion of the specific food allergen were: 16/27 for almond, 17/21 for apple, 36/44 for hazelnut, 16/23 for peach, 11/28 for peanut, and 22/31 for walnut. In total, forty SPTs (23%) were performed in patients who experienced no symptoms at all after consumption of the specific food allergen. Sixteen SPTs (9%) were performed in patients who could not answer the question as to whether they experienced symptoms after consumption of a specific food allergen, because they were on a strict diet free from the food for a long time, caused by, e.g., a positive sIgE in the past during routine testing. The results of any symptoms ever experienced after consumption of the specific food allergen gathered from the questionnaire are shown in Table 2.

Table 2. Symptoms per food allergen.

Symptoms Per Food Allergen							
	Almond	**Apple**	**Hazelnut**	**Peach**	**Peanut**	**Walnut**	**Total**
n (%)	27 (16) *n* (%)	21 (12) *n* (%)	44 (25) *n* (%)	23 (13) *n* (%)	28 (16) *n* (%)	31 (18) *n* (%)	174 (100) *n* (%)
Consuming/ no symptoms	8 (30)	3 (14)	7 (16)	6 (26)	13 (46)	3 (10)	40 (23)
NA (strict diet)	3 (11)	1 (5)	1 (2)	1 (4)	4 (14)	6 (19)	16 (9)
Symptoms:							
GI/OAS	12 (44)	12 (57)	23 (52)	12 (52)	5 (18)	14 (45)	78 (45)
Skin	0	0	1 (2)	0	1 (4)	0	2 (1)
Lung	1 (4)	0	0	0	0	0	1 (1)
GI/OAS + Skin	1 (4)	1 (5)	2 (5)	1 (4)	0	1 (3)	6 (3)
GI/OAS + Lung	2 (7)	3 (14)	8 (18)	3 (13)	2 (7)	5 (16)	23 (13)
GI/OAS + Skin + Lung	0	1 (5)	2 (5)	0	2 (7)	2 (6)	7 (4)
Skin + Lung	0	0	0	0	1 (4)	0	1 (1)

SPT = skin prick test; *n* = number; NA = not applicable because of patient on a strict diet free from the food allergen; GI = gastro-intestinal symptoms; OAS = oral allergy syndrome; skin = skin symptoms; lung = respiratory symptoms.

3.2. Skin Prick Tests

One hundred and seventy-four SPTs were performed with the six included food allergens: 27 (16%) with almond, 21 (12%) with apple, 44 (25%) with hazelnut, 23 (13%) with peach, 28 (16%) with peanut, and 31 (18%) with walnut.

The mean HEP index with ALK food allergen extracts vs. HM extracts was; 0.96 vs. 0.51 for almond, 0.47 vs. 0.38 for apple, 1.40 vs. 1.61 for hazelnut, 0.11 vs. 0.83 for peach, 0.86 vs. 1.13 for peanut, and 0.42 vs. 0.39 for walnut, respectively. P-values for the comparison of the number of positive SPTs (≥ 3mm) were: 0.5 for almond, 0.5 for apple, 1.0 for hazelnut, <0.001 for peach, 0.63 for peanut, and 1.0 for walnut. The SPT-HEP results and ICC of the six food allergens are shown in Table 3.

Table 3. Skin prick test results per food allergen.

SPT Results Per Food Allergen								
		Almond	**Apple**	**Hazelnut**	**Peach**	**Peanut**	**Walnut**	**Total**
ALK	Positive ≥3 mm	24	19	41	4	21	14	123
	%	89	90	93	17	75	45	71
	Mean HEP index	0.96	0.47	1.40	0.11	0.86	0.42	
	Range HEP index	0–4.05	0–1.14	0–18.85	0–1.42	0–5.17	0–2.57	
HM	Positive ≥3 mm	22	17	42	20	23	14	138
	%	81	81	95	87	82	45	79
	Mean HEP index	0.51	0.38	1.61	0.83	1.13	0.39	
	Range HEP index	0–1.37	0–1.44	0–11.44	0–1.91	0–6.07	0–2.31	
ICC		0.36	0.74	0.79	0.17	0.66	0.78	
95% CI for ICC		0 to 0.65	0.47 to 0.89	0.65 to 0.88	0 to 0.49	0.39 to 0.82	0.59 to 0.89	
p-value HEP		0.03	<0.0001	<0.0001	0.015	<0.0001	<0.0001	
Strength of ICC		weak	moderate	strong	weak	moderate	strong	

SPT = Skin Prick Test; *n* = number; HEP = Histamine Equivalent Prick test; ALK = Allergy Laboratories Kopenhagen, Denmark; HM = homemade; ICC = intra-class correlation coefficient.

The differences in the SPT-HEP results with both extracts are also depicted in Figure 1.

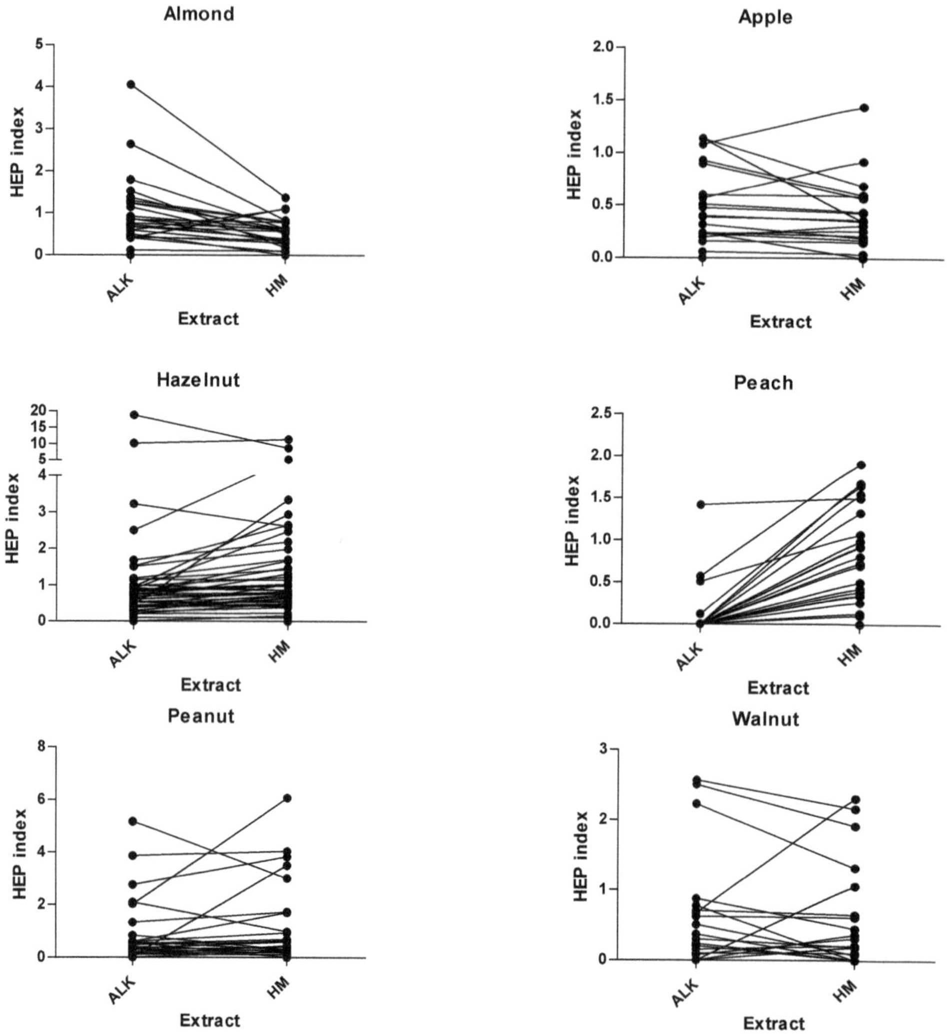

Figure 1. SPT-HEP results for ALK-HM regarding the 6 food allergens: almond, apple; hazelnut; figure peach; peanut; walnut.

The ICCs between the SPT-HEP results for both extract methods, HM and commercial, were all significant; 0.36 (weak) for almond, 0.74 (moderate) for apple, 0.79 (strong) for hazelnut, 0.17 (weak) for peach, 0.66 (moderate) for peanut, and 0.78 (strong) for walnut.

3.3. Proportion of Agreement of SPT-HEP Results with Symptoms

Sensitization in relation to food-specific symptoms (proportion of agreement) and the confidence interval (CI) for ALK vs. HM extracts was calculated: for almond, 0.75 (CI 0.58–0.92) and 0.67 (CI 0.48–0.86), respectively; apple, 0.75 (CI 0.56–0.94) and 0.75 (CI 0.56–0.94), respectively; hazelnut, 0.84 (CI 0.73–0.95) and 0.81 (CI 0.70–0.93), respectively; peach, 0.36 (CI 0.16–0.56) and 0.77 (CI 0.60–0.95), respectively; peanut, 0.50 (CI 0.30–0.70) and 0.54 (CI 0.34–0.74), respectively; and walnut, 0.52 (CI 0.32–0.72) and 0.56 (CI 0.37–0.75), respectively. The sensitization in relation to the symptoms and CI of all six food allergen extracts is shown in Figure 2.

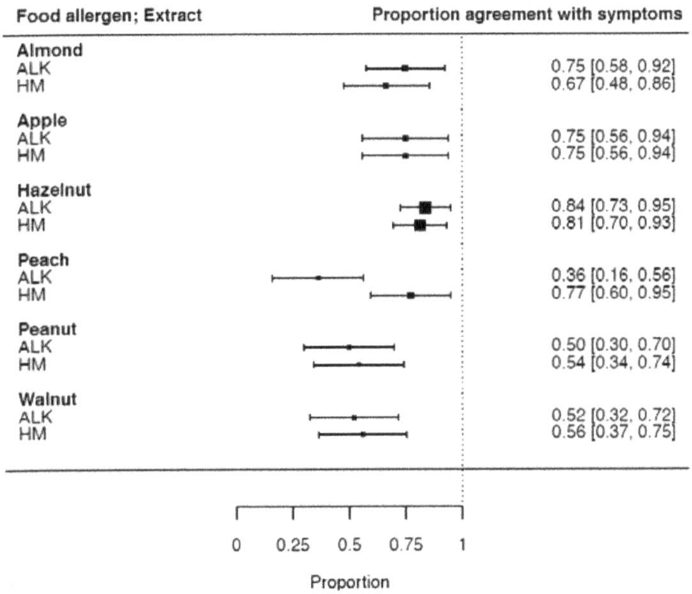

Figure 2. SPT-HEP results of ALK and HM extracts in relation to symptoms (proportion of agreement) and CI of 6 food allergens. ALK = Allergy Laboratories Kopenhagen, Denmark; HM = homemade.

3.4. Serum-Specific IgE Measurements

Serum-specific IgE measured by ImmunoCAP™ (monoplex, Thermo Fisher Scientific, Uppsala, Sweden) for almond was positive in 10/27 cases. Specific IgE measured by ISAC (multiplex, Thermo Fisher Scientific, Uppsala, Sweden) was positive in all 42 sera. Specific IgE was positive in 18/20 for Mal d1 (apple), 40/42 for Cor a1 (hazelnut), 20/22 for Pru p1 (peach), 19/28 for Ara h8 (peanut), and 3/30 for Jug r 1 (walnut). The median and range of all sIgE measurements are shown in Table 4.

Table 4. Serum specific IgE measurements.

Serum Specific IgE Measurements					
		n =	Positive *n* =	Mean ISU	Range ISU
Almond *		27	10	0.60 *	0.0–2.98 *
Apple	Mal d1	20	18	15.38	0–64.1
Hazelnut	Cor a1	42	40	7.42	0–31.2
	Cor a8	42	2	1.34	0–54.4
	Cor a9	42	2	0.30	0–9.83
	Cor a14	42	3	3.03	0–105.6
Peach	Pru p1	22	20	7.50	0–60.9
	Pru p3	22	1	0.23	0–4.67
Peanut	Ara h2	28	5	1.56	0–20.3
	Ara h6	28	5	1.15	0–14.4
	Ara h8	28	19	2.88	0–14.0
	Ara h9	28	3	3.39	0–8.29
Walnut	Jug r1	30	3	2.41	0–65
	Jug r3	30	0	0	0

n = number; * = measured by ImmunoCAP™ Thermo Fisher Scientific, Uppsala, Sweden

It appeared that other allergen components were only positive in a few cases; hazelnut Cor a8 and peach Pru p3 (both lipid transfer proteins [LTP]) were only positive in two and one patients, respectively, while major 2S albumins hazelnut Cor a 14 and peanut Ara h2 and Ara h6 were only positive in three, five and five cases, respectively. Proportion of agreement calculations were not feasible for these allergen components due to low power.

The proportion of agreement of specific IgE measurements in relation to symptoms and CI was calculated: for almond: 0.29 (CI 0.11–0.47); apple (Mal d1; PR10): 0.79 (CI 0.61–0.97); hazelnut (Cor a1; PR10): 0.80 (CI 0.68–0.93); peach (Pru p1; PR–10): 0.73 (CI 0.54–0.91); peanut (Ara h8; PR10): 0.71 (CI 0.53–0.89); and for walnut (Jug r1; 2S albumine): 0.21 (CI 0.05–0.37).

3.5. Accuracy of Sensitization Measurements in Relation to Reported Symptoms

Sensitivity and specificity measurements as well as the PPV, NPV and LR of SPT results in comparison to the reported symptoms were obtained (Table 5.) The mean sensitivity of SPT HM extracts and ALK extracts was 0.84 and 0.73, respectively. The mean specificity of SPT HM extracts and ALK extracts was 0.38 and 0.37, respectively.

Table 5. Accuracy of sensitization measurements in relation to reported symptoms.

Accuracy of Sensitization Measurements in Relation to Reported Symptoms							
Extract		**Sensitivity**	**Specificity**	**PPV**	**NPV**	**LR+**	**LR−**
Almond	ALK	1.00	0.25	0.73	1.00	1.33	0.00
	HM	0.88	0.25	0.70	0.50	1.17	0.50
	sIgE Almond	0.44	0.20	0.25	0.38	0.56	2.78
Apple	ALK	0.88	0.00	0.83	0.00	0.88	NA
	HM	0.82	0.33	0.88	0.25	1.24	0.53
	Mal d1	0.88	0.00	0.88	0.00	0.88	NA
Hazelnut	ALK	0.97	0.25	0.85	0.67	1.30	0.11
	HM	0.97	0.13	0.83	0.50	1.11	0.23
	Cor a1	0.82	0.50	0.97	0.13	1.64	0.36
Peach	ALK	0.19	0.83	0.75	0.28	1.13	0.98
	HM	0.94	0.33	0.79	0.67	1.41	0.19
	Pru p1	0.75	0.50	0.94	0.17	1.50	0.50
Peanut	ALK	0.82	0.23	0.47	0.60	1.06	0.79
	HM	0.91	0.23	0.50	0.75	1.18	0.39
	Ara h2	1.00	0.65	0.36	1.00	2.86	0.00
	Ara h8	0.53	0.71	0.82	0.38	1.85	0.66
Walnut	ALK	0.50	0.67	0.92	0.15	1.50	0.75
	HM	0.50	1.00	1.00	0.21	NA	0.50
	Jug r1	1.00	0.14	0.10	1.00	1.16	0.00

ALK = Allergy Laboratories Kopenhagen, Denmark; HM = homemade; PPV = positive predictive value; NPV = negative predictive value; LR+ = positive likelihood ratio; LR− = negative likelihood ratio; NA = not applicable.

4. Discussion

In this study, we compared SPT results for HM food allergen extracts with results for commercially available extracts, in patients with reported food-specific allergic symptoms, e.g., OAS, GI symptoms, skin symptoms, and/or respiratory symptoms, after ingestion of the suspected food. We performed SPTs in 54 patients, using both the HM extract and the commercially available extract of the same food allergen, within the same patient, at the same time. We found a strong correlation between both extract methods for hazelnut and walnut, moderate correlation for peanut and apple, and weak correlation for almond and peach. This indicates that these HM food allergen extracts are a good to moderate alternative in the absence of standardized commercially available extracts.

We found comparable sensitivity and specificity results for HM and ALK food allergen extracts. As expected, the mean sensitivity was high (0.84 and 0.73, respectively), but the specificity is considerably low for both extracts (0.38 and 0.37, respectively). Asero et al. described this low specificity earlier for food allergen extracts [14]. In particular, the hazelnut extract performed poorly in both extracts. One reason might be that we compared the results with doctor-diagnosed allergies without performing DBPCFCs. Another reason might be that most patients are sensitized to Cor a1, and consequently we lose these labile proteins during extraction. Interestingly, the sensitivity of the HM peach extract (0.94) performed very well, in contrast to the ALK extract, which has a sensitivity of 0.19. These results are in line with previous studies with fruit allergens, which point out that, in commercially available extracts of fruit and vegetables, the proteins may be destroyed during the manufacturing process [6]. A review by Foong and Santos in 2020 established higher sensitivity and specificity of SPTs with fresh fruit and vegetables, compared to commercial extracts, and acknowledged their importance in patients with pollen sensitization [15].

The considerable differences between SPT-HEP results of the HM and the commercially available extract of almond (mean HEP 0.51, range 0–1.37 vs. mean HEP 0.96, range 0–4.05, respectively) and the proportion of agreement for almond (0.67 vs. 0.75) must be seen in perspective. The sIgE measurements in relation to symptoms (proportion of agreement) and CI for almond (0.29 (CI 0.29–0.47)) were low. However, the perception of the patients with symptoms due to almond consumption can be argued; being sensitization to almond is often followed by a negative food challenge. In a cohort study by Arends et al., 189 almond challenges among a group of Dutch children were analyzed. A positive SPT with almond was found in 148 children (78%); 97/101 double blind placebo-controlled food challenges (DBPCFC) were negative [16].

In the 28 SPTs we performed with the peanut extract, we found 21 (75%) positive SPT($\geq$3mm) results with the commercially available extract vs. 23 (82%) with the HM extract (p-value 0.63). Thirteen patients (46%) could consume peanut without experiencing allergic symptoms. These outcomes were established in earlier studies; in 2005, Mortz et al. investigated the prevalence of peanut sensitization in an unselected population of adolescents and evaluated the clinical relevance of a positive sIgE or SPT to peanut, and the possible correlation between peanut and pollen sensitization. In a group of Danish adolescents, a peanut sensitization evaluated by ImmunoCAP™ and SPT of 5.8% and 3.4%, respectively, was found, while the point of prevalence of a peanut allergy, confirmed by oral challenge, was estimated to be 0.5%. Most peanut-sensitized adolescents had atopic diseases; intermittent allergic rhinitis was seen in 58–74%. The possibility of correlation between peanut and pollen (grass) sensitization was suggested [17]. Food challenge is still the gold standard for diagnosing food allergies, including suspected reaction to peanut [17–20].

HM food allergen extracts are prepared by standardized protocols. The HM allergen extracts of nuts and peanut are in all cases defatted during pre-processing. The removal of fat and oils, which are able to cause false positive type IV skin reactions, and other small particles, e.g., minerals, improves the exposure of allergenic proteins and extraction efficiency, and removes components that are insoluble in water [12,21].

Defatted and dried HM allergen material (dry powder) can be stored at −20 °C, which improves the long-term stability. De Jong et al. showed good stability results with the same method (HM), comparing fresh, 3-month-old, and 6-month-old extracts. In this earlier study, batch-to-batch comparisons with coriander, hazelnut, peach, and sesame seed gave coefficients of variation of 39%, 33%, 37%, and 26%, respectively. Overall, pair wise comparison of dose response SPT results with the four different HM extracts using 5%, 10%, and 20% were significant in all cases [12]. Secondly, the HM extracts appeared to be safe, as no adverse events occurred in the 2004 study, as well as in the current study. Finally, the method of preparing HM extracts is clearly extremely cost-effective. An analyst can prepare the material in the hospital laboratory, using food from the local grocery, and even more rare sources (e.g., new food sources such as seaweed, tropical fruits such as papaya and new legumes such as lentils) can be extracted easily at a low cost.

There are some limitations to this study: first, most patients included in this study suffer from an inhalant allergy (sensitization birch pollen: 94%, grass pollen: 74%). We did not specifically select these patients, but as we performed the study in a peripheral hospital (second line), the population differs from, e.g., an Academic Center. Consequently, the suspected food allergy in these patients was most likely caused by cross reactivity, which could be confirmed by a high percentage of sensitization to several PR-10-specific allergens (Cor a 1, Ara h 8, Pru p 1, Mal d 1). This might cause some bias, as we therefore did not test the allergen extracts in patients with a primary food allergy. The low sIgE found for LTP proteins (Cor a8, Ara h9, Jug r3, and Pru p3) confirms the population of the included patients. Consequently, the proportion of agreement for walnut Jug r1 is low (0.24). Finally, in a peripheral hospital, we did not perform the gold standard for the diagnosis of food allergy; the DBPCFC. Comparing SPT results with suspected food allergy is not in accordance to the guidelines, but must be seen as a first step in the diagnosis of a food allergy [2].

5. Conclusions

In this study, we found that the SPT-HEP results of the HM extracts are comparable with the SPT-HEP results of commercially available extracts for hazelnut and walnut and moderately comparable for apple and peanut. We recommend further studies with HM extracts of food allergens in another population, e.g., children and patients with a primary food allergy. Furthermore, we also recommend the characterization and identification of allergenic proteins in HM food allergen extracts. Commercial food allergen extracts will be less available in the near future, caused by new European government regulations Developing and validating educational tools on how to produce suitable and reproducible HM food allergen extracts will increase the establishment of vertical and horizontal networks between Academic Centers of excellence, allergy specialists, and primary health care practitioners [22]. These developments will increase the knowledge, quality, and use of HM food allergens extracts, and might be one step forward in the complex diagnosis of food allergies.

Author Contributions: Conceptualization: S.T.; Data curation: S.T., M.E.v.S.; Formal analysis: S.T., F.E.v.B., M.W.J.S., N.W.d.J.; Investigation: S.T., M.B.-v.Z., C.d.G.-i.'t.V.; Methodology: S.T., N.W.d.J.; Project administration: S.T.; Resources: S.T.; Software: S.T., N.W.d.J.; Supervision: N.W.d.J.; Validation: S.T., N.W.d.J.; Visualization: S.T.; Writing—original draft: S.T.; Writing—review and editing: S.T., F.E.v.B., M.B.-v.Z., C.d.G.-i.'t.V., M.E.v.S., P.L.A.v.D., M.S.v.M., M.W.J.S., N.W.d.J. All authors have read and agreed to the published version of the manuscript.

Funding: This research received no external funding.

Institutional Review Board Statement: Medical ethical approval was obtained for this study at 01-02-2018, trial number MEC-2017-486, NL61899.078.17. The study was conducted in accordance with the Declaration of Helsinki, and approved by the Ethics Committee of the Erasmus Medical Centre (protocol code HoMaFA study; MEC-2017-486, NL61899.078.17; date of approval 01-02-2018).

Informed Consent Statement: Informed consent was obtained from all subjects involved in the study. Written informed consent has been obtained from the patient(s) to publish this paper.

Data Availability Statement: The original database is not available online.

Conflicts of Interest: The authors declare no conflict of interest.

Abbreviations

ACE	Academic Center of Excellence
ALK	Allergy Laboratories Kopenhagen
CI	confidence interval
DBPCFC	double blind placebo controlled food challenge
EAACI	European Academy of Allergy and Clinical Immunology
e.g.,	exempli gratia
F	female
FN	false negative
FP	false positive
HEP	Histamine Equivalent Prick test
HM	homemade
ICC	intra-class correlation coefficient
IgE	immunoglobulin E
ISAC	immuno-solid phase allergen chip
LR	likelihood ratio
LTP	lipid transfer proteins
M	male
Mm	millimetre
n	number
NA	not applicable
NPV	negative predictive value
Nr	number
OAS	oral allergy syndrome
OFC	oral food challenge
PPV	positive predictive value
Resp	respectively
Sens	sensitivity
sIgE	specific immunoglobulin E
Spec	specificity
SPT	skin prick test
TN	true negative
TP	true positive
vs.	versus

References

1. Grabenhenrich, L.B. Epidemiologische Daten zur Nahrungsmittelallergie in Europa [The epidemiology of food allergy in Europe]. *Bundesgesundheitsblatt Gesundh. Gesundh.* **2016**, *59*, 745–754. [CrossRef]
2. Sampson, H.A.; van Wijk, R.G.; Bindslev-Jensen, C.; Sicherer, S.; Teuber, S.S.; Burks, A.W.; Dubois, A.E.J.; Beyer, K.; Eigenmann, P.A.; Spergel, J.M.; et al. Standardizing double-blind, placebo-controlled oral food challenges: American Academy of Allergy, Asthma & Immunology—European Academy of Allergy and Clinical Immunology PRACTALL consensus report. *J. Allergy Clin. Immunol.* **2012**, *130*, 1260–1274. [CrossRef] [PubMed]
3. Gomes-Belo, J.; Hannachi, F.; Swan, K.; Santos, A. Advances in Food Allergy Diagnosis. *Curr. Pediatr. Rev.* **2018**, *14*, 139–149. [CrossRef] [PubMed]
4. Klimek, L.; Hoffmann, H.J.; Kalpaklioglu, A.F.; Demoly, P.; Agache, I.; Popov, T.A.; Muraro, A.; Schmid-Grendelmeier, P.; Bonini, S.; Bonertz, A.; et al. In-vivo diagnostic test allergens in Europe: A call to action and proposal for recovery plan—An EAACI position paper. *Allergy* **2020**, *75*, 2161–2169. [CrossRef] [PubMed]
5. Skypala, I.J.; Venter, C.; Meyer, R.; Dejong, N.W.; Fox, A.T.; Groetch, M.; Elberink, J.N.O.; Sprikkelman, A.B.; Diamandi, L.; Vlieg-Boerstra, B.J.; et al. The development of a standardised diet history tool to support the diagnosis of food allergy. *Clin. Transl. Allergy* **2015**, *5*, 7–10. [CrossRef]
6. Gupta, M.; Nadeau, K.C. Diagnosis of Food Allergy. *Immunol. Allergy Clin. N. Am.* **2018**, *38*, 39–52. [CrossRef] [PubMed]
7. Kattan, J.D.; Sicherer, S.H. Optimizing the Diagnosis of Food Allergy. *Immunol. Allergy Clin. N. Am.* **2015**, *35*, 61–76. [CrossRef] [PubMed]
8. Soares-Weiser, K.; Takwoingi, Y.; Panesar, S.S.; Muraro, A.; Werfel, T.; Hoffmann-Sommergruber, K.; Roberts, G.; Halken, S.; Poulsen, L.; Van Ree, R.; et al. The diagnosis of food allergy: A systematic review and meta-analysis. *Allergy* **2014**, *69*, 76–86. [CrossRef] [PubMed]

9. Van Der Valk, J.P.M.; Van Wijk, R.G.; Hoorn, E.; Groenendijk, L.; Groenendijk, I.M.; De Jong, N.W. Measurement and interpretatio of skin prick test results. *Clin. Transl. Allergy* **2015**, *6*, 8. [CrossRef] [PubMed]
10. De Jong, N.W.; Terlouw, S.; van Boven, F.E.; van Maaren, M.S.; Schreurs, M.W.J.; van den Berg-Somhorst, D.B.P.M.; Esser, D Bastiaan-Net, S. Birch Pollen Related Pear Allergy: A Single-Blind Oral Challenge TRIAL with 2 Pear Cultivars. *Nutrients* **202** *13*, 1355. [CrossRef] [PubMed]
11. Heinzerling, L.; Mari, A.; Bergmann, K.C.; Bresciani, M.; Burbach, G.; Darsow, U.; Durham, S.; Fokkens, W.; Gjomarkaj, M Haahtela, T.; et al. The skin prick test—European standards. *Clin. Transl. Allergy* **2013**, *3*, 3. [CrossRef] [PubMed]
12. De Jong, N.W. *Reproducibility and Stability of 'in House Manufactured' Extracts Used in the Diagnosis of IgE mediated allergy*; 2004; pp 45–53. ISBN 90-77595-73-2.
13. Koo, T.K.; Li, M.Y. A Guideline of Selecting and Reporting Intraclass Correlation Coefficients for Reliability Research. *J. Chirop Med.* **2016**, *15*, 155–163. [CrossRef] [PubMed]
14. Asero, R.; Ballmer-Weber, B.K.; Beyer, K.; Conti, A.; Dubakiene, R.; Fernandez-Rivas, M.; Hoffmann-Sommergruber, K.; Lidholm J.; Mustakov, T.; Elberink, J.N.G.O.; et al. IgE-Mediated food allergy diagnosis: Current status and new perspectives. *Mol. Nut Food Res.* **2006**, *51*, 135–147. [CrossRef] [PubMed]
15. Foong, R.; Santos, A.F. Biomarkers of diagnosis and resolution of food allergy. *Pediatr. Allergy Immunol.* **2021**, *32*, 223–233 [CrossRef] [PubMed]
16. Arends, N.J.T.; de Groot, N.E.H.; Emons, J.A.M.; Brand, H.K.A.; Verhoeven, D.; van Veen, L.N.; de Jong, N.W. Almond allergy in a cohort of Dutch atopic children: Results from 189 oral food challenges with almond. *Linical Transl. Allergy* **2017**, *7* (Suppl. 1) OP07.
17. Mortz, C.G.; Andersen, K.E.; Bindslev-Jensen, C. The prevalence of peanut sensitization and the association to pollen sensitization in a cohort of unselected adolescents—The Odense Adolescence Cohort Study on Atopic Diseases and Dermatitis (TOACS) *Pediatr. Allergy Immunol.* **2005**, *16*, 501–506. [CrossRef]
18. Emmett, S.; Angus, F.; Fry, J.; Lee, P. Perceived prevalence of peanut allergy in Great Britain and its association with other atopic conditions and with peanut allergy in other household members. *Allergy* **1999**, *54*, 380–385. [CrossRef] [PubMed]
19. De Martino, M.; Novembre, E.; Cozza, G.; De Marco, A.; Bonazza, P.; Vierucci, A. Sensitivity to tomato and peanut allergens in children monosensitized to grass pollen. *Allergy* **1988**, *43*, 206–213. [CrossRef] [PubMed]
20. Tariq, S.M.; Stevens, M.; Matthews, S.; Ridout, S.; Twiselton, R.; Hide, D.W. Cohort study of peanut and tree nut sensitisation by age of 4 years. *BMJ* **1996**, *313*, 514–517. [CrossRef] [PubMed]
21. David, N.A.; Penumarti, A.; Burks, A.W.; Slater, J.E. Food allergen extracts to diagnose food-induced allergic diseases: How they are made. *Ann. Allergy Asthma Immunol.* **2017**, *119*, 101–107. [CrossRef] [PubMed]
22. Muraro, A.; Werfel, T.; Hoffmann-Sommergruber, K.; Roberts, G.; Beyer, K.; Bindslev-Jensen, C.; Cardona, V.; Dubois, A.; Dutoit, G.; Eigenmann, P.; et al. EAACI Food Allergy and Anaphylaxis Guidelines: Diagnosis and management of food allergy. *Allergy* **2014**, *69*, 1008–1025. [CrossRef] [PubMed]

nutrients

Article

Birch Pollen Related Pear Allergy: A Single-Blind Oral Challenge TRIAL with 2 Pear Cultivars

Nicolette W. de Jong [1,*], Severina Terlouw [1], Frank E. van Boven [1], M.S. van Maaren [1], Marco W.J. Schreurs [2], Dianne B.P.M. van den Berg-Somhorst [3], Diederik Esser [3] and Shanna Bastiaan-Net [3]

1 Internal Medicine, Allergology & Clinical Immunology, Erasmus MC, University Medical Centre, 3015 GD Rotterdam, The Netherlands; s.terlouw@erasmusmc.nl (S.T.); f.boven@erasmusmc.nl (F.E.v.B.); m.vanmaaren@erasmusmc.nl (M.S.v.M.)

2 Department of Immunology, Laboratory Medical Immunology, Erasmus MC, University Medical Centre, 3015 GD Rotterdam, The Netherlands; m.schreurs@erasmusmc.nl

3 Wageningen Food & Biobased Research, Wageningen University & Research Centre, 6708 WE Wageningen, The Netherlands; dianne.somhorst@wur.nl (D.B.P.M.v.d.B.-S.); diederik.esser@wur.nl (D.E.); shanna.bastiaan@wur.nl (S.B.-N.)

* Correspondence: n.w.dejong@erasmusmc.nl

Abstract: Approximately 70% of birch pollen allergic patients in Europe experience hypersensitivity reactions to Immunoglobulin E (IgE) cross-reactive food sources. This so-called pollen-food syndrome (PFS) is defined by allergic symptoms elicited promptly by the ingestion of fruits, nuts, or vegetables in these patients. So far, in the literature, less attention has been given to Bet v 1 cross-reactive symptoms caused by pear (*Pyrus communis*). In the Netherlands, pears are widely consumed. The primary objective of this study was to measure the type and severity of allergic symptoms during pear challenges in birch pollen allergic patients, with a positive history of pear allergy, using two different pear varieties. Fifteen patients were included, skin prick test (SPT), prick-to-prick test (PTP), specific Immunoglobulin E (sIgE), and single-blind oral challenges were performed with two pear (*Pyrus communis*) varieties: the 'Cepuna' (brand name Migo®) and the 'Conference' pears. All patients were sensitized to one or both pear varieties. A total of 12 out of 15 participants developed symptoms during the 'Cepuna' food challenge and 14/15 reacted during the 'Conference' challenge. Challenges with the 'Cepuna' pears resulted in less objective symptoms (n = 2) in comparison with challenges with 'Conference' pears (n = 7). Although we did not find significance between both varieties in our study, we found a high likelihood of fewer and less severe symptoms during the 'Cepuna' challenges. Consequently selected pear sensitized patients can try to consume small doses of the 'Cepuna' pear outside the birch pollen season.

Keywords: birch pollen; allergy; Bet v 1; OAS; pear; oral challenge

Citation: de Jong, N.W.; Terlouw, S.; van Boven, F.E.; van Maaren, M.S.; Schreurs, M.W.J.; van den Berg-Somhorst, D.B.P.M.; Esser, D.; Bastiaan-Net, S. Birch Pollen Related Pear Allergy: A Single-Blind Oral Challenge TRIAL with 2 Pear Cultivars. *Nutrients* **2021**, *13*, 1355. https://doi.org/10.3390/nu13041355

Academic Editor: Margarida Castell Escuer

Received: 25 February 2021
Accepted: 15 April 2021
Published: 18 April 2021

1. Introduction

In the general population in Europe, the prevalence of birch pollen sensitization ranges from approximately 8 to 16% [1]. The prevalence of sensitization to Bet v 1, a PR-10 allergen, and the major allergen of birch pollen, is notably high among European patients with pollen allergies. In a study of 260 patients with tree pollen allergy in Germany, 92% were sensitized to Bet v 1 [2], and in a retrospective study of 854 patients with birch pollen sensitization in Italy, sensitization to Bet v 1 ranged from 53% to 95%, depending on the region [3]. Approximately 70% of birch pollen allergic patients experience hypersensitivity reactions to IgE cross-reactive food sources [4]. This so-called pollen-food syndrome (PFS) is defined by allergic symptoms elicited promptly by the ingestion of fruits, nuts, or vegetables in patients with seasonal allergic rhinoconjunctivitis (SAR) [5]. Patients are sensitized to pollen allergenic molecules highly cross-reacting with their homologs in the offending foods [6,7]. Symptoms of PFS are often restricted and isolated to the oral cavity and include labial and oropharyngeal pruritus, paraesthesia, and angioedema of the oral mucosa tongue, lips, palate, and oropharynx, or laryngeal tightness, which altogether are labeled as oral allergy syndrome (OAS). Gastrointestinal symptoms and, rarely, life-threatening wheezing and anaphylaxis, occur in less than 10% of patients [8]. Bet v 1, a PR-10 type of protein, is the most prevalent cause of cross-sensitization. The PR-10 related reactions are mainly in response to *Rosaceae* fruits (i.e., apples) and nuts (i.e., hazelnut). The most frequently described cross-reacting fruits are apple, peach, cherry, and apricot, but a wide range of fruits contain Bet v 1 homologs. So far, in the literature, less attention has been given to Bet v 1 cross-reactive symptoms caused by pear (*Pyrus communis*). Most larger prevalence studies in Europe did not include pear while in most apple allergic patients pear also causes symptoms [9]. One systematic review by Zuidmeer et al. [10] found a study on pear allergy [11] reporting 0.3% pear allergy in Germany. Furthermore, Rodriquez et al. [12] performed skin prick tests in 26 patients in Spain with adverse reactions to *Rosaceae* fruits and 18 patients with positive SPT for apple appeared to be positive for pear as well (69%). In the Netherlands, pears are widely consumed. Each household consumes an average of 4.7 kilos of pear per year and pear is in a solid third place in the Dutch fresh fruit top 10 [13]. In the Erasmus MC Rotterdam, SPT with pear is regularly positive in birch pollen allergic patients but clinical relevance is often unclear. Diagnosis of pear allergy has to be confirmed by a double-blind placebo-controlled food challenge. The primary objective of this study was to measure type and severity of allergic symptoms during pear challenges in birch pollen allergic patients, with a positive history of pear allergy, using two different pear varieties.

2. Materials and Methods

2.1. Patients

Adult patients visiting the outpatient clinic of the department of Allergology of the Erasmus MC with a doctor's diagnosed birch pollen allergy and a positive history of pear allergy were asked to participate in the study. The patients were approached from August 2019 and inclusion started on 1 October 2019 till 1 February 2020. Medical ethical approval was received in August 2019; registered as METC NL70165.078.09. The purpose was to perform the study just outside the 'birch pollen season' (February to May) to circumvent that participating patients could not stop their anti-histamines, and/or preventing possible bias in patients having more symptoms during that season.

2.2. Pears

Two pear (*Pyrus communis*) varieties 'Cepuna' (brand name Migo®) and 'Conference' pear were tested according to a normal consumer simulation consisting of simulating refrigerated transport to the supermarket and consumer storage in a fruit bowl. Pears were acclimatized in advance (via 'chambreren') by GKE NV and delivered ready for consumption every week by courier. Pears delivered were stored in the refrigerator. The evening before inclusion, the number of pears needed was brought to room temperature. Before use, the pears were rinsed under the tap with water as in a home situation.

2.3. Skin Test

Skin prick test (SPT) and prick-to-prick (PTP) tests were performed with both pear varieties on the forearm during the first day of the oral challenge, with pear juice and fresh pears respectively, next to 2 positive controls (histamine), birch pollen extract, and a negative control (ALK-Abello; Almere, the Netherlands). The difference between SPT and PTP is that in the PTP test, the needle is first pricked into the fresh intact unpeeled pear near the stalk [14] and the juice sticking to the needle is subsequently transferred into the skin of the participant. The SPT was performed by applying a drop of whole fresh pear juice on the skin of the volar aspect of the forearm. Twenty minutes after the skin tests, the contours of the wheal were encircled with a fine-tip pen and transferred to a record sheet by translucent tape [15]. Subsequently, the surface was measured with an area scanner and compared with the positive control which gives the HEP index score as described by van der Valk et al. [15]. No threshold values have yet been defined for the SPT and PTP HEP index values for pear allergy. SPT and PTP tests were considered positive when $\geq$3 mmØ [16,17].

2.4. Specific Serum IgE

Specific Immunoglobulin E (sIgE) antibody concentration was measured in blood serum. Specific serum IgE for pear allergen extract (f94), birch pollen allergen component Bet v 1 PR10 (t215), peach allergen component Pru p 3 LTP (f420), and grass pollen allergen component Phl p 12 profilin (g212) were measured by fluorescence immunoassay (FEIA) using the ImmunoCAP™ test system, according to the manufacturer's instructions (Thermo Fisher Scientific, Upsala, Sweden). sIgE concentration was considered positive when >0.35 KU/L.

2.5. Single Blind Oral Challenges

As validated recipes for double-blind placebo-controlled food challenges with pear do not exist, we performed single-blind oral pear challenges. The oral pear challenges consisted of two-day admissions in the outpatient clinic, in which the pear 'Cepuna' was tested on one day and the pear 'Conference' on the other day. For each participant, the order in which each pear variety was tested was randomized. The single-blind oral food challenge consisted of five doses of pear, including the peel as described by Rodriguez et al. [12]. The doses were calculated using the 'voedingswaardetabel.nl' site. According to this site, pears, in general, contain 0.5 g protein/100 g pear. Following the PRACTALL guidelines, we challenged 10 μg, 30 μg, 100 μg, 300 μg, and 1000 μg protein, which resulted in the following dose series: 2, 6, 20, 60, and 200 g of pear. The person that weighed the pear doses and prepared the challenge was not the same person as the one who gave the pear pieces to the patient. So, the nurse was not blinded but was ignorant about which variety was given to the patient. The symptoms during the challenge were recorded according to the PRACTALL guidelines, [18] and scored as mild, moderate, or severe. The challenge was stopped as soon as the participant responded three consecutive times with subjective symptoms to a certain dose and was stopped immediately in case the patient reacted with objective symptoms, as based on the reference of Sampson et al. [18]. This means that not every patient consumed the last doses, dose four and dose five. To minimize sensory perception between the pear varieties, patients were blindfolded and wore a nose clip during the oral challenge (single-blind). After the challenge, participants remained in the clinic for 2 h to monitor possible reactions. Twenty-four hours after the challenge the patients were contacted by telephone to register possible late reactions. Subjective symptoms were recorded as itching in the mouth and on the lips, in the ears, nose, or eyes, and nausea. Objective symptoms are seen as the more serious symptoms and consist of itchy skin or red skin (urticaria), wheezing, and laryngeal symptoms.

2.6. IgE-Immunoblotting/SDS Page/Electrophorese

Pears were cut into an upper and a lower part. The upper quarter of the pear, without the inner core, was cut into pieces and snap-frozen in liquid nitrogen. From the bottom part, only the peeled skin (still containing a little bit of flesh) was collected and combined with the upper quarter of the pear to obtain an equal amount of flesh versus peel material. Per variety, 10 pears were sampled to obtain a representative sample batch. Subsequently, the sampled material was ground under liquid nitrogen using an IKA mill and the acquired powder was stored at −80 °C until further use. Total protein was extracted using the method described by Vieths et al. [19] with slight adjustments. Since pears are high in polyphenol content, and these can interfere with protein isolation [20], ground pear samples were homogenized in acetone/dry ice and incubated overnight while stirring and cooled by dry ice. Precipitates were washed twice with acetone/dry ice, and once with acetone/diethyl ether/dry ice (1:1, v/v, −60 °C). Subsequently, precipitates were filtered (Whatman®; 595$\frac{1}{2}$, ø240 mm, Pfullingen, Germany), lyophilized, and stored a −20 °C until protein extraction. Total protein extracts were obtained by extraction witl 0.001 M potassium phosphate (mix K2HPO4 and KH2PO4) buffer pH7.4, containing 0.15 M NaCl by overnight incubation at 4 °C on a stirring weal (2 g of acetone powder/30 ml extraction buffer). The next day, samples were centrifuged (4 °C, 60 min, 4700 rpm) anc the protein supernatants were concentrated using 3 kDa Amicon concentrators (Merck Millipore; Tullagreen, Ireland). The protein concentration was determined using Bradforc assay (Thermo Fisher Scientific Inc., Madison, WI, USA) according to the manufacturer's instructions using bovine serum albumin (BSA) as a protein standard. IgE-Immunoblotting was performed as described previously [21]. In brief, 20 µg of pear protein concentrate was separated by SDS PAGE on Bolt™ 4–12% Bis-Tris Plus gels next to a Precision Plus Protein Dual Xtra Standard molecular weight marker (Bio-rad, Hercules, CA, USA) according to the manufacturer's instructions (Invitrogen, Carlsbad, CA, USA) and either stained by Simply Blue safe stain (Thermo Fisher Scientific Inc.) or transferred to a 0.2 µm nitrocellulose membrane (LKB, Bromma, Sweden) by Tris-glycine buffer (25 mM Tris, 190 mM glycine 0.1% SDS, 20% methanol) for 36 min at 70 V using a Criterion blotter (Bio-rad). The transfer was verified using the MemCode Reversible Protein Stain Kit (Thermo Fisher Scientific Inc.). After blocking in 3% BSA, blots were incubated overnight with 1:5 diluted patient serum (2 mL). The first and secondary antibodies used were the polyclonal rabbit anti-IgE antibody from Dako (1:1000; Glostrup, Denmark) and AP-conjugated polyclonal goat anti-rabbit antibody from Sigma Aldrich (1:20,000; Saint Louis, MO, USA). Blots were stained for 30 min in 20 mL NBT/BCIP staining solution (Sigma Aldrich; St. Louise, MO USA). Imaging and analysis of antibody binding were performed using a Universal Hood III and Image Lab 4.1 software (Bio-Rad; Hercules, CA, USA).

2.7. Statistics

One of the objectives was to compare both pear varieties based on allergenicity. Calculating Bayes factors compared the differences in the pear outcomes. The Bayes factor (BF) is a likelihood ratio of a null hypothesis and an alternative [22]. Evidence for the alternative hypothesis (H1) was set as BF > 3 (moderately), BF > 10 (strongly), BF > 30 (very strong), and BF > 100 (extremely), and evidence for the null hypothesis (H0) was set as BF < 1/3. Dependencies between pear outcomes were analyzed by the Fisher exact test The classification concerns included the difference in the amount of pear consumed and the various scores. All calculations were performed with R, in the Fisher exact test, $p < 0.05$ is considered to be statistically significant.

3. Results

3.1. Patients

Based on the medical history of patients registered at Erasmus MC, a total of 74 patients with birch pollen allergy were approached, of which 17 were included (20%). Thirty-one patients did not want to participate despite previous symptoms while eating pear, while 28 patients had tree pollen allergies without symptoms when consuming pear or other fruit. Of the 17 included patients, two dropped out: one patient was negative in SPT and PTP on both pear varieties tested, and one patient did not attend the second visit. Finally, fifteen patients were included in the study, 80% of which were female. The average age was 37 years (range 20–64 years). Eleven patients did not consume pear. Four patients indicated eating, very occasionally, processed pears (heated, cooked). Of the eleven patients who did not eat pears, 10 patients had eliminated pears from their diet for >3 years, and one patient less than 3, but more than 2 years from their diet (Table 1).

3.2. SPT/PTP

The SPT HEP index for 'Cepuna' pear and 'Conference' pear was on average 0.20 (range 0–0.58) and 0.22 (range 0–0.83), respectively. The prick-to-prick (PTP) test HEP index with 'Cepuna' and 'Conference' pear averaged 0.81 (range 0–2.57) and 0.61 (range

0.13–2.37), respectively. SPT was negative (<3 mm Ø) in 9 cases (60%) for both pear varieties. PTP was negative with 'Cepuna' in 2 cases (13%) and with 'Conference' in 1 case (6%). All patients were sensitized to pear in at least one of the tests (SPT, PTP, or sIgE). SPT with birch pollen was positive in 14/15 patients with an average HEP of 1.28 (Table 2). A test of difference produced a BF of 29 (BF > 10 = strongly) in favor of a PTP being larger than SPT in 'Cepuna' as well as 'Conference'. The mean of SPT and PTP did not differ between 'Cepuna' and 'Conference' (BF = 1.35). SPT and PTP were not associated with the challenge outcomes for 'Cepuna' as well as 'Conference' pears (p = 0.13 to 1.0 resp.).

Table 1. Characteristics of the patients.

		n	%
Numbers included		15	
Patients	Female	12	80%
	Mean age/range	37	18–65
Inhalant allergy	HDM	11	73%
	GP	11	73%
	BP	15	100%
Food allergy	>1 allergen	1	7%
	>2 allergens	1	7%
	>3 allergens	13	87%
Current Pear consumption	No	11	73%
	Yes (processed)	4	27%
History of Symptoms	Itchy mouth	14	93%
	Skin	0	0%
	Wheeze	3	20%
Anti-allergic medication use		12	80%

HDM: house dust mite; BP: birch pollen; GP: grass pollen.

Table 2. Results of the SPT, PTP, sIgE, and open single-blind challenges.

	SPT	PTP	SPT	PTP	SPT	sIgE	sIgE	Single Blind Oral Challenge			
	Cepuna		Conference		Bp	Pear	Bp	Cepuna		Conference	
nrs	HEP	HEP	HEP	HEP	KU/L	KU/L	KU/L	pos/neg	dose*	pos/neg	dose*
1	neg	0.90	neg	0.25	0.61	0.6	19.2	pos	3	pos	1
2	neg	0.40	neg	0.28	1.32	2.1	12.4	pos	1	pos	1
3	0.58	2.57	0.83	1.17	5.06	0.4	16.2	pos	1	pos	5
4	0.50	1.21	neg	2.37	2.42	16.7	102.5	pos	1	pos	1
5	neg	0.51	neg	0.88	0.16	0.8	29.6	neg	neg	pos	1
6	neg	0.30	neg	0.38	1.02	neg	11.3	neg	neg	pos	2
7	0.25	1.09	0.46	0.77	1.52	0.7	15.6	pos	1	pos	2
8	neg	0.71	neg	0.24	neg	neg	8.8	pos	4	pos	3
9	neg	neg	0.14	0.34	0.61	NA	NA	pos	2	neg	neg
10	neg	1.33	neg	0.14	0.62	8.6	3	pos	1	pos	1
11	neg	neg	neg	neg	1.43	5.1	46.3	pos	1	pos	1
12	0.38	1.32	0.37	0.89	1.18	neg	51.5	neg	neg	pos	1
13	neg	0.32	neg	0.35	0.71	2.1	25	pos	5	pos	2
14	0.22	0.38	0.21	0.37	1.20	neg	6.7	pos	1	pos	1
15	0.24	0.53	0.20	0.69	0.94	0.6	neg	pos	5	pos	1

SPT: skin prick test; PTP: prick-to-prick test; sIgE: serum immune globulin E; HEP: histamine equivalent prick index; pos: positive; Neg: < 3 mmØ; dose*: lowest dose during the pear challenge that the patient reacted to; NA: not available. Bp: birch pollen.

3.3. *sIgE*

One patient refused to give blood (nr 9), so 14 sera for sIgE measurements were available (Table 2). The sIgE serum concentration for pear was on average 2.91 KU/L (range 0–16.7 KU/L) and was negative (<0.35 KU/L) in four cases (29%). These patients may be solely sensitized to Pyr c 1, and it is not clear whether it is in the f94 pear extract. For Bet v 1 (PR10) sIgE, the average was 24.86 KU/L (range 0–102.5 KU/L) and negative in one case (7%). Furthermore, sIgE to Pru p 3 (LTP) was positive in 2 patients, nr 1 and 10 with values of 2.1 KU/L and 91.6 KU/L respectively. SIgE to Phl p 12 (profilin) was also positive in 3 cases: nrs 5, 13 and 15: 4.8; 0,7 and 27.2 KU/L respectively. SIgE serum concentration was not associated with SPT, PTP, and challenges for both the 'Cepuna' and 'Conference' pear (p = 0.15 to 1.0).

3.4. *Pear Challenge*

Twelve out of fifteen participants (80%) developed symptoms during the 'Cepuna' food challenge. Three participants could eat the whole 'Cepuna' pear without symptoms (nrs 5, 6, and 12). Fourteen out of fifteen participants (93%) developed symptoms during the 'Conference' food challenge, in which only one participant (nr 9) could eat the whole pear without symptoms. None of the patients showed a late reaction (24 h after the food challenge) after either challenge. The BF of a reduced number of positive challenges was 8 for 'Cepuna' pear, and 0,4 for 'Conference' pear (Table 2).

Challenges with the 'Cepuna' pears resulted in less objective symptoms (two patients) in comparison with challenges with 'Conference' pears (seven patients) (BF = 4192). Most of the scores were assessed as mild (score 1). During the 'Cepuna' challenge, four patients scored moderate (score 2) for itchy mouth (nrs 3, 8, 9, and 14) and one patient scored moderate (score 2) for wheeze and larynx symptoms (nr 4). During the 'Conference' challenge, three patients scored moderate (score 2) for itchy mouth (nrs 7, 12, and 15) and one patient for nose and/or ears symptoms (nr 7). One patient scored severe (score 3) for itchy mouth (blisters) (nr 6) (Table 3).

Table 3. Symptoms during pear challenges.

	'Cepuna'		**'Conference'**	
PT nr	**Subjective**	**Objective**	**Subjective**	**Objective**
1	IM		IM	SK, LA
2	IM, IE		IM, IE	LA
3	IM/2, IN		IM	
4	IM	WH/2, LA/2	IM	
5	——		IM	
6	——		IM/3	
7	IM, IE, IN		IE, IN/2	WH, LA
8	IM/2		IM	
9	IM/2		——	
10	IM	LA	IM	LA
11	IM, IN		IM, IN	SK, LA
12	——		IM/2, IN	WH
13	IM		IM, N	
14	IM/2, IN, N		IM, IN, N	LA
15	IM		IM/2	

Score: All patients score mild/1, except when given/2 for moderate, or/3 for severe. IM: itchy mouth; IE: itchy ears; IN: itchy nose and ears; N: nausea; SK: skin; WH: wheeze; LA: laryngeal.

Overall, four patients were treated with antihistamine for their allergic reaction (nrs 1, 2, 3, and 4) because they asked for it. The wheeze and larynx symptoms were mild and consequently, no adrenaline or corticosteroids were administered.

3.5. Immunoblotting

For Western blot analysis, 15 µg of total protein concentrates were separated on SDS PAGE under reducing conditions and transferred to membranes that were subsequently incubated with the serum of each of the participating patients (except for pt nr 9).

The blots incubated with patient sera indicate that most pear allergic individuals carry IgE antibodies against relatively moderate molecular weight (MW) proteins (between 25 and 75 kDa) while a few patients also react to proteins in the small MW range of 15–25 kDa (Figure 1 and Supplementary Table S1).

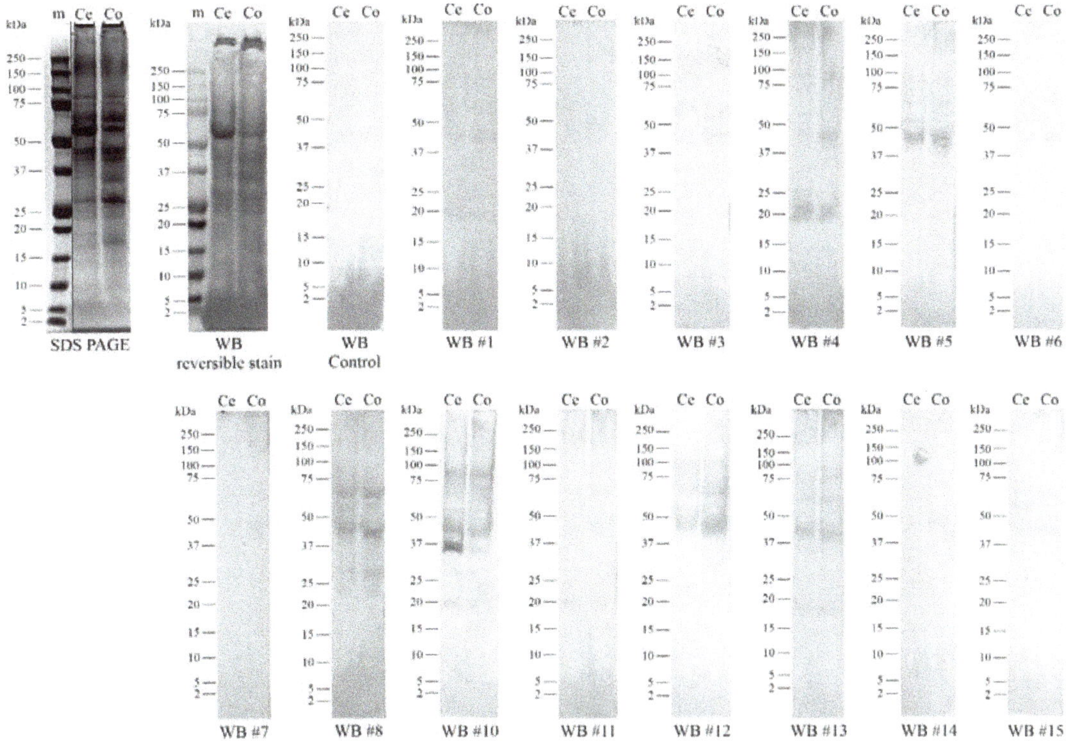

Figure 1. Western blots (WB) of 'Cepuna' and 'Conference' pear total protein concentrate using the patient serum. The WB control blot was exposed to buffer instead of serum. Patient numbering is indicated by #.

All patients, except patient nr 15, seem to have IgE antibodies that bind (in more or less intensity) a protein band with an estimated MW of ~45 kDa. The majority of patient sera bound to protein bands with an estimated MW of ~19 kDa and ~55 kDa (Table S1 Supplementary).

Patients nrs 5, 6, and 12 did not show an allergic response to the consumption of 'Cepuna' pear while responding to 'Conference' pear upon the first or second dose. However, in Western blot, these patient sera recognized the ~45 kDa band with almost equal intensity for both pear varieties, suggesting that IgE binding to this protein band might be clinically irrelevant.

Up till today, four allergens are officially identified in pear: Pyr c 1.0101, Pyr c 3.0101, Pyr c 4.0101, and Pyr c 5.0101, which represent the protein allergen types Bet v 1-like, non-specific lipid transfer protein (nsLTP), profilin and isoflavone reductase, respectively (Table 4). In apple, another protein allergen type is known, named thaumatin, which might also be present in pear but has not yet been identified.

Table 4. Pear (Pyrc) and apple (Mald) allergens described in the Allergome database (http://www.allergome.org/index.php, accessed on 10 October 2020).

Allergen Name	Protein Type	kDa Based on AA Sequence	kDa without Signal Peptide
Pyr c 1.0101	Bet v 1-like	17,581	
Pyr c 3.0101	nsLTP	11,463	9125
Pyr c 4.0101	Profilin	14,064	
Pyr c 5.0101	Isoflavone reductase	33,823	
Mal d 2.0101	Thaumatin	25.7	23,211

The estimated ~35 kDa band, recognized by patient sera nr 10 and nr 14, might be Pyr c 5.0101 or another isoflavone reductase isoform, based on the similarity of MW. The Bet v-like allergen in pear, Pyr c 1.0101, might represent the ~19 kDa band on gel, while profilin (Pyr c 4.0101) might represent the estimated ~15 kDa band, slightly bound by patient serum nr 6. The birch pollen sIgE measurements back up the hypothesis that the ~19 kDa band might represent Pyr c 1.0101, as patient nr 4, who showed the highest amount of birch pollen sIgE (102.5 kU/L), also showed the highest band intensity on the blot. Serum IgE from patient nr 8 seems to bind a ~2 kDa protein band, which might represent a thaumatin type of protein, as identified as an allergen in apple (Mal d 2.0101). The identity of the high MW bands bound by the patient sera IgE is impossible to estimate but could be identified in the future by performing an LC-MS/MS analysis.

4. Discussion

In this study, we describe a group of birch pollen allergic patients, suffering from oral allergy symptoms during the consumption of pear. This Bet v 1 (PR-10) related fruit allergy is hardly described. According to Beyer et al. [23], the foods that most frequently elicit allergic reactions in birch pollen allergic patients were apple (78%), carrot (52%), and peach (49%). However, in the same study, pear comes close to these numbers at 36%. This study in Germany focused on food allergy-related quality of life (FAQL) in birch pollen-associated food allergy (FA) symptoms. The mean food allergy Quality of Life Questionnaire—Adult form (FAQLQ-AF) score was 3.7. This shows the extent of everyday impairment in this group of patients with food allergies.

Pear is becoming popular in Europe and consumed widely because of its nutritional benefits [24]. We included patients who previously reported symptoms after consumption of pear and in most cases, this could be confirmed in the single-blind oral challenges. Four patients had a negative challenge to one of both varieties, of which two patients appeared to have low to negative sensitization profiles. At the same time, all patients reacted to one of both pears, irrespective of their sensitization pattern. Remarkably, most patients experienced subjective symptoms directly after consuming dose one (10 µg pear). Patients with OAS recognized these symptoms; however, challenges were continued with the next doses, as subjective symptoms should occur on three consecutive doses to be positive, following the PRACTAL guidelines. Nevertheless, v. Erp et al. showed that subjective symptoms are significantly associated with disagreement when assessed by different clinical experts [25]. So, variability in the interpretation of food challenge outcomes exists, especially when objective symptoms are absent.

Overall, in this study, besides subjective symptoms, objective symptoms also occurred in seven cases, but we cannot neglect that subjective symptoms are prone to interpretation bias by the patient as well as the researcher. Furthermore, a shortcoming in this study might be the open challenges. As earlier described, up to 12.9% of placebo reactions can appear during food challenges and this is usually captured by performing double-blind placebo-controlled food challenges (DBPCFC) [26]. Unfortunately, double-blind challenges with pear were not possible, as validated recipes with e.g., the right matrix do not exist. To capture this problem the patients were blindfolded and used a nose clip, and the nurse who provided the pear doses was not the same as the nurse who assessed the symptoms.

Nevertheless, we are aware the study is most likely somewhat biased by this phenomenon and that taste preference could have played a role in the assessment of the symptom scores.

To statistically compare the several types of data we used the Bayes factor (BF), which is a weighted average likelihood factor of a particular hypothesis. The BF of a reduced number of positive challenges was 8 for 'Cepuna' pear, and 0.4 for 'Conference' pear and challenges with the 'Cepuna' pears resulted in less objective symptoms in comparison with challenges with 'Conference' pears (BF = 4192 (extremely strong)). In both cases, the data are in favor of the 'Cepuna' pear in comparison to the 'Conference' pear. Particularly, the likelihood of a difference is extreme in the case of objective symptoms.

SPT was negative in nine cases with both pear varieties, while PTP was only negative in two cases. In two of those negative SPTs, a negative challenge confirmed the negative sensitization, but the other seven cases were positive in the oral challenge. A test of difference produced a BF of 29 in favor of a PTP being larger than SPT in 'Cepuna' as well as 'Conference'. So far, these comparisons of skin tests with pear have never been made. PTP is widely accepted as a reliable tool for measuring sensitization to fruits in patients with OAS [27]. Our results also tie in well with the study of Vlieg-Boerstra et al. who concluded that SPT was not useful to assess the allergenicity of 68 apple cultivars [28]. So, in this study, we again confirmed that PTP with fresh fruit is the best method to be used in the diagnosis of fruit food allergy. The study was performed in birch pollen allergic patients and cross-sensitization to PR-10 allergens are most likely. Symptoms caused by these PR-10 allergens are often subjective and mild (OAS) [29]. Unfortunately, we could only slightly confirm the presence of these PR10 antibodies in the sera of the patients in immunoblot, as binding at 17.5–19 kDa was only evident in a few patients. This might be caused by the extraction method of the pears, which was not described in earlier literature, and apple protocols were used as an alternative [19], which might have resulted in an underrepresentation of Pyr c 1 in our extract [18]. Such is also often the case in commercial diagnostic fruit extracts including f94, which might also explain why four patients were negative for the pear f94 sIgE measurements while being positive for birch pollen sIgE (Table 2). Another explanation could be that most of these patients recognize PR-10 conformational epitopes, which are (partly) destroyed under reducing SDS PAGE conditions. In follow-up studies, native or non-reducing conditions might be considered to study Pyr c 1 IgE binding in more detail. Furthermore, it might be possible that the patients are sensitized to proteins in the pear with a higher molecular weight, e.g., Pyr c 5.0101 as in some cases, binding in immunoblot is present to high molecular weight proteins. We did not quantify sIgE to this 33.8 kDa protein. However, as these patients are all birch pollen allergic, their pear allergy is most likely caused by cross-reactivity to PR10 allergens [4]. Although sensitized to PR-10 allergens, several patients in this study also reacted with objective symptoms, which are usually seen as more severe. This is quite remarkable. These objective symptoms are often caused by non-PR-10 allergens e.g., nsLTPs. In our study, only one patient (nr 10) was sensitized to nsLTP (Pru p 3) allergens with a high sIgE value of 91.6 KU/L. This patient reacted on dose 1 with subjective as well as objective symptoms (laryngeal symptoms) to both pear varieties. In contrast, in immunoblot, we could not detect IgE binding at an MW band migrating at around 11.4 kDa (nsLTP). Le et al. [30] found comparable results in a Dutch population suffering from birch pollen-related apple allergy. Of the 14 patients, only one was positive for nsLTP (Pru p 3) sIgE. Although anaphylaxis did not occur in our study with pear challenges, especially for the 'Conference' pear we found seven patients with skin symptoms, wheeze, and laryngeal symptoms. The wheeze and larynx symptoms were mild and consequently, no adrenaline or corticosteroids were administered; nevertheless, PR-10 pear proteins appear to be able to cause these objective symptoms after pear consumption [31]. On Western blot, most patient sera bound 'Cepuna' and 'Conference' protein bands in equal intensity, making it difficult to correlate these IgE binding results to the single-blind oral challenge results or the PTP, SPT, or sIgE measurements. Although the protein content in both pear varieties differed only slightly (0.315% ± 0.007% and 0.330% ± 0.006% for 'Cepuna' and

'Conference', respectively (DUMAS, Nf6.25); $p < 0.05$), the protein yield differed by a factor of 1.6 (2.8 mg versus 4.5 mg respectively), which might indicate that proteins in 'Cepuna' are more difficult to extract or less bio-available. This difference in protein availability might perhaps also occur when pears are eaten, but proving this would require further research. Differences in allergenicity may also be influenced by matrix components other than allergen content, e.g., the polyphenol content [32], which was not tested for in this study. We are aware that the applied enzyme-based immunoblot detection technique limits the sensitivity of the overall detection signal which could have restricted our data analysis especially in the case of low KU/L sIgE titers. To circumvent these limitations, future studies could consider X-ray film using radio-labeled antibodies to increase detection sensitivity. In addition, gradient gels specifically for low MW protein separation can be considered to increase sensitivity in the 7–35 kDa range, given that the known pear allergens Pyr c 1 to 5 run in this range (Table 4). We compared two pear varieties. This study was industry initiated and their hypothesis was that the 'Cepuna' pear was less allergenic. They regularly received signals from allergic patients in the Netherlands who experienced less to no symptoms during consumption of this pear. There was a very strong likelihood of fewer symptoms during 'Cepuna' pear challenges and even negative in three cases. In addition, the 'Cepuna' pear caused objective symptoms in only two patients versus seven patients during the 'Conference' challenge. The burden of pollen-related food allergy is often underestimated in patients with a multi-fruit allergy, and therefore it is of the highest interest to find one fruit that can be consumed. Fruits contain all kinds of nutrients and vitamins that are indispensable in the daily diet. Food allergic patients are interested in having low-allergen food available and want to eat the food they are allergic to [33]. Kootstra et al. [34] compared different apple cultivars and found that more than half (53%) of the patients ($n = 15$) could consume the 'Santana' apple without symptoms ($p = 0.02$) [34].

5. Conclusions

Although we did not find a significant difference in symptoms during single-blind oral challenges between both pear varieties in our study, we found a very high likelihood of fewer symptoms during the 'Cepuna' challenges. Consequently selected patients can try to consume small doses of the 'Cepuna' pear outside the birch pollen season.

Supplementary Materials: The following are available online at https://www.mdpi.com/article/10.3390/nu13041355/s1. Table S1: Indicative molecular weight of protein bands bound by IgE antibodies in patient sera; protein bands corresponding to the Western blots in Figure 1. Ce: 'Cepuna' pear; Co: 'Conference' pear; WB: Western blot; the number of x's indicate the (by eye) estimated intensity of the coloration. Serum nr 9 was not available for Western blotting.

Author Contributions: Funding acquisition: N.W.d.J., D.E., and S.B.-N.; clinical trial: N.W.d.J., S.T., F.E.v.B., M.W.J.S., and M.S.v.M.; protein isolation and quantification: D.B.P.M.v.d.B.-S.; immunoblotting: S.B.-N.; writing—original draft: N.W.d.J., S.B.-N.; review and editing: N.W.d.J., S.T., F.E.v.B., M.S.v.M., M.W.J.S., D.B.P.M.v.d.B.-S., D.E., and S.B.-N. All authors have read and agreed to the published version of the manuscript.

Funding: This research was funded by GKE NV. Experimental design and implementation have not been influenced by the funding organization. The views expressed in this manuscript are those of the authors and do not necessarily reflect the position or policy of the funding organization.

Institutional Review Board Statement: Medical ethical approval was received in August 2019; registered as METC NL70165.078.09.

Informed Consent Statement: Written informed consent has been obtained from the patient(s) to publish this paper.

Data Availability Statement: The original database ie not available online.

Acknowledgments: We thank A. Boudewijn and G.H.A. de Jong for their assistance in the protein isolation procedures.

Conflicts of Interest: The authors declare no conflict of interest.

References

1. Biedermann, T.; Winther, L.; Till, S.J.; Panzner, P.; Knulst, A.; Valovirta, E. Birch pollen allergy in Europe. *Allergy* **2019**, *74*, 1237–1248. [CrossRef]
2. Canis, M.; Gröger, M.; Becker, S.; Klemens, C.; Kramer, M.F. Recombinant marker allergens in diagnosis of patients with allergic rhinoconjunctivitis to tree and grass pollens. *Am. J. Rhinol. Allergy* **2011**, *25*, 36–39. [CrossRef]
3. Ciprandi, G.; Comite, P.; Mussap, M.; De Amici, M.; Quaglini, S.; Barocci, F.; Marseglia, G.L.; Scala, E. Profiles of Birch Sensitization (Bet v 1, Bet v 2, and Bet v 4) and Oral Allergy Syndrome Across Italy. *J Investig. Allergol. Clin. Immunol.* **2016**, *26*, 244–248. [CrossRef]
4. Geroldinger-Simic, M.; Zelniker, T.; Aberer, W.; Ebner, C.; Egger, C.; Greiderer, A.; Prem, N.; Lidholm, J.; Ballmer-Weber, B.K.; Vieths, S.; et al. Birch pollen-related food allergy: Clinical aspects and the role of allergen-specific IgE and IgG4 antibodies. *J. Allergy Clin. Immunol.* **2011**, *127*, 616–622. [CrossRef] [PubMed]
5. Katelaris, C.H. Food allergy and oral allergy or pollen-food syndrome. *Curr. Opin. Allergy Clin. Immunol.* **2010**, *10*, 246–251. [CrossRef]
6. Valenta, R.; Niederberger, V. Recombinant allergens for immunotherapy. *J. Allergy Clin. Immunol.* **2007**, *119*, 826–830. [CrossRef] [PubMed]
7. Werfel, T.; Asero, R.; Ballmer-Weber, B.K.; Beyer, K.; Enrique, E.; Knulst, A.C.; Mari, A.; Muraro, A.; Ollert, M.; Poulsen, L.K.; et al. Position paper of the EAACI: Food allergy due to immunological cross-reactions with common inhalant allergens. *Allergy* **2015**, *70*, 1079–1090. [CrossRef] [PubMed]
8. Mari, A.; Ballmer-Weber, B.K.; Vieths, S. The oral allergy syndrome: Improved diagnostic and treatment methods. *Curr. Opin. Allergy Clin. Immunol.* **2005**, *5*, 267–273. [CrossRef] [PubMed]
9. Burney, P.G.; Potts, J.; Kummeling, I.; Mills, E.N.; Clausen, M.; Dubakiene, R.; Barreales, L.; Fernandez-Perez, C.; Fernandez-Rivas, M.; Le, T.M.; et al. The prevalence and distribution of food sensitization in European adults. *Allergy* **2014**, *69*, 365–371. [CrossRef]
10. Zuidmeer, L.; Goldhahn, K.; Rona, R.J.; Gislason, D.; Madsen, C.; Summers, C.; Sodergren, E.; Dahlstrom, J.; Lindner, T.; Sigurdardottir, S.T.; et al. The prevalence of plant food allergies: A systematic review. *J. Allergy Clin. Immunol.* **2008**, *121*, 1210–1218. [CrossRef] [PubMed]
11. Zuberbier, T.; Edenharter, G.; Worm, M.; Ehlers, I.; Reimann, S.; Hantke, T.; Roehr, C.C.; Bergmann, K.E.; Niggemann, B. Prevalence of adverse reactions to food in Germany-a population study. *Allergy* **2004**, *59*, 338–345. [CrossRef]
12. Rodriguez, J.; Crespo, J.F.; Lopez-Rubio, A.; De La Cruz-Bertolo, J.; Ferrando-Vivas, P.; Vives, R.; Daroca, P. Clinical cross-reactivity among foods of the Rosaceae family. *J. Allergy Clin. Immunol.* **2000**, *106*, 183–189. [CrossRef] [PubMed]
13. Meest Geconsumeerde Fruitsoorten. Resultaten van VCP 2012–2014. *Rijksinst. voor Volksgezond.* **2018**. Available online: https://www.rivm.nl/sites/default/files/2018-11/2018-01%20memo%20fruit%20top%20tien.pdf (accessed on 10 October 2020).
14. Vlieg-Boerstra, B.J.; van de Weg, W.E.; van der Heide, S.; Dubois, A.E. Where to prick the apple for skin testing? *Allergy* **2013**, *68*, 1196–1198. [CrossRef] [PubMed]
15. van der Valk, J.P.; Gerth van Wijk, R.; Hoorn, E.; Groenendijk, L.; Groenendijk, I.M.; de Jong, N.W. Measurement and interpretation of skin prick test results. *Clin. Transl. Allergy* **2015**, *6*, 1–5. [CrossRef]
16. Dreborg, S. Allergen standardisation and skin test: EAACI position paper. *Allergy* **1993**, *48*, 49–82. [CrossRef]
17. Dreborg, S. Skin testing. The safety of skin tests and the information obtained from using different methods and concentrations of allergen. *Allergy* **1993**, *48*, 473–475. [CrossRef]
18. Sampson, H.A.; Gerth van Wijk, R.; Bindslev-Jensen, C.; Sicherer, S.; Teuber, S.S.; Burks, A.W.; Dubois, A.E.; Beyer, K.; Eigenmann, P.A.; Spergel, J.M.; et al. Standardizing double-blind, placebo-controlled oral food challenges: American Academy of Allergy, Asthma & Immunology-European Academy of Allergy and Clinical Immunology PRACTALL consensus report. *J. Allergy Clin. Immunol.* **2012**, *130*, 1260–1274.
19. Vieths, S.; Schöning, B.; Petersen, A. Characterization of the 18-kDa apple allergen by two-dimensional immunoblotting and microsequencing. *Int. Arch. Allergy Immunol.* **1994**, *104*, 399–404. [CrossRef]
20. Suleria, H.A.R.; Barrow, C.J.; Dunshea, F.R. Screening and Characterization of Phenolic Compounds and Their Antioxidant Capacity in Different Fruit Peels. *Foods* **2020**, *9*, 1206. [CrossRef] [PubMed]
21. Bastiaan-Net, S.; Reitsma, M.; Cordewener, J.H.G.; van der Valk, J.P.M.; America, T.; Dubois, A.E.J.; Gerth van Wijk, R.; Savelkoul, H.F.J.; de Jong, N.W.; Wichers, H.J. IgE Cross-Reactivity of Cashew Nut Allergens. *Int. Arch. Allergy Immunol.* **2019**, *178*, 19–32. [CrossRef] [PubMed]
22. Makowski, D.; Ben-Shachar, M.S.; Chen, S.H.A.; Lüdecke, D. Indices of Effect Existence and Significance in the Bayesian Framework. *Front. Psychol.* **2019**, *10*, 2767. [CrossRef]
23. Beyer, S.; Franke, A.; Simon, J.C.; Treudler, R. Measurement of health-related quality of life in adult patients with birch pollen-associated food allergy. *J. Dtsch. Dermatol. Ges.* **2016**, *14*, 397–404. [CrossRef]
24. Reiland, H.; Slavin, J. Systematic Review of Pears and Health. *Nutr. Today* **2015**, *50*, 301–305. [CrossRef]
25. van Erp, F.C.; Knulst, A.C.; Meijer, Y.; Gabriele, C.; van der Ent, C.K. Standardized food challenges are subject to variability in interpretation of clinical symptoms. *Clin. Transl. Allergy* **2014**, *4*, 1–6. [CrossRef]

26. Vlieg-Boerstra, B.J.; van der Heide, S.; Bijleveld, C.M.; Kukler, J.; Duiverman, E.J.; Dubois, A.E. Placebo reactions in double-blind placebo-controlled food challenges in children. *Allergy* **2007**, *62*, 905–912. [CrossRef] [PubMed]
27. Vlieg-Boerstra, B.J.; van de Weg, W.E.; van der Heide, S.; Skypala, I.; Bures, P.; Ballmer-Weber, B.K.; Hoffmann-Sommergruber, K Zauli, D.; Ricci, G.; Dubois, A.E. Additional indications for the low allergenic properties of the apple cultivars Santana and Elise *Plant Foods Hum. Nutr.* **2013**, *68*, 391–395. [CrossRef] [PubMed]
28. Vlieg-Boerstra, B.J.; van de Weg, W.E.; van der Heide, S.; Kerkhof, M.; Arens, P.; Heijerman-Peppelman, G.; Dubois, A.E Identification of low allergenic apple cultivars using skin prick tests and oral food challenges. *Allergy* **2011**, *66*, 491–498 [CrossRef] [PubMed]
29. Turner, P.J.; Dawson, T.C.; Skypala, I.J.; Fox, A.T. Management of pollen food and oral allergy syndrome by health care professionals in the United Kingdom. *Ann. Allergy Asthma Immunol.* **2015**, *114*, 427–428. [CrossRef]
30. Le, T.M.; van Hoffen, E.; Lebens, A.F.; Bruijnzeel-Koomen, C.A.; Knulst, A.C. Anaphylactic versus mild reactions to hazelnut and apple in a birch-endemic area: Different sensitization profiles? *Int. Arch. Allergy Immunol.* **2013**, *160*, 56–62. [CrossRef] [PubMed]
31. Skypala, I.J. Can patients with oral allergy syndrome be at risk of anaphylaxis? *Curr. Opin. Allergy Clin. Immunol* **2020**, *20*, 459–464 [CrossRef] [PubMed]
32. Russell, W.R. Phenolic acid content of fruits commonly consumed and locally produced in Scotland. *Food Chem.* **2009**, *115*, 100–104 [CrossRef]
33. Miles, S. Attitudes towards low-allergen food in food allergic consumers. *Nutr. Food Sci.* **2004**, *35*, 220–228. [CrossRef]
34. Kootstra, H.S.; Vlieg-Boerstra, B.J.; Dubois, A.E. Assessment of the reduced allergenic properties of the Santana apple. *Ann Allergy Asthma Immunol.* **2007**, *99*, 522–525. [CrossRef]

nutrients

MDPI

Article

Improving Successful Introduction after a Negative Food Challenge Test: How to Achieve the Best Result?

Joyce Emons *, Marije van Gunst, Olivia Liem, Lonneke Landzaat and Nicolette Arends

Division Erasmus MC, Department of Pediatrics, Division of Respiratory Medicine and Allergology, University Medical Center Rotterdam, 3015 CN Rotterdam, The Netherlands; m.vangunst@erasmusmc.nl (M.v.G.); o.liem@erasmusmc.nl (O.L.); l.landzaat@erasmusmc.nl (L.L.); n.arends@erasmusmc.nl (N.A.)

* Correspondence: j.a.m.emons@erasmusmc.nl

Received: 27 July 2020; Accepted: 1 September 2020; Published: 7 September 2020

Abstract: Oral food challenges (OFC) confirm or exclude the presence of a food allergy. The outcome can be positive (allergic symptoms), inconclusive, or negative (no symptoms). In the case of a negative OFC, parents and children are advised to introduce the challenged food allergen into their diet. However, previous studies showed difficulties in a successful introduction at home. The aim of this prospective non-randomized intervention study is to evaluate the effect of a new strategy with more guidance regarding the dietary introduction after a negative food challenge test. We compared two cohorts: an historical (retrospective) control group of 157 children, previously described, who did not receive any special advice after a negative OFC, versus a new cohort consisting of 104 children, who were guided according to our new strategy of written introduction schemes, food diaries, and several phone calls. In the historical control group, introduction was successful in 56%, partially successful in 16%, and 28% failed to introduce at home. After introduction of our new strategy, complete introduction was found in 82%, 11% had partially introduced, and only 8% failed to introduce the allergen. In conclusion, comprehensive advice and dietary recommendation after a negative OFC results in an increase in successful home introduction. Therefore, more attention, guidance, and follow-up of children and parents are desirable after a negative OFC.

Keywords: oral food challenge; successful introduction; children; food allergy; allergy; cow's milk; hens egg; peanut; hazelnut

1. Introduction

Food allergy is a well-known worldwide health problem. Prevalence numbers vary from 1–11% with patient self-reported food allergy up to 35% [1–3]. The most common food allergens in young children are cow's milk (2.5%), egg (1.3%), peanut (0.8%), wheat (0.4%), soy (0.4%), and tree nuts (0.2%) [4]. The gold standard for the diagnosis of food allergy is an oral food challenge test (OFC). Besides diagnosing an allergy, OFCs are also frequently performed to examine whether tolerance is developed in children who have a history of food allergy. After a positive outcome of a food challenge, a specific diet avoiding the culprit allergen is advised in order to prevent allergic reactions. This diet has a high impact on the quality of life of allergic children and their parents and deserves medical attention and guidance in order to avoid dietary shortage, malnutrition, or excessive avoidance behavior [5,6].

After a negative outcome of a food challenge test, it is recommended that children should (re)introduce the investigated allergen into their diet to improve dietary management and consequently to improve their quality of life. However, usually fewer consultations and follow-ups take place after a negative OFC. Recurrence of allergy is described in patients with a peanut allergy. These patients passed a food challenge test but failed to consume peanut frequently and had a recurrence of their allergy [7–9]. In addition, there is increasing evidence that atopic children who avoid allergenic foods for which they are sensitized are at increased risk of developing an acute allergy with the possibility of a severe allergic reaction in such cases [10]. Therefore, unnecessary elimination diets should be avoided as much as possible. For these reasons, a negative OFC can only be considered successful if it is followed by a successful introduction in the diet. Unfortunately, failure of introduction is reported frequently in the literature due to several reasons [9,11–14]. Reasons for (re)introduction failure are: symptoms during introduction, aversion of the food, fear of the child or parents, habit of not eating the food, other allergies, positive challenge test in patients or parents' experience, and allergy in the family [11].

The aim of this study is to evaluate whether new comprehensive advice and a written allergen-specific introduction protocol can increase the rate of a successful allergen introduction after a negative OFC for cow's milk, hen's egg, peanut, or hazelnut.

2. Materials and Methods

This prospective non-randomized intervention study was conducted between 16 March and 18 May at the Erasmus Medical Center, Sophia Children's Hospital in Rotterdam, The Netherlands. Approval of the Dutch medical ethical committee was received (MEC-2016-597). There were 104 children aged 0–18 years with a negative OFC to cow's milk, hen's egg, peanut, and hazelnut included in the study. OFCs were either open or double-blinded placebo-controlled. Outcomes were assessed and compared before and after the intervention.

2.1. Food Challenges and Intervention

Van der Valk et al. conducted a retrospective study in the same population and clinic from 2008–2013 [11]. A total of 188 negative OFCs were performed in 157 children. None of the children and parents received any special advice after their negative challenge test. The percentage of successful introductions after negative OFCs and reasons of introduction failure were examined. Since this investigated historical group is similar to our enrolled group of children, these results were used as a baseline prior to our intervention.

For this interventional study, parents and children were asked to participate after a negative OFC with one of the following allergens: cow's milk, hen's egg, peanut, or hazelnut (study group). OFC were either open, where the child received an unmasked food (the suspected allergen), or double-blinded (DBPCFC) with the allergen hidden and processed in a matrix. The matrix used for egg, peanut, and hazelnut was gingerbread; for cow's milk the matrix was soymilk, rice milk, or the hydrolyzed formula the child was using at that time. In the DBPCFC the child received on one day the placebo and the other day the suspected allergen. Blinding was guaranteed for the physician, the nurse, and the patient. Blinding was broken 24 h after the challenge. The food challenge test consisted of a six-step doses regime with increasing dosages every 20–30 min of 1, 3, 10, 30, 100, 300, and 1000 mg protein equivalent. Cumulatively, these dose were comparable to 50 mL of cow's milk, one fifth of hen's egg, seven peanuts, or ten hazelnuts. The challenge was discontinued and scored positive when objective allergic symptoms occurred, or subjective allergic symptoms occurred twice on two successive administrations of the challenge material. Objective symptoms and signs were defined as angioedema, urticaria, significant increase in eczema, rash, vomiting, diarrhea, rhinoconjunctivitis, stridor, coughing, wheezing, hoarseness, collapse, tachycardia, and hypotension. Subjective symptoms were defined as exacerbation of generalized itch (in the case of atopic eczema), abdominal pain, nausea and/or cramps, oral allergy symptoms, itchy throat or sensation of throat swelling, difficulty in swallowing, and 'other' symptoms such as drowsiness and irritability. Patients were observed for at least 1 h after the last dosage before discharge.

After inclusion, children and parents received a written step-wise introduction protocol concerning the challenged allergen. The protocol contained a list of several products containing the food allergen with stepwise advice on how to introduce carefully and in a well-controlled way. Additionally, parents were asked to fill in a food diary in order to assess the amount and frequency of the introduced allergen (see Files S1–S4).

The food diary was returned after 6 weeks and evaluated by a telephone consultation with the parents. A questionnaire of 40 questions was carried out during this consultation (see File S5).

2.2. *Success of Introduction*

Level of introduction was categorized into 3 groups: complete introduction, partial introduction, and failed introduction. Complete introduction was defined as regular (at least once a week) unlimited intake of the pure allergen. Partial introduction was defined as consuming small amounts of allergen in pure or processed products. Children with a failed introduction did not succeed in introduction and were still avoiding the tested allergen.

2.3. *Questionnaires*

The questionnaire contained a total of 40 questions (File S5). The first part concerned the patient and their family characteristics. The middle part of the questionnaire contained questions regarding symptoms before, during, and after the challenge test. The last part of the questionnaire focused on the successful or failed introduction of the investigated food and the parental experience of the new introduction protocol.

2.4. *Data Analysis*

Rate of successful dietary introduction was compared between the control group and the study group. Data were collected and processed in IBM SPSS Statistics (Version 25, North Castle, New York, USA). The data were analyzed by means of frequencies, differences, and coherence. Differences in introductions between the control group and the research group were analyzed using a chi-square test. Multivariable logistic regression analysis was performed to study the effect of the intervention and several covariables on the introduction of the allergen. Significance was defined as a p-value < 0.05.

3. Results

3.1. *Population*

A total of 104 children participated in the current study and 157 children in the control group. No patients were lost to follow-up. Baseline characteristics are presented in Table 1.

In the control group, a total of 188 negative food challenge tests were analyzed for either cow's milk, chicken's egg, peanut, or hazelnut. In the study group, a total of 104 food challenge tests were analyzed and performed in the period. Most challenge tests were DBPCFC (73%) and a minority (27%) were open. Patient's characteristics for both groups are shown in Table 1. The majority of the children (86% and 87%) were sensitized to the tested food allergen (sIgE detectable or positive SPT). Almost half of the patients (41%) had never consumed the allergen before, 41% of the patients had IgE mediated symptoms in their history, 10% had non-IgE mediated symptoms, and 8% did not consume the allergen for a longer period and a sensitization was found. For cow's milk allergy in the study group, only 45% of patients were sensitized, 27% of patients had symptoms of an IgE mediated cow's milk allergy, and 73% had non-IgE mediated allergy in their history. In both control and study groups, most children (94% and 90%) had other features of the atopic syndrome (asthma, rhinoconjunctivitis, eczema). In the study group there were more peanut challenge tests and less cow's milk tests. Patients were on average a little younger and fewer patients had eczema.

Table 1. Characteristics of the control and study group.

	Control Group *n* (%)	Study Group *n* (%)	*p* Value
Total number	188	104	
Boys	112 (60%)	69 (66%)	$p = 0.31$
Girls	76 (40%)	34 (33%)	
Age (year)	7.5 (5.5–11.3) *	5.0 (3.0–8.0)	$p \leq 0.01$
Atopy			
Asthma	63 (40%)	39 (39%)	$p = 1$
Rhinoconjunctivitis	74 (47%)	53 (55%)	$p = 0.26$
Eczema	136 (87%)	74 (74%)	$p = 0.02$
No atopic characteristics	10 (6%)	9 (9%)	$p = 0.41$
Positive Sensitization (SPT/ IgE)	156 (86%)	84 (87%)	$p = 0.98$
Food challenge test			
DBPCFC	146 (78%)	76 (73%)	$p = 0.46$
Open	42 (22%)	28 (27%)	
Tested allergen			
Cow's milk	41 (22%)	11 (11%)	$p = 0.04$
Egg	39 (21%)	24 (23%)	
Peanut	82 (20%)	32 (31%)	
Hazelnut	70 (37%)	37 (36%)	

* = median (range), SPT = Skin prick test, IgE = Immunoglobulin E, DBPCFC = Double-blind placebo-controlled food challenge.

3.2. Success and Failure of Introduction

A significant improvement of successful introduction was found in the study group compared to the control group (Table 2, $p < 0.01$ and Figure 1). In this study group, failure of introduction was highest in hen's egg (17%), followed by hazelnut (8%), peanut (3%), and no failure was seen with the introduction of milk after a negative challenge test. In the control group, the highest failure of introduction was seen for peanut (61%), followed by hazelnut (52%), followed by cow's milk (32%) and the lowest failure of introduction was seen with egg (26%). Reasons for introduction failure are depicted in Table 3.

Table 2. Success of introduction.

	Control Group *N* = 188 (%)	Study Group *N* = 104 (%)	
Successful introduction	106 (56%)	85 (82%)	$p < 0.01$
Partly introduction	30 (16%)	11 (11%)	
Failed introduction	52 (28%)	8 (8%)	

After introduction of our new strategy, dietary introduction was not successful in only eight children: four patients with hen's egg, three patients with hazelnut, and one patient with peanut. Egg introduction failed in two cases because of stomach ache and vomiting after eating boiled egg (challenge test was with baked egg), one patient failed due to fear after an anaphylactic reaction to another allergen and the fourth one was unsuccessful because of social issues in the family. Hazelnut introduction failed because of parental interpretation of subjective symptoms during the challenge test that they believed to be caused by the hazelnut (twice) or social issues in the family. The reason for failure of peanut introduction was fear.

Multivariable logistic regression analysis showed that gender, age, asthma, rhinoconjunctivitis, type of challenge test, and sensitization were not associated with a higher success of introduction. The intervention of our new strategy again was significant in this analysis ($p = 0.0001$). Eczema was found to be associated with successful introduction with an odds ratio of 4.1 (95% CI 1.4, 11.9) ($p = 0.009$), but also the patients with no atopic features were more successful in this analysis (OR 8.1; 95% CI 1.2, 52.2; $p = 0.00288$).

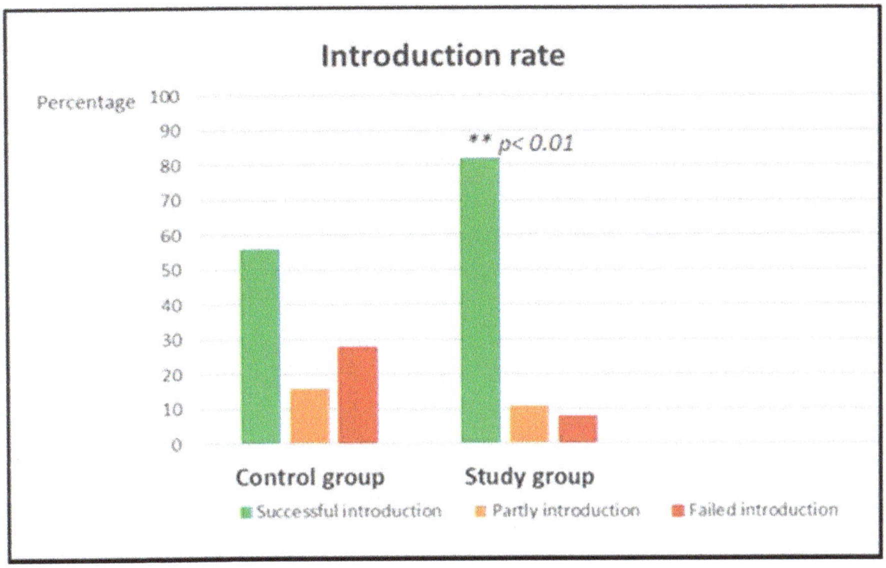

Figure 1. Percentages of success rate of introduction. ** $p < 0.01$.

Table 3. Reasons of failed introduction.

	Control Group $N = 52$ (%)	Study Group $N = 8$ (%)	
Symptoms at introduction	12 (23%)	4 (50%)	$p = 0.15$
Aversion of the food	11 (21%)	0	
Symptoms during OFC	2 (4%)	1 (10%)	
Fear for reaction (child)	7 (14%)	2 (20%)	
Dietary habit of avoidance	6 (13%)	0	
Fear for reaction (parents)	5 (10%)	0	
Other/unknown	9 (15%)	1 (10%)	

OFC: Oral food challenges.

3.3. *Parental Experience Regarding the New Introduction Protocol*

Most parents (74%) reported in the questionnaire that the introduction protocol was clear, informative, and helpful. Furthermore, they reported that this approach contributed to the introduction of the investigated allergen. Two thirds of the parents (65%) reported that the diary was also helpful with introduction.

4. Discussion

This is the first study showing that intensive guidance of allergen introduction after a negative challenge test results in a higher rate of successful introduction at home. In this prospective non-randomized intervention study, a written introduction protocol was used for cow's milk, hen's egg, peanut, and hazelnut. Together with the use of a diary and regular phone call appointments, introduction of allergens was significantly improved.

The 82% success rate in this study is higher compared to most other studies in the literature for these allergens. Eigenmann et al. reported 74.6% of successful introduction in 73 patients with a negative food challenge test of several allergens including milk, egg, and peanut [12]. In a Dutch study by van Erp et al., 68% of 103 children with a negative peanut challenge test failed to introduce peanut at home [9]. Whether an introduction is successful also seems to depend on the type of allergen tested, with milk giving the best outcome. Flammarion et al. studied the frequency of recurrent reactions during the introduction of cow's milk and its consequences for daily life in 67 children and reported a successful introduction in 83% of patients [13]. An additional study that investigated introduction after a negative cow's milk challenge test reported successful introduction of 80% [14]. In our study milk was best introduced as well, with 100% successful introduction in this group. Perhaps the nutritional importance of daily intake for milk is the explanation for this or less fear for introduction compared to peanut and tree nuts. In addition, in the Dutch diet a lot of dairy products are used like cheese and yoghurt. Comparative studies with high successful rates for milk introduction were performed in France with also a high dairy intake and the Netherlands. Perhaps in Asian and African countries, where the dairy intake is lower, these percentages would be lower.

All the above studies provide little to no information about the given advice and guidance by the medical staff or dietician after a negative challenge test. Schrijvers et al. studied the effect of personal follow-up and follow-up by phone after a negative cow's milk challenge test in Dutch children. They found an increase of 22% in both personal and follow-up by phone approaches (91%) compared to follow-up by phone alone (69%) [15]. No additional written advice was given in this study. We hypothesize that the tailored approach for each patient contributed to the success of introduction. The importance of introduction was highlighted for each patient. Patients were able to ask questions at several time points. They were reminded of the introduction in the extra contact moments and in addition in the diary that needed to be completed at home. The food diaries contained examples of food products that helped the parents in the selection of other products in case of food aversion, picky eaters, and dietary habits. The diaries are easy and a good way for the medical staff to check the amount and frequency of introduction with possible symptoms occurring that might influence introduction.

In addition to the written advice in the protocol and diary, there were also two telephonic consultations in follow-up of the challenge test with the medical staff including a dietician in some of the cases. Additional contact with a dietician resulted in a more successful introduction. An extra telephonic consultation is a good way to check whether introduction has succeeded, to help with problems, and to remind parents to introduce the allergen regularly into the child's diet.

In both groups the most important cause for a failed introduction were symptoms occurring during introduction at home. This might be caused by a false negative challenge test (due to desensitization during the challenge test) or by symptoms occurring after ingestion of higher dosages or less heated products in the case of milk and egg in which it is known that heating decreases their allergenic potential. Challenge tests for egg were performed with baked (well heated) egg and therefore less heated egg at home could still cause an allergic reaction. Milk OFCs were done with pure non-heated milk. It is important to evaluate these reactions with a pediatric allergist. In a few cases a re-challenge may be necessary for the culprit food. Other causes for a failed introduction were increase in eczema, fear from children or parents, complaints during the food challenge test, or no clear excuse was reported, but parents reported that there was simply no time. The same reasons were reported in other studies as well [9,11]. All the factors described can easily be clarified and addressed if this is acknowledged by the medical staff. In particular, fear is known to be present in a high percentage of allergic patients and this can have a large effect on quality of life. When this is recognized, it can be discussed, introductory steps can be taken more slowly, and psychological help can be offered when necessary.

Patient-related characteristics like asthma, gender, ethnicity, or age did not influence the rate of introduction in this study. This is in contrast to the study of Eigenmann et al. that reported more successful introduction in boys [12]. Another Dutch study regarding cow's milk introduction after a negative challenge test supported our findings and did not find an association between age and gender on the rate of successful introduction as well [16]. Eczema was found to be associated with more success of introduction. This is surprising since eczema is a chronic disease with frequent exacerbations in time and known by clinicians to complicate the introduction process. Parents confuse eczema with allergy symptoms. It is important for the medical staff to treat eczema properly and aim for optimal control with a dermatologist in consultation when necessary. However, in this study it was not found to be an important risk factor.

The number of patients in this study and in the subgroups consisted of a relatively small number of children, which might have affected the results. However, previously described comparative studies are even smaller. The current study is a prospective non-randomized intervention study and is compared to a previous retrospective study performed in the same hospital. The studies were not blinded and not placebo controlled. Since both groups are from the same hospital with the same medical staff and same food challenge protocols, it is likely that the intervention was the main cause of this increase in successful introduction.

Finally, our advice is to implement this new strategy in more Dutch Centers where allergic patients are treated and challenge tests are performed to test its national effectiveness. Furthermore, protocols can be translated and adapted to international dietary habits for other countries. In order to keep up to date, it may be possible to realize a digital protocol and/or app to advise patients and parents.

5. Conclusions

Dietary introduction after a negative food challenge is not always successful. Extra comprehensive advice and dietary recommendation from the medical staff results in a significant increase of allergen introduction into the diet. More guidance is advised for follow-up after negative food challenge tests.

Supplementary Materials: The following are available online at http://www.mdpi.com/2072-6643/12/9/2731/s1. File S1: Diary Negative Food challenges egg, File S2: Diary Negative Food challenges hazelnut, File S3: Diary Negative Food challenges milk, File S4: Diary Negative Food challenges peanut, File S5: Questionnaire English.

Author Contributions: J.E.: writing original draft preparation, inclusion patients, supervising OFC, advising patients after negative OFC, analysis and interpretation results. M.v.G.: writing original draft preparation, performing OFC, collecting data and analysis, telephone calls. O.L.: inclusion patients, performing OFC, advising patients after negative OFC. L.L.: inclusion patients, performing OFC, advising patients after negative OFC. N.A.: conceptualization, methodology, inclusion patients, supervising OFC, advising patients after negative OFC, analysis, and interpretation results. All authors have read and agreed to the published version of the manuscript.

Funding: No external funding was received for this research.

Acknowledgments: We acknowledge all the nurses performing the OFC in addition to the dietician who guided some of the patients extensively.

Conflicts of Interest: The authors declare no conflict of interest.

References

1. Rona, R.J.; Keil, T.; Summers, C.; Gíslason, D.; Zuidmeer, L.; Sodergren, E.; Sigurdardottir, S.T.; Lindner, T.; Goldhahn, K.; Dahlström, J.; et al. The prevalence of food allergy: A meta-analysis. *J. Allergy Clin. Immunol.* **2007**, *120*, 638–646. [CrossRef] [PubMed]
2. Savage, J.; Johns, C.B. Food allergy: Epidemiology and natural history. *Immunol. Allergy Clin. N. Am.* **2015**, *35*, 45–59. [CrossRef] [PubMed]
3. Venter, C.; Arshad, S.H. Epidemiology of food allergy. *Pediatr. Clin. N. Am.* **2011**, *58*, 327–349. [CrossRef] [PubMed]
4. Sicherer, S.H.; Sampson, H.A. 9. Food allergy. *J. Allergy Clin. Immunol.* **2006**, *117* (Suppl. S2), S470–S475. [CrossRef] [PubMed]
5. Avery, N.J.; King, R.M.; Knight, S.; Hourihane, J.O. Assessment of quality of life in children with peanut allergy. *Pediatr. Allergy Immunol.* **2003**, *14*, 378–382. [CrossRef] [PubMed]

6. Cummings, A.J.; Knibb, R.; King, R.M.; Lucas, J.S. The psychosocial impact of food allergy and food hypersensitivity in children, adolescents and their families: A review. *Allergy* **2010**, *65*, 933–945. [CrossRef] [PubMed]
7. Busse, P.J.; Nowak–Wegrzyn, A.; Noone, S.A.; Sampson, H.A.; Sicherer, S.H. Recurrent peanut allergy. *N. Engl. J. Med.* **2002**, *347*, 1535–1536. [CrossRef] [PubMed]
8. Fleischer, D.M.; Conover-Walker, M.K.; Christie, L.; Burks, A.W.; Wood, R.A. Peanut allergy: Recurrence and its management. *J. Allergy Clin. Immunol.* **2004**, *114*, 1195–1201. [CrossRef] [PubMed]
9. van Erp, F.C.; Boot, J.; Knulst, A.C.; Pasmans, S.G.M.; Van Der Ent, C.K.; Meijer, Y. Reintroduction failure after negative peanut challenges in children. *Pediatr. Allergy Immunol.* **2014**, *25*, 580–585. [CrossRef] [PubMed]
10. Tran, M.; Lefebvre, D.L.; Dai, D.; Dharma, C.; Subbarao, P.; Lou, W.; Azad, M.B.; Becker, A.B.; Mandhane, P.J.; Turvey, S.E.; et al. Timing of food introduction and development of food sensitization in a prospective birth cohort. *Pediatr. Allergy Immunol.* **2017**, *28*, 471–477. [CrossRef] [PubMed]
11. van Der Valk, J.P.M.; Van Wijk, R.G.; Vergouwe, Y.; De Jong, N.W. Failure of introduction of food allergens after negative oral food challenge tests in children. *Eur. J. Pediatr.* **2015**, *174*, 1093–1099. [CrossRef] [PubMed]
12. Eigenmann, P.A.; Caubet, J.C.; Zamora, S.A. Continuing food-avoidance diets after negative food challenges. *Pediatr. Allergy Immunol.* **2006**, *17*, 601–605. [CrossRef] [PubMed]
13. Flammarion, S.; Santos, C.; Romero, D.; Thumerelle, C.; Deschildre, A. Changes in diet and life of children with food allergies after a negative food challenge. *Allergy* **2010**, *65*, 797–798. [CrossRef] [PubMed]
14. Dambacher, W.M.; De Kort, E.H.; Blom, W.M.; Houben, G.F.; De Vries, E. Double-blind placebo-controlled food challenges in children with alleged cow's milk allergy: Prevention of unnecessary elimination diets and determination of eliciting doses. *Nutr. J.* **2013**, *12*, 22. [CrossRef] [PubMed]
15. Schrijvers, M.; Hendriks, T.; Mars, H.; Bertrams-Maartens, I.A. The effect of follow-up after a negative double-blinded placebo-controlled cow's milk challenge on successful reintroduction. *Eur. J. Pediatr.* **2019**, *178*, 1113–1117. [CrossRef] [PubMed]
16. Van De Ven, C.A.; Herpertz, I.; Van Lente, L.; Van Der Meulen, G.N.; Kamps, A.W.A. Successful introduction of milk after a negative double-blind placebo-controlled food challenge test is independent of the total dose and milk product used during the challenge test. *Immun. Inflamm. Dis.* **2020**, *8*, 292–298. [CrossRef] [PubMed]

nutrients

MDPI

Article

Peptide Release after Simulated Infant In Vitro Digestion of Dry Heated Cow's Milk Protein and Transport of Potentially Immunoreactive Peptides across the Caco-2 Cell Monolayer

Hannah E. Zenker [1], Harry J. Wichers [2,3], Monic M. M. Tomassen [2], Sjef Boeren [4], Nicolette W. De Jong [5] and Kasper A. Hettinga [1,*]

1 Food Quality & Design Group, Wageningen University & Research Centre, 6708 WG Wageningen, The Netherlands; hannah.zenker@wur.nl

2 Wageningen Food & Biobased Research, Wageningen University & Research Centre, 6708 WG Wageningen, The Netherlands; harry.wichers@wur.nl (H.J.W.); monic.tomassen@wur.nl (M.M.M.T.)

3 Laboratory of Food chemistry, Wageningen University & Research Centre, 6708 WG Wageningen, The Netherlands

4 Laboratory of Biochemistry, Wageningen University & Research Centre, 6708 WE Wageningen, The Netherlands; sjef.boeren@wur.nl

5 Internal Medicine, Department of Allergology & Clinical Immunology, Erasmus Medical Centre, 3000 CA Rotterdam, The Netherlands; n.w.dejong@erasmusmc.nl

* Correspondence: kasper.hettinga@wur.nl

Received: 13 July 2020; Accepted: 13 August 2020; Published: 18 August 2020

Abstract: Dry heating of cow's milk protein, as applied in the production of "baked milk", facilitates the resolution of cow's milk allergy symptoms upon digestion. The heating and glycation-induced changes of the protein structure can affect both digestibility and immunoreactivity. The immunological consequences may be due to changes in the peptide profile of the digested dry heated milk protein. Therefore, cow's milk protein powder was heated at low temperature (60 °C) and high temperature (130 °C) and applied to simulated infant in vitro digestion. Digestion-derived peptides after 10 min and 60 min in the intestinal phase were measured using LC-MS/MS. Moreover, digests after 10 min intestinal digestion were applied to a Caco-2 cell monolayer. T-cell epitopes were analysed using prediction software, while specific immunoglobin E (sIgE) binding epitopes were identified based on the existing literature. The largest number of sIgE binding epitopes was found in unheated samples, while T-cell epitopes were equally represented in all samples. Transport of glycated peptide indicated a preference for glucosyl lysine and lactosyl-lysine-modified peptides, while transport of peptides containing epitope structures was limited. This showed that the release of immunoreactive peptides can be affected by the applied heating conditions; however, availability of peptides containing epitopes might be limited.

Keywords: cow's milk protein; peptides; Caco-2 cell; immunogenicity; allergenicity; glycation

1. Introduction

Cow's milk protein consists of two major protein fractions, casein and whey protein, and is an important protein source in infant nutrition. Heating and glycation of cow's milk protein (MP) has been shown to alter its digestibility and immunogenicity. Dry heating, as applied in this study, is not commonly used in the dairy industry; however, it has an important role in mimicking the heat treatment when MP powder is baked into a muffin or waffle. These products are often referred to as "baked milk" and have been shown to accelerate the resolution of cow's milk allergy symptoms in allergic children [1]. Under these heating conditions (low aw-level, high sugar content, high temperature), protein aggregation and modifications via the Maillard reaction (MR) are favored [2]. The MR is the reaction between primary amino-groups of proteins, peptides and amino acids and the reactive carbonyl group of reducing sugars, for instance lactose. During the early stage of the MR, the initial condensation to the Amadori product occurs followed by a rearrangement to lactosyl lysine or glucosyl lysine. In the advanced stage of the MR, a pool of different advanced glycation end products (AGEs) are formed [3]. Amongst these, N^{ε}-carboxymethyllysine (CML) has been used as a marker for the advanced stage of the MR and is one of the most abundant AGEs in processed dairy products [4,5]. The extent of the MR can affect the digestibility and immunoreactivity of MP. With respect to digestibility, the effect of thermal processing of milk and dairy products on peptide generation during gastrointestinal digestion in vitro and in vivo has been subject to several studies [6–11]. Moreover, it was shown that lysine blockage via the MR affects peptide size distribution after simulated infant in vitro digestion of infant formula [12] and that glycation of isolated milk proteins changes the composition of the peptides in digestion [13]. Heating and glycation can also affect immunogenicity and allergenicity of MP [14]. For example, binding of specific immunoglobulin E (sIgE) to either isolated MP or MP in mixture has been shown to decrease for extensively glycated milk proteins, possibly related to a masking effect on epitopes [15–17]. However, the effects of glycation and protein aggregation under the applied heating conditions are difficult to disentangle and can both affect sIgE binding [18]. Corzo-Martínez et al. [19] also showed that impaired digestibility may increase the residual allergenicity after in vitro digestion, when comparing heat-glycated and unheated β-lactoglobulin. Differences in the peptide profiles after ingestion of dry heated MP vs. unheated MP could affect the immunological response by differential preservation or glycation induced-modification of linear sIgE binding epitopes. Moreover, the generation of peptides carrying a glycation structure can modulate the inflammatory response by binding to the receptors for AGEs on antigen presenting cells [14,20]. Binding of AGEs to AGE receptors has particularly been shown for protein-bound CML and pyrroline [21,22], while for peptide bound AGEs this was only demonstrated for CML [20]. The availability of AGE-modified peptides to the gastrointestinal immune system by means of translocation across the epithelial barrier is an important determinant in the immunological response to a foreign antigen. The metabolic transit of AGEs has been shown in previous literature on the excretion of CML and pyrroline in urine [23,24]. Moreover, translocation across the Caco-2 cell monolayer has been shown for lactosylated and CML-modified dipeptides [25]. As reviewed by O'Hagan et al. [26], the literature reports that small quantities of intact proteins, other macromolecules, and intact antigens can pass the intestinal epithelial layer in vivo. Furthermore, the identification of cow's milk derived peptides, ranging from 6 to 17 amino acids, in human milk has recently been described, indicating their absorption via the gastrointestinal tract [27]. However, to our knowledge the transport of food derived glycated peptides larger than two amino acids has not yet been investigated. Transport of larger glycated peptides can be an important factor for the binding of AGE-modified peptides as it has been suggested that CML is more abundant in fractions of in vitro digests that are larger than 1 kDa [28]. Therefore, transport of larger AGE-modified peptides could also contribute to the pool of dietary derived AGEs. This could be crucial as it has been shown with the example of CML that binding to AGE receptors is dependent on the concentration in which the CML is present in the vicinity of the receptors [22]. In this study, the peptide profiles of low temperature (LT) and high temperature (HT) heated MP after simulated infant in vitro digestion was compared to that of non-treated milk (NT). The most abundant glycation induced post translational modifications (PTMs), including CML and pyrroline as potential AGE receptor ligands, were monitored before and after transport across a Caco-2 cell monolayer. Special attention was given to sIgE binding epitopes and T-cell epitopes to assess immunomodulatory potential of the digest on the peptide level.

2. Materials and Methods

2.1. Chemicals

Dulbecco's Modified Eagles Medium supplemented with high glucose, HEPES, L-glutamine (42,430,082), both with and without phenol red as well as trypsin-EDTA (0.25%) with phenol red, and HyClone™ Fetal Bovine Serum were purchased from Thermo Fisher Scientific (Waltham, MA, USA). All other chemicals were obtained from Merck KGaA (Darmstadt, Germany).

2.2. Preparation of Milk Powders and Heat Treatment

Liquid raw cow's MP concentrate was obtained from FrieslandCampina (Wageningen, The Netherlands) and was composed of micellar casein (MCI88 liquid) and whey protein (acid WPC80 liquid) in the ratio 80:20. After the addition of lactose in the ratio 1:1.5 (protein/lactose), the solutions were freeze dried.

Heat treatment was applied at two different temperatures and durations as described elsewhere [29]. Briefly, for LT heated MP (LT-MP), the powder was heated for three weeks at 60 °C (a_w 0.23) and for HT heated MP (HT-MP) the powder was heated for 10 min at 130 °C (a_w 0.23). An unheated part of the milk powder was used as heating control (NT-MP).

2.3. Infant In Vitro Digestion

Simulated infant in vitro digestion was conducted in duplicate and was based on the protocol by Ménard et al. [30] with adaptations specific for the type of product described elsewhere. Compared to the adult digestion model, pH in the gastric phase (GP) was higher, while pH in the intestinal phase (IP) was lower. At the same time, enzyme concentrations were lower compared to the adult model [29]. Briefly, protein concentration of the meal was set to 1.2%. Digestion in the GP was conducted for 60 min with a pepsin activity set to 268 U/mL and at pH 5.3, but without the use of gastric lipase, as the milk powder contained <1% fat. Digestion in the IP was conducted for 60 min using pancreatin adjusted for its trypsin activity set to 16 U/mL digest and at pH 6.6. Samples were taken after 10 min and 60 min in the IP and stopped by the addition of 0.5 mM Pefabloc in the ratio Pefabloc/digest of 1/20 (*v/v*).

2.4. Caco-2 Cell Culture

The Caco-2 cell line was purchased from the American Type Culture Collection (Manassas, VA, USA), cultured in Dulbecco's Modified Eagles Medium (DMEM) supplemented with 10% heat inactivated fetal bovine serum (FBS Hyclone) at 37 °C and in a humidified atmosphere containing 5% CO_2. Cells were sub-cultured weekly upon confluence 85–95% using trypsination. Caco-2 cells were used from passage 30–40 and seeded into 24-well trans-wells (Greiner Bio-One, Kremsmünster, Austria) at a concentration of 0.225×10^6 cells/mL in DMEM with 10% heat inactivated FBS. The medium was changed (apical (150 µL) and basolateral (700 µL)) every two–three days and cells were used for the transport experiment after 21 days of incubation. Before transport experiments, the transepithelial electrical resistance (TEER) value was measured and only wells with a TEER value higher than 750 $\Omega \cdot cm^2$ were used.

2.5. Transport across the Caco-2 Cell Monolayer

Digest of one of the in duplicate in vitro digestions were diluted 1:1 with DMEM without phenol red, supplemented with 0.1% penicillin-streptomycin (10,000 U/mL) and applied to the apical side of the Caco-2 cell monolayer. TEER was measured at 37 °C using a Millicell-ERS 'Ω Meter (Millipore, Molshein, France) and samples were incubated for 2 h with Caco-2 cells at 37 °C and 5% CO_2 saturation. TEER was then measured and samples were taken from the basolateral side. Each sample was applied in duplicate. Samples were kept at −20 °C until further analysis.

2.6. Peptide Analysis

Digests after 10 min in the IP contained 3.6 mg/mL, 3.7 mg/mL, 3.6 mg/mL protein in NT-MP, LT-MP, and HT-MP, respectively. Digest after 60 min in the IP contained 4.2 mg/mL, 3.5 mg/mL, 3.8 mg/mL protein in NT-MP, LT-MP, and HT-MP, respectively. Samples were mixed 1:1 with trichloroacetic acid (20%) and centrifuged (10 min, 3500× *g*, 4 °C). The supernatants were cleaned using an in-house stage tip following a protocol described by Dingess et al. [31]. All samples were concentrated to compensate for the dilution during trichloroacetic acid precipitation.

Peptides were analysed on a Thermo nLC 1000 system (Thermo, Waltham, MA, USA) coupled to a LTQ orbitrap XL (Thermo Fisher Scientific, Breda, The Netherlands) for peptides in the in vitro digest, or Q Exactive HF-X X (Thermo Fisher Scientific, Breda, The Netherlands) for peptides on the basolateral side, as well as glycated peptides. Each sample was measured once. Chromatographic separation was conducted over a 0.10 × 250 mm ReproSil-Pur 120 C18-AQ 1.9 µm beads analytical column. A gradient consisting of acetonitrile in water spiked with 0.1% formic acid was used. Acetonitrile increased from 9% to 34% within 50 min using a flow rate of 0.5 µL/min. Full scan positive mode spectra (FTMS) were measured in the Orbitrap between *m*/*z* 380 and 1400 using high resolution (60,000). Collision-induced dissociation (LTQ) or Higher-energy collisional dissociation (Q Exactive HF-X) fragmentation was applied using an isolation width of 2 *m*/*z* and 1.2 *m*/*z*, respectively and 30 % and 24% normalized collision energy, respectively. MSMS scans were recorded in the data dependent mode for 2–3 2–5+charged peaks in the MS scan. For glycated peptides measured by the Q Exactive HF-X, a stepped collision energy (sCE 20-30-40) was used based on the method published by Liu et al. [32]. LC-MS/MS runs were processed using the MaxQuant version 1.6.3.4 with the Andromeda search engine [33]. Digestion mode was set to "unspecific". A fixed modification was set for the formation of propionamide on cysteines, while variable modifications were set for acetylation of the peptide N-terminus, deamidation of asparagine and glutamine, and oxidation of methionine.

Peptides were identified using a bovine database from Uniprot (https://www.uniprot.org) that includes all the bovine milk proteins observed by Boggs et al. [34] (PRIDE PXD003011) in combination with a database for common contaminants. For peptide identification in MaxQuant with unspecific enzyme cleavage, peptides with a minimum length of 8 amino acids and maximum peptide length of 25 amino acids were identified to limit false identifications. Both peptide and protein false discovery rates were set to 1%. Post translational modifications were included for lactosylation (+324 Da), hexose modification (+162 Da), N^{ε}-carboxymethyllysine modification (+58 Da), and pyrroline modification (+108 Da). For simplicity glucosyl lysine was used to refer to the hexose modification, although other hexoses could also result in this mass shift. Phosphorylated and glycated peptides were not included in the quantitation during the MaxQuant search. Due to the limited number of measurements, as well as the limitations in obtaining quantitative data from glycated peptides, data were reported as peptide count.

2.7. Data Analysis

Data were filtered for peptides derived from the six major milk proteins: α_{s1}-casein, α_{s2}-casein, β-casein, κ-casein, β-lactoglobulin, and α-lactalbumin. All peptides with score >80 were used for the overall peptide profiles, while for the sIgE binding epitopes and T-cell epitopes, only peptides with a score >100 were used. For total peptide count per sample in the digest, each duplicate digestion was filtered separately for non-modified peptides (intensity >0) and for phosphorylated and glycated peptides (identification by matching and/or by MS/MS). For all further analysis, only peptides identified in both duplicate digestions of the same heat treatment were reported.

2.8. sIgE Binding Epitope Identification

sIgE binding epitopes were identified by comparison of digestion-derived peptide sequences with sIgE binding epitopes as reviewed previously [35]. Peptides were reported as potential sIgE binding epitopes if their sequence matched ≥80% of the sequence of a known sIgE binding epitope.

2.9. T-Cell Epitope Prediction

T-cell epitopes were predicted using IEDB MHC Class II Binding Prediction software (http://tools.iedb.org/mhcii/, 02.06.2020) where an MHC class II allele reference set was obtained from (https://help.iedb.org/hc/en-us/articles/114094151851, 02.06.2020). The default method "IEDB recommended 2.2" was used for T-cell epitope predictions. All peptides within the size range 15–24 amino acids, which was previously reported as the size range for T-cell epitopes, were applied to the prediction software [36]. Peptides were reported as potential T-cell epitopes following the recommendations of the prediction tool, where each peptide reaching a percentile rank <10.0% can be considered as potential T-cell epitope.

3. Results

3.1. Identification of Peptides in In Vitro Digests

Peptides released upon infant in vitro digestion after 10 and 60 min in the IP were analysed using LC-MS/MS. Only peptides derived from the six major MPs, α_{s1}-casein, α_{s2}-casein, β-casein, κ-casein, β-lactoglobulin, and α-lactalbumin were considered in the data analysis. Dry heating of MP decreased the number of peptides released upon infant in vitro digestion (Figure 1), where HT heating resulted in even less peptides than LT heating after 10 min (315 ± 36 vs. 369 ± 26 peptides) and 60 min (207 ± 1 vs. 246 ± 7 peptides) in the IP.

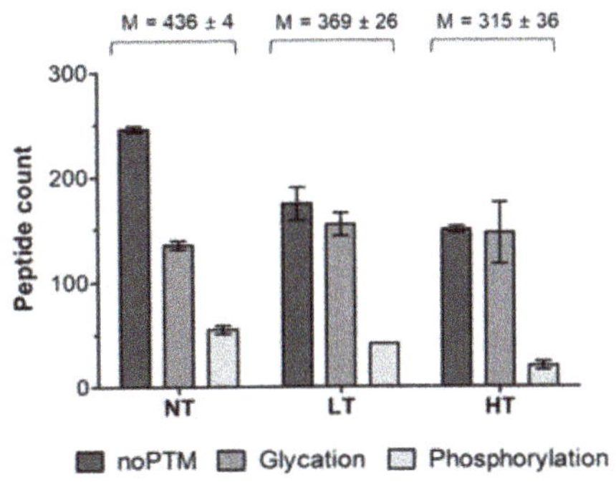

(a)

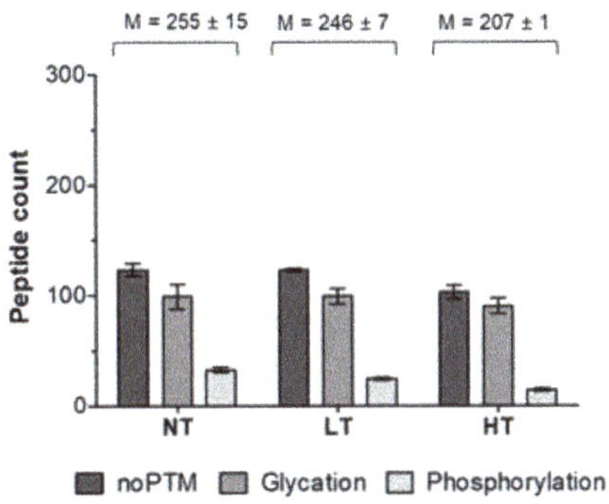

(b)

Figure 1. Total count of non-modified, glycated and phosphorylated digestion-derived peptides derived from cow's milk protein. Samples were non-treated (NT), heated at low temperatures (LT), and heated at high temperature (HT) (**a**) after 10 min in the intestinal phase and (**b**) after 60 min in the intestinal phase. Number of peptides without post translational modification (noPTM), glycated, and phosphorylated peptides were compared. The minimum length for identification was eight amino acids. Error bars represent the standard deviation of duplicate digestions. The mean (M) of the total count of peptides per treatment and digestion time point ± standard deviation for duplicate digestions is shown above the bars.

Differences in the modification state of the peptides (non-modified vs. glycated vs. phosphorylated peptides) were higher after 10 min than after 60 min in the IP. Heated samples showed comparable levels of glycated and non-modified peptides, while the NT-MP sample had two-fold more non-modified peptides than glycated peptides, after 10 min in the IP. At the same time, the number of phosphorylated peptides was 4.6-fold lower in HT-MP compared to NT-MP after 10 min in the IP, while LT-MP only showed a 1.6-fold decrease. This trend continued until 60 min in the IP, however to a lesser extent. Most peptides after 10 min in the IP were derived from β-casein, followed by β-lactoglobulin and α_{s1}-casein, while a smaller number of peptides originated from α_{s2}-casein, followed by κ-casein, and α-lactalbumin (Supplementary Materials: Figure S1a). This trend did not change after 60 min in the IP (Supplementary Materials: Figure S1b).

In line with the number of peptides per protein, peptides generated after 10 min in the IP covered large parts of the protein sequences of β-casein, β-lactoglobulin, and α-caseins (Figure 2a).

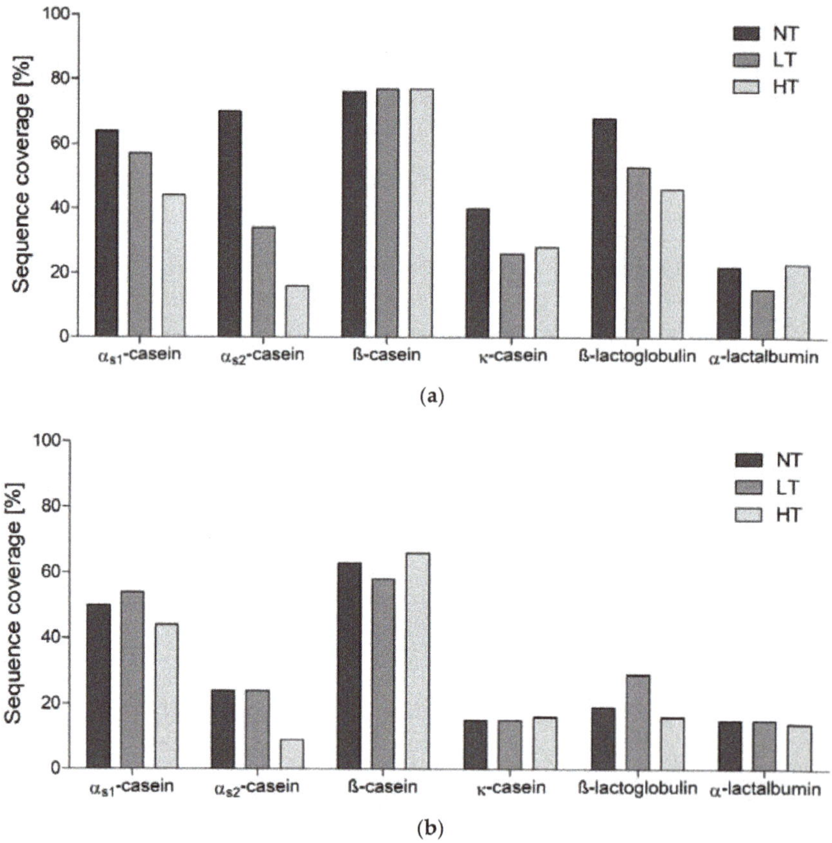

Figure 2. Sequence coverage of the six main milk proteins by digestion-derived peptides with and without posttranslational modification derived from in vitro digests of cow's milk protein. Samples were non-treated (NT), heated at low temperature (LT), and heated at high temperature (HT) (**a**) after 10 min in the intestinal phase and (**b**) after 60 min in the intestinal phase.

Sequence coverage was higher for NT-MP compared to LT-MP and HT-MP. This difference was highest for the α-caseins and β-lactoglobulin, while β-casein showed no changes. After 60 min in the IP, only α_{s2}-casein showed remarkably lower sequence coverage in HT-MP compared to NT-MP and LT-MP (Figure 2b). Due to the larger differences observed after 10 min in the IP, and the possibility of an immune response already at this stage of digestion, we focused mainly on the samples from 10 min in the IP.

In Figure 3, peptide sequence alignment is shown for peptides generated after 10 min in the IP. Independent from the heat treatment, all proteins showed specific regions that were similarly covered in all samples, but with different numbers of peptides generated in the same region.

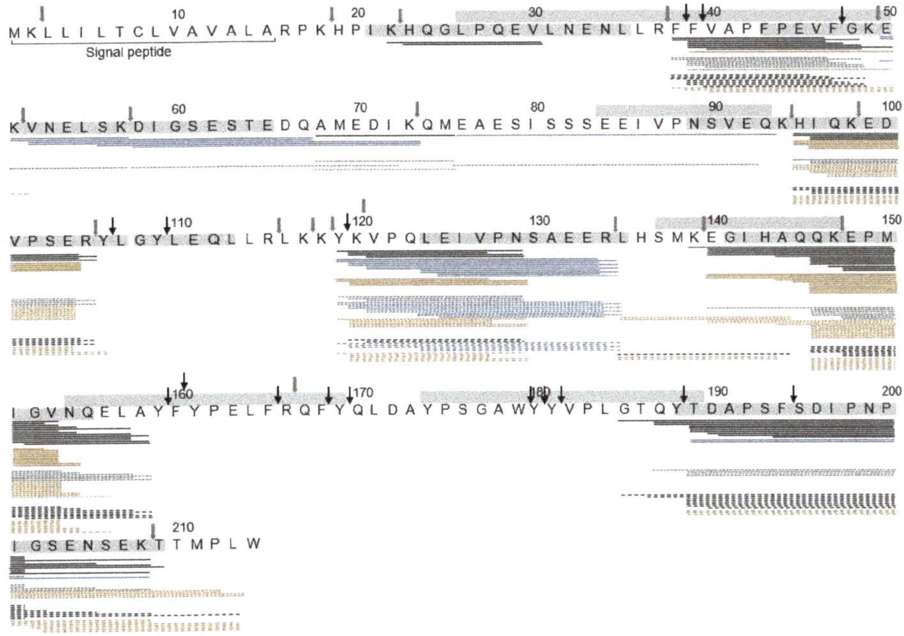

(a)

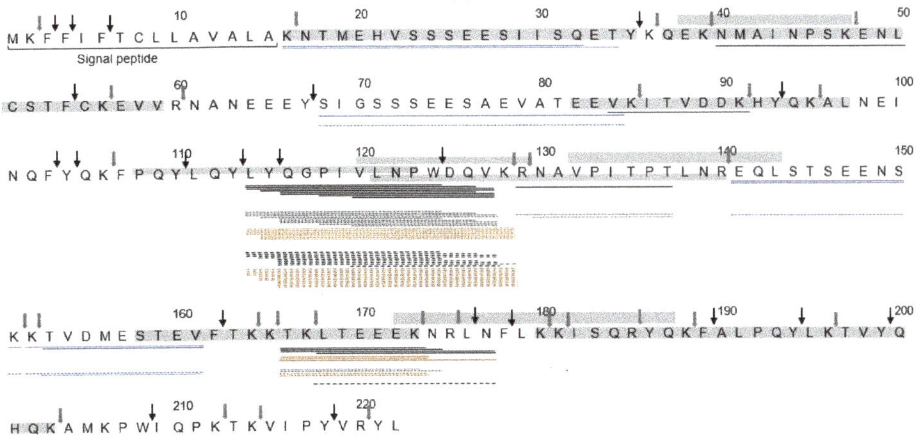

(b)

Figure 3. *Cont.*

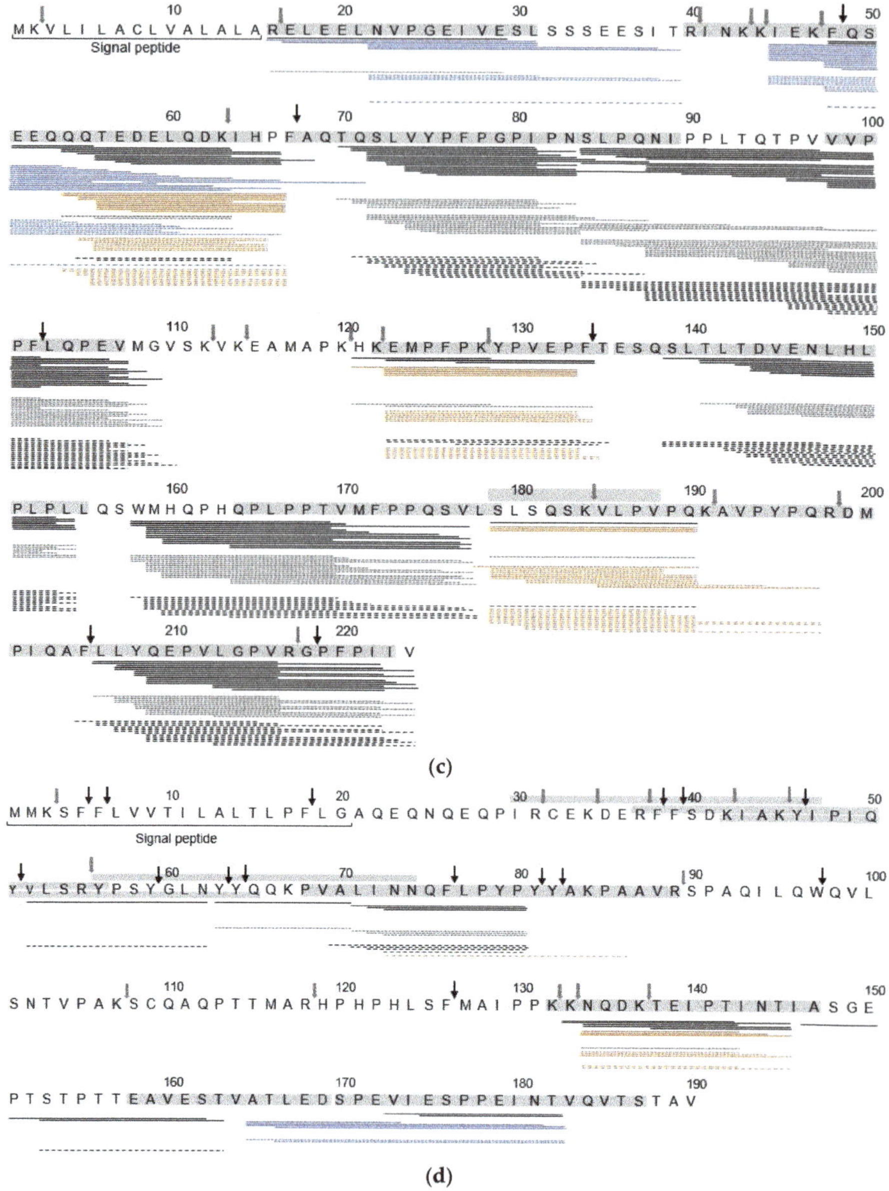

Figure 3. *Cont.*

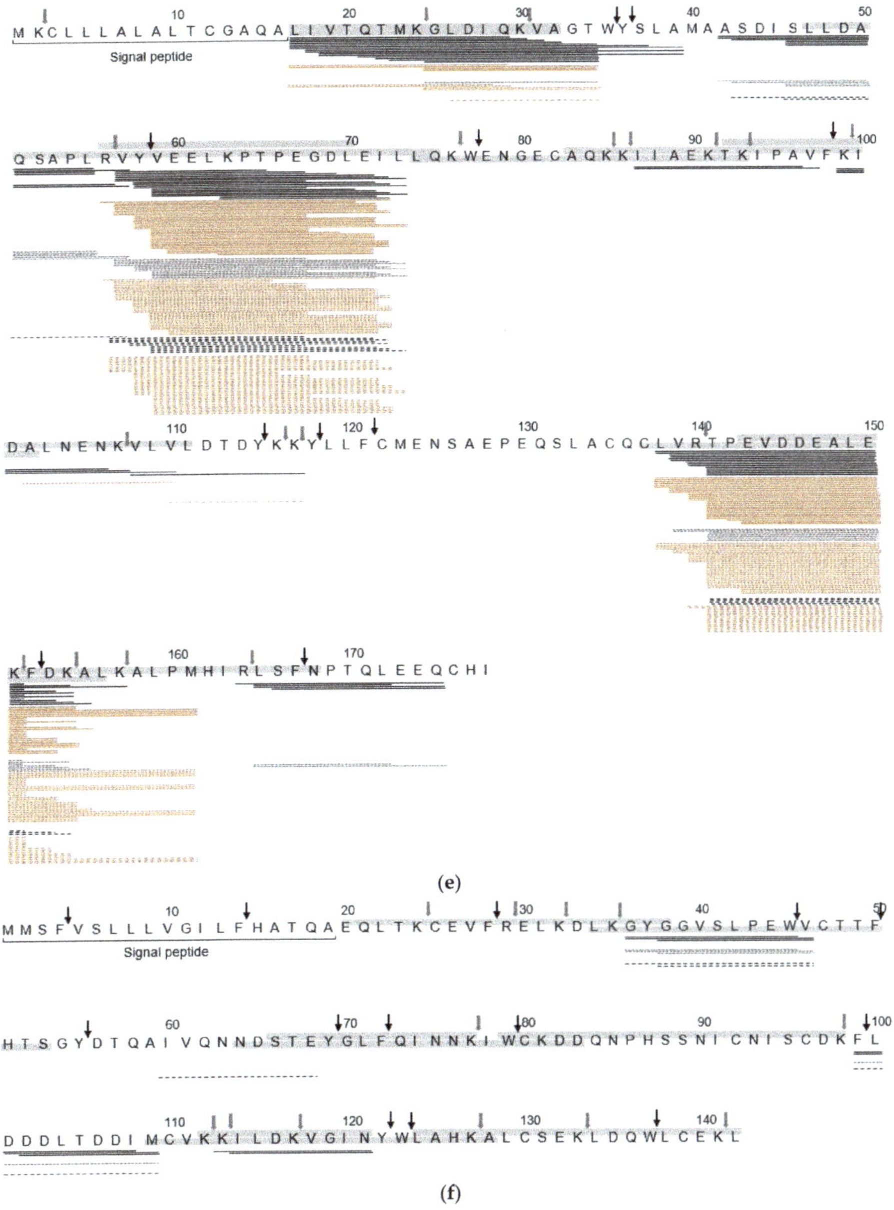

(e)

(f)

Figure 3. Sequence alignment of digestion-derived peptides identified after 10 min in the intestinal phase. Peptides derived from (**a**) α_{s1}-casein, (**b**) α_{s2}-casein, (**c**) β-casein, (**d**) κ-casein, (**e**) β-lactoglobulin, (**f**) α-lactalbumin identified after simulated infant in vitro digestion of non-treated cow's milk protein (full line), heated at low temperature (dotted line), and heated at high temperature (dashed line). Glycated peptides (orange), phosphorylated peptides (blue), trypsin cleavage sites (thick grey down arrow), chymotrypsin cleavage sites (thin black down arrow). Trypsin and chymotrypsin cleavage sites were determined using Expasy Bioinformatics Resource Portal (https://web.expasy.org/peptide_cutter/ last visited 08.06.2020).

In all proteins, the most differences in peptide distribution along the protein sequence were observed between HT-MP and the other samples. The peptides derived from β-lactoglobulin (Figure 3e) came from three main regions (f17–39, f56–73, and f138–161) and only minor differences were observed between heat treatments in f56–73, while in f17–39 and f138–161 much fewer peptides were found in HT-MP compared to NT-MP and LT-MP. In contrast, peptides derived from α_{s2}-casein and β-casein where distributed all over the amino acid chain. Interestingly, most differences in caseins were observed as a result of the absence of phosphorylated peptides in HT-MP. In α_{s1}-casein (Figure 3a), the region between f52–92 was mostly covered by phosphorylated peptides in NT-MP and LT-MP, while this region did not lead to the formation of peptides in HT-MP. Moreover, the number of peptides covering the same sequence part f52–92 was much lower in LT-MP than NT-MP. Similar observations, where the number of phosphorylated peptides was lower in at least one of the heated samples compared to NT-MP, were also made for α_{s2}-casein f16–32, f68–84, f141–161(Figure 3b), β-casein f22–39 and f48–66 (Figure 3c), and κ-casein f165–182 (Figure 3d). In contrast, glycated peptides only had a minimal effect on differences in sequence coverage when comparing samples. In α_{s1}-casein (Figure 3a), the regions f135–138 and f210–213 and in β-casein (Figure 3c) the region 191–197 were covered in LT-MP and HT-MP as a result of the presence of glycated peptides. The presence of glycated peptides, however, affected the number of peptides which arise from specific areas of the proteins. This was especially seen for the region f140–155 of α_{s1}-casein in NT-MP, f113–128 of α_{s2}-casein, and f179–197 of β-casein in both LT-MP and HT-MP.

While progressing intestinal digestion, only small changes were observed in the peptide alignment along the protein sequence of α_{s1}-casein, β-casein, κ-casein, and α-lactalbumin (Supplementary Materials: Figure S2). In contrast, β-lactoglobulin showed two resistant areas f57–73 and f139–154 as well as α_{s1}-casein at f119–134.

3.2. Identification of sIgE Epitopes and T-Cell Epitopes in the In Vitro Digest

sIgE binding epitopes were identified by comparison with known epitopes from the literature (Table 1) [35]. Peptides were reported as potential sIgE binding epitopes when at least 80% of the peptide sequence matched a known sIgE epitope sequence. Peptides derived from β-lactoglobulin contained 18 sIgE epitopes, followed by 16 derived from α_{s1}-casein, 14 from β-casein, 3 from α_{s2}-casein, and 1 from α-lactalbumin and κ-casein, respectively. The majority of sIgE epitopes were found in peptides derived from NT-MP; however, up to 69% of α_{s1}-casein derived sIgE epitopes and 77% of β-casein derived sIgE epitopes were also found in either one or both heated samples. The peptides α_{s1}-casein f189–213, α_{s2}-casein f116–128, and β-casein f96–110 were only found in heated samples; however, their length only differed by a maximum of four amino acids from a similar peptide found in NT-MP and those four amino acids were not covering an additional sequence part of the sIgE binding epitope.

Additionally, glycated peptides that matched sequence parts of sIgE binding epitopes were identified after 10 min in the IP (Table 2). Most of such peptides were found in β-lactoglobulin; however, only five of them were exclusively found in heated samples and covered similar sequence parts as peptides that were also found in NT-MP. Glycated peptides matching the sequence of an sIgE binding epitope from α_{s2}-casein and β-casein were exclusively found in heated samples. For α_{s1}-casein, three out of four glycated peptides with sequence homology to an sIgE binding epitope were only found in HT-MP.

Table 1. sIgE binding epitopes [1] identified in digestion-derived peptides after 10 min in the intestinal phase. Peptides were identified in cow's milk protein, non-treated (NT), heated at low temperature (LT), and heated at high temperature (HT), after simulated infant in vitro digestion and derived from casein (cn), β-lactoglobulin (lg), and α-lactalbumin (lac). Peptides matching exactly the sIgE binding epitope sequence are indicated (*). Amino acids (AAs) position indicates the position within the proteins including the signal peptide. Peptides carrying a post translational modifications (PTM) are marked with phosphorylation (Phos), whereas phosphorylated serine (S) and threonine (T) residues are highlighted in bold and underline.

Protein	Sample	Peptide Sequence	AAs Position	sIgE Epitope AAs Position	PTM	
α_{s1}-cn	NT, LT, HT	VNEL**S**KDIG**S**E**ST**EDQ	52–67	54–63	Phos	
	NT, LT, HT	VNEL**S**KDIG**S**E**ST**EDQAMEDIK	52–73	54–63	Phos	
	NT, LT, HT	KVPQLEIVPN**S**AEE	120–133	124–135	Phos	
	NT, LT, HT	KVPQLEIVPN**S**AEER	120–134	124–135	Phos	
	NT, LT	VPQLEIVPN**S**AEER	121–134	124–135	Phos	
	NT, LT, HT	LEIVPN**S**AEE	124–133	124–135	Phos	
	NT, LT, HT	LEIVPN**S**AEER	124–134	124–135	Phos	
	NT	EIVPN**S**AEER	125–134	124–135	Phos	
	NT	KEGIHAQQKEPMIGV	139–153	137–147	N/A	
	NT	EGIHAQQKEPMIGV	140–153	141–155	N/A	
	NT, HT	GTQYTDAPSFSDIPNPI	185–201	186–200	N/A	
	NT, LT, HT	QYTDAPSFSDIPNPI	187–201	186–200	N/A	
	NT	QYTDAPSFSDIPNPIGSENSEK	187–208	188–209	N/A	
	NT, LT, HT	TDAPSFSDIPNPIGSENSEK	189–208	188–209	N/A	
	NT,LT,HT	TDAP**S**F**S**DIPNPIG**S**EN**S**EK	189–208	188–209	Phos	
	NT	TDAPSFSDIPNPIGSENSEKT	189–209	188–209	N/A	
α_{s2}-cn	NT	KNTMEHV**SSS**EE**S**IISQ	16–32	16–35	Phos	
	NT	KNTMEHV**SSS**EE**S**IISQET	16–34	16–35	Phos	
	HT	QGPIVLNPWDQVK	116–128	120–129	N/A	
β-cn	NT, LT	RELEELNVPGEIVE	16–29	16–31	N/A	
	NT	RELEELNVPGEIVE**S**L	16–31	16–31	Phos	*
	NT	ELEELNVPGEIVE**S**L	17–31	16–31	Phos	
	NT	TEDELQDKIHPFA	56–68	60–69	N/A	
	NT, LT, HT	SLVYPFPGPIPNS	72–84	70–85	N/A	
	NT, LT, HT	PVVVPPFLQPE	96–106	98–107	N/A	
	NT, LT, HT	PVVVPPFLQPE	96–107	98–107	N/A	
	NT, LT, HT	PVVVPPFLQPEVMG	96–109	98–107	N/A	
	LT, HT	PVVVPPFLQPEVMGV	96–110	98–107	N/A	
	NT, LT, HT	VVPPFLQPE	98–106	98–107	N/A	
	NT, LT, HT	VVPPFLQPEV	98–107	98–107	N/A	*
	NT, LT, HT	EMPFPKYPVEPF	123–134	122–135	N/A	
	NT, LT	QPLPPTVMFPPQS	164–176	164–179	N/A	
	NT, LT, HT	QPLPPTVMFPPQSV	164–177	164–179	N/A	
κ-cn	NT	KNQDKTEIPTINT	133–145	132–147	N/A	
β-lg	NT	LIVTQTMKGLDIQ	17–29	17–32	N/A	
	NT	LIVTQTMKGLDIQKV	17–31	17–32	N/A	
	NT	LIVTQTMKGLDIQKVA	17–32	17–32	N/A	
	NT	LIVTQTMKGLDIQKVAGT	17–34	17–32	N/A	
	NT	LIVTQTMKGLDIQKVAGTWYS	17–37	17–32	N/A	
	NT	LIVTQTMKGLDIQKVAGTWYSLA	17–39	17–32	N/A	
	NT	IVTQTMKGLDIQKVAGT	18–34	17–32	N/A	
	NT	IVTQTMKGLDIQKVAGTWYSLA	18–39	17–32	N/A	
	NT	VTQTMKGLDIQKVAGT	19–34	17–32	N/A	
	NT	VTQTMKGLDIQKVAGTWYSLA	19–39	17–32	N/A	
	NT, LT, HT	VYVEELKPTPEGDLE	57–71	56–70	N/A	
	NT, LT, HT	VYVEELKPTPEGDLEI	57–72	56–70	N/A	
	NT, LT	VYVEELKPTPEGDLEIL	57–73	56–70	N/A	
	NT, LT, HT	YVEELKPTPEGDLE	58–71	56–70	N/A	
	NT, LT, HT	YVEELKPTPEGDLEI	58–72	56–70	N/A	
	NT, LT	YVEELKPTPEGDLEIL	58–73	56–70	N/A	
	NT	LVRTPEVDDEALEK	138–151	136–150	N/A	
	NT	LVRTPEVDDEALEKFDK	138–154	137–156	N/A	
α-lac	NT	KILDKVGIN	113–121	112–121	N/A	

[1] Peptides were reported as sIgE binding epitopes if their sequence contained at least 80% of the sequence of an sIgE binding epitope.

Table 2. sIgE binding epitopes [1] identified in glycated digestion-derived peptides after 10 min in the intestinal phase. Peptides were identified in cow's milk protein, non-treated (NT), heated at low temperature (LT), and heated at high temperature (HT), after simulated infant in vitro digestion and derived from casein (cn) and β-lactoglobulin (lg). Amino acids (AAs) position indicates the position within the proteins including the signal peptide. Peptides containing post translational modification (PTM) to lactosyl lysine (Lac), glucosyl lysine (Gluc), N^{ε}-carboxymethyllysine (CML), and pyrroline (Pyr) on lysine (K) are indicated, and modified K residues are highlighted in bold and underlined with modification site probability given in brackets if multiple options were identified.

Protein	Sample	Peptide Sequence	AAs Position	sIgE Epitope AAs Position	PTM
α_{s1}-cn	NT, LT	EGIHAQQ**K**EPMIGV	140–153	141–155	Lac
	HT	TDAPSFSDIPNPIGSENSE**K**	189–208	188–209	Lac
	HT	TDAPSFSDIPNPIGSENSE**K**TTMPL	189–213	188–209	Gluc
	HT	TDAPSFSDIPNPIGSENSE**K**TTMPL	189–213	188–209	Lac
	NT, LT	EGIHAQQ**K**EPMIGV	140–153	141–155	Lac
α_{s2}-cn	LT, HT	LYQGPIVLNPWDQV**K**	114–128	120–129	Lac
	LT, HT	LYQGPIVLNPWDQV**K**	114–128	120–129	Gluc
	LT, HT	LYQGPIVLNPWDQV**K**	114–128	120–129	CML
	LT, HT	LYQGPIVLNPWDQV**K**	114–128	120–129	Pyr
	LT, HT	YQGPIVLNPWDQV**K**	115–128	120–129	Lac
	LT, HT	YQGPIVLNPWDQV**K**	115–128	120–129	Gluc
	LT, HT	YQGPIVLNPWDQV**K**	115–128	120–129	CML
	HT	QGPIVLNPWDQV**K**	116–128	120–129	Lac
	HT	QGPIVLNPWDQV**K**	116–128	120–129	Gluc
	HT	QGPIVLNPWDQV**K**	116–128	120–129	CML
	HT	QGPIVLNPWDQV**K**	116–128	120–129	Pyr
β-cn	LT, HT	EMPFP**K**YPVEPF	123–134	122–135	Lac
	LT, HT	EMPFP**K**YPVEPF	123–134	122–135	Gluc
	HT	SLSQS**K(1)**VLPVPQ**K(1)**AVPYPQ	179–197	182–199	Lac
β-lg	LT, HT	LIVTQTM**K**GLDIQ	17–29	17–32	Lac
	NT	LIVTQTM**K(1)**GLDIQ**K(1)**VAGT	17–34	17–32	Lac
	LT	RVYVEEL**K**PTPEGDLE	56–71	56–70	Lac
	NT	RVYVEEL**K**PTPEGDLEI	56–72	56–70	Lac
	NT	VYVEEL**K**PTPEGDLE	57–70	56–70	Lac
	NT, LT, HT	VYVEEL**K**PTPEGDLE	57–71	56–70	Lac
	NT, LT, HT	VYVEEL**K**PTPEGDLE	57–71	56–70	Gluc
	NT, LT, HT	VYVEEL**K**PTPEGDLE	57–71	56–70	CML
	NT, LT, HT	VYVEEL**K**PTPEGDLE	57–71	56–70	Pyr
	NT, LT, HT	VYVEEL**K**PTPEGDLEI	57–72	56–70	Lac
	NT, LT, HT	VYVEEL**K**PTPEGDLEI	57–72	56–70	Gluc
	NT, LT, HT	YVEEL**K**PTPEGDLE	58–71	56–70	Lac
	NT, LT, HT	YVEEL**K**PTPEGDLE	58–71	56–70	Gluc
	NT, LT, HT	YVEEL**K**PTPEGDLE	58–71	56–70	CML
	NT, LT, HT	YVEEL**K**PTPEGDLE	58–71	56–70	Pyr
	LT, HT	YVEEL**K**PTPEGDLEI	58–72	56–70	Lac
	LT, HT	YVEEL**K**PTPEGDLEI	58–72	56–70	Gluc
	LT, HT	YVEEL**K**PTPEGDLEI	58–72	56–70	CML
	NT, HT	YVEEL**K**PTPEGDLEIL	58–73	56–70	Lac
	NT	LVRTPEVDDEALE**K(1)**FD**K(1)**	138–154	137–156	Lac
	NT	LVRTPEVDDEALE**K(1)**FD**K(1)**	138–154	137–156	Pyr
	NT, LT	LVRTPEVDDEALE**K(1)**FD**K(1)**AL**K(1)**ALPM	138–161	137–156	Lac
	NT, LT	LVRTPEVDDEALEKFD**K(0.8)**AL**K(0.2)**ALPM	138–161	137–156	Gluc
	NT, LT	LVRTPEVDDEALE**K(1)**FDKALKALPM	138–161	137–156	CML
	NT, LT	LVRTPEVDDEALE**K(1)**FD**K(1)**ALKALPM	138–161	137–156	Pyr

[1] Peptides were reported as sIgE binding epitopes if their sequence contained at least 80% of the sequence of an sIgE binding epitope.

T-cell epitopes were predicted using IEDB MHC Class II Binding Prediction software. All modified and non-modified peptides that were predicted as potential T-cell binding epitopes are shown in Table 3.

Table 3. Potential T-cell epitopes identified after 10 min in the intestinal phase. Peptides were identified as potential T-cell epitopes using IEDB MHC Class II Binding Prediction software (http://tools.iedb.org/mhcii/). Digestion-derived peptides identified from cow's milk protein, non-treated (NT), dry heated at low temperature (LT), and dry heated at high temperature (HT) applied to simulated infant in vitro digestion, derived from casein (cn) and β-lactoglobulin (lg). Peptides matching exactly the sIgE binding epitope sequence are indicated (*). Amino acids (AAs) position indicates the position within the proteins including the signal peptide. Unmodified peptides and peptides with post translational modifications (PTM), via phosphorylation (Phos), as well as modification to glucosyl lysine (Gluc), lactosyl lysine (Lac), N^{ε}-carboxymethyllysine (CML), and pyrroline (Pyr) were reported. Modified amino acids are highlighted in bold and underlined.

Protein	Sample	Sequence	HLA-Allele	AAs Position	PTM	Perc. Rank
α_{s1}-cn	NT, LT	EAESI**SSS**EEIVPNSVEQ	HLA-DQA1*03:01/DQB1*03:02; HLA-DQA1*04:01/DQB1*04:02	76–93	Phos	2.5 5.3
	NT, LT	YKVPQLEIVPN**S**AEE	HLA-DRB1*04:05; HLA-DQA1*04:01/DQB1*04:02; HLA-DQA1*03:01/DQB1*03:02	119–133	Phos	1.9 4.1 5.8
	NT, LT, HT	YKVPQLEIVPN**S**AEER	HLA-DRB1*04:05; HLA-DQA1*04:01/DQB1*04:02; HLA-DQA1*03:01/DQB1*03:02	119–134	Phos	2.2 5.9 6.5
	NT, LT, HT	KVPQLEIVPN**S**AEER	HLA-DRB1*04:05; HLA-DQA1*04:01/DQB1*04:02; HLA-DQA1*03:01/DQB1*03:02	120–134	Phos	1.9 5.6 5.9
α_{s2}-cn	NT, LT, HT	SIGSSSEESAEVATEEV	HLA-DQA1*04:01/DQB1*04:02; HLA-DQA1*03:01/DQB1*03:02	68–84	n.a.	0.14 0.18
	LT, HT	LYQGPIVLNPWDQV**K**	HLA-DRB1*13:02	114–128	Gluc	9.7
	LT, HT	LYQGPIVLNPWDQV**K**	HLA-DRB1*13:02	114–128	Lac	9.7
	LT, HT	LYQGPIVLNPWDQV**K**	HLA-DRB1*13:02	114–128	CML	9.7
	LT, HT	LYQGPIVLNPWDQV**K**	HLA-DRB1*13:02	114–128	Pyr	9.7
β-cn	HT	SLTLTDVENLHLPLP	HLA-DPA1*03:01/DPB1*04:02	139–153	N/A	6.3
β-lg	NT	VTQTMKGLDIQKVAGT	HLA-DRB4*01:01	19–34	N/A	7.9
	NT	ASDISLLDAQSAPLRV	HLA-DRB4*01:01; HLA-DRB1*01:01; HLA-DRB1*13:02; HLA-DRB1*12:01 HLA-DQA1*03:01/DQB1*03:02	42–57	N/A	4.0; 6.6; 7.9; 8.2; 9
	HT	SDISLLDAQSAPLRV	HLA-DRB4*01:01; HLA-DRB1*01:01; HLA-DRB1*12:01; HLA-DRB1*13:02; HLA-DRB1*09:01	43–57	N/A	3.3 4.4 6.2 6.3 7.2

Overall, 13 potential T-cell epitopes were found in the digest, with most epitopes deriving from α_{s1}-casein and α_{s2}-casein, followed by β-lactoglobulin, and β-casein. In the digest of NT-MP, 7 T-cell epitopes were found, of which five were also found in at least one of the heated samples. LT and HT heating resulted in the release of nine T-cell epitopes, respectively, with six solely found in heated samples. Of these, 40% were found in the digest of LT-MP and HT-MP were also glycated.

3.3. *Peptides Identified at the Basolateral Compartment of the Caco-2 Cell Monolayer*

To study the epithelial transport, in vitro digests sampled after 10 min in the IP were applied to a Caco-2 cell monolayer. The number of peptides found in the basolateral compartment for each sample decreased with heating intensity. Observed were 181, 129, and 121 peptides in NT-MP, LT-MP, and HT-MP, respectively. Moreover, most peptides were derived from α_{s1}-casein, β-casein, and β-lactoglobulin (data not shown). Independent from the heat treatment, the majority of peptides were found in the size range between 8–10 and 11–13 amino acids (Figure 4). Compared to the composition in the digest before transport, relatively higher numbers of peptides in the size range between 8–10 and 11–13 were found (Figure 4 and Supplementary Materials: Figure S3). Interestingly, peptides up to 24 amino acids long were identified on the basolateral side of the Caco-2 cell monolayer, however at low numbers.

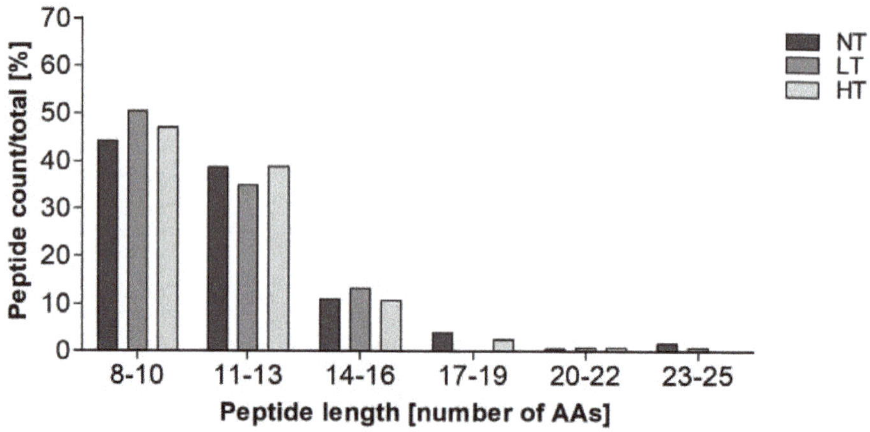

Figure 4. Peptide length distribution on the basolateral side of digestion-derived peptides, sampled after 10 min in the intestinal phase from simulated infant in vitro digests of cow's milk protein, non-treated (NT), heated in the presence of lactose at low temperature (LT), and high temperature (HT), expressed as peptide count relative to the total number (NT: 181, LT: 129, HT: 121) of peptides in one sample.

In NT-MP, less glycated peptides were found on the basolateral side (37%), compared to LT-MP (50%) and HT-MP (56%). This relative number of glycated peptides on the basolateral side increased in all samples compared to the digest before transport (35 ± 0%, 47 ± 1%, 49 ± 5% for NT-MP, LT-MP, and HT-MP, respectively). In all samples, the majority of those glycated peptides was modified to lactosyl lysine, followed by modification to glucosyl lysine, pyrroline and CML. Interestingly, the largest increase on the basolateral side was observed for the relative number of lactosyl lysine with 3%, 5%, and 5% increase and glucosyl lysine modified peptides with 5%, 7%, and 8% increase in NT-MP, LT-MP, and HT-MP, respectively (Supplementary Materials: Figure S3). This effect was larger in heated samples than in NT. HT-MP also showed 5% higher relative numbers of CML-modified peptides on the basolateral side, while the relative numbers of pyrroline-modified peptides shown were comparable to the digest.

3.4. *sIgE Binding Epitopes on the Basolateral Side of the Caco-2 Cell Monolayer*

Peptides identified at the basolateral side that carried at least 80% of the sequence of a known sIgE epitope are shown in Table 4. Similar to the observations in the digest (Table 2), most epitopes were found in peptides derived from β-lactoglobulin; however, only two of these peptides were unmodified. Moreover, only 19% of the glycated and non-glycated sIgE binding epitopes found in the digest (Tables 1 and 2) were also found on the basolateral side (Table 4). Contrastingly, on average the total number of glycated and non-modified peptides found on the basolateral side corresponded to 49% of the number of glycated and non-modified peptides in the digest. Two of the peptides containing a sIgE epitope derived from α_{s1}-casein (f52–67 and f123–133) were not identified in the digests before the Caco-2 cell experiment (Table 1). However, these peptides could derive from other precursor peptides (e.g., phosphorylated f56–67 and f124–133). Additionally, a peptide derived from β-lactoglobulin (f57–71) in non-modified and glycated form was previously found in all samples, while identification on the basolateral side was only possible in NT-MP.

Table 4. Digestion-derived peptides covering sIgE binding epitope sequences [1], identified on the basolateral side of the Caco-2 cell monolayer. Peptides were generated after digestion of cow's milk protein, non-treated (NT), heated at low temperature (LT), and heated at high temperature (HT), in an infant in vitro model. Amino acids (AAs) position indicates the position within the proteins including the signal peptide. Peptides derived from α_{s1}-casein (α_{s1}-cn), β -casein (β-cn), and β-lactoglobulin (β-lg). Peptides with and without post translational modification (PTM) to lactosyl lysine (Lac), glucosyl lysine (Gluc), N^{ε}-carboxymethyllysine (CML), and pyrroline (Pyr) are shown. Digestion-derived peptides covering the exact sequence of a sIgE binding epitope are indicated with *.

Protein	Sample	Peptide Sequence	AAs Position	sIgE Epitope AAs Position	PTM	
α_{s1}-cn	NT, LT	VNELSKDIGSESTEDQ	52–67	54–63	N/A	
	NT, LT	KVPQLEIVPNSAEE	120–133	124–135	N/A	
	NT, LT	QLEIVPNSAEE	123–133	124–135	N/A	
	NT, LT, HT	LEIVPNSAEER	124–134	124–135	N/A	
β-cn	NT, LT	PVVVPPFLQPEV	96–107	98–107	N/A	
	NT, LT, HT	VVPPFLQPE	98–106	98–107	N/A	
	NT, LT, HT	VVPPFLQPEV	98–107	98–107	N/A	*
β-lg	NT, LT, HT	VYVEELKPTPEGDLE	57–71	56–70	N/A	
	NT	VYVEEL**K**PTPEGDLE	57–71	56–70	CML	
	NT	VYVEEL**K**PTPEGDLE	57–71	56–70	Lac	
	NT	VYVEEL**K**PTPEGDLE	57–71	56–70	Pyr	
	NT, HT	YVEELKPTPEGDLE	58–71	56–70	N/A	
	NT, LT, HT	YVEEL**K**PTPEGDLE	58–71	56–70	CML	
	NT, LT, HT	YVEEL**K**PTPEGDLE	58–71	56–70	Lac	
	NT, LT, HT	YVEEL**K**PTPEGDLE	58–71	56–70	Pyr	

[1] Peptides were reported as sIgE binding epitopes if their sequence contained at least 80% of the sequence of an sIgE binding epitope.

From the T-cell epitopes identified in the digest, only one T-cell epitope derived from α_{s1}-casein was found in HT-MP (f68–84). This data suggested an overall low passage of T-cell epitopes and sIgE binding epitope.

4. Discussion

4.1. Heat Treatment Dependent Differences in Peptide Profiles

Dry heated MP at LT and HT was subjected to simulated infant in vitro digestion and peptides were identified after 10 min and 60 min in the IP. As most differences between heat treatments in the digests were observed after 10 min in the IP, at which time the mucosal immune system in the gastrointestinal tract may already encounter antigens, the focus was on this digestion time point. Heat treatment of MP resulted in 15% and 28% less peptides released upon digestion after 10 min in the IP in LT-MP and HT-MP compared to NT-MP, while after 60 min in the IP only 3% and 19% less peptides were observed in LT-MP and HT-MP, respectively (Figure 1). The absence of peptides can be a result of both increased and impaired hydrolysis [7]. However, in our previous study we showed that HT dry heating impairs hydrolysis after 10 and 60 min in the IP suggesting that the absence of peptides results from decreased hydrolysis [29]. It is also possible that a larger pool of different linear and crosslinking MRPs can result in lower number of peptides as only the most abundant modifications were monitored. At the same time, the relative number of glycated peptides identified in the digest of heated samples was higher compared to NT-MP (Figure 1a). This is in line with the levels of CML and pentosidine that were reported previously for the samples used in this study which increased with increases in heating temperature [29]. Most of the peptides were modified to glucosyl lysine and lactosyl lysine (Supplementary Materials: Figure S4) and already a large proportion of lactoyslated peptides was observed in NT-MP. The comparison of peptide intensities, however, indicated that the quantities of glycated peptides in the heated samples were higher than in NT-MP (Supplementary Materials: Figure S5). This was in agreement with the findings of Milkovska-Stamenova et al. [37], who found 50 lactosylation sites in raw milk which increased to only 70–80 in ultra-high temperature treated milk. At the same time, quantification of the glycated peptides in their study showed much lower levels in raw milk compared to processed dairy products.

Most peptides were derived from α_{s1}-casein, β-casein, and β-lactoglobulin (Supplementary Materials: Figure S1). In the two casein examples, this is probably related to the relatively higher concentration compared to the other proteins. For β-lactoglobulin, this is also related to the larger number of glycated peptides, of which each was counted as a separate peptide (Figure 3e). A heat treatment dependent decrease of sequence coverage after 10 min in the IP was especially observed for the two α-caseins and β-lactoglobulin (Figure 2a). For β-lactoglobulin, this originated from the absence or low number of peptides in the regions f17–39, f88–116, and f164–174 (Figure 3e). β-lactoglobulin, as a globular protein, is more sensitive to heating induced structural modifications compared to casein [38,39]. The regions f17–39 and f88–116 are rich in lysine residues, explaining the impairment of peptide generation especially in heated samples from this area via tryptic hydrolysis. The region f164–175 is located on the outside of the globular protein, partly incorporated in a β-strand and α-helix structure, which makes it rather easily accessible for digestive enzymes. However, it has been shown that upon heating in solution a α-β transition occurs, contributing to the aggregation of β-lactoglobulin via hydrophobic interactions [40]. This could explain the absence of peptides in the f164–175 region, as HT heating promotes the aggregation of β-lactoglobulin, but not in LT-MP and NT-MP. For α-caseins, a heating dependent decrease of sequence coverage was mainly reflected by the absence of phosphorylated peptides in HT-MP (Figure 3a,b) and is in line with the lower number of phosphorylated peptides (Figure 1a). Dephosphorylation has been reported upon heating in solution of caseinate at HT (140 °C) [41]. Both, hydrolysis of phosphoserine as well as β-elimination may induce dephosphorylation of casein. Michael addition subsequently to β-elimination and subsequent Michael addition results in protein crosslinking [41], which may also explain the lower number of peptides in HT-MP from sequence parts that are more phosphorylated. A study from Wada et al. [42] showed that dephosphorylation could decrease digestibility of heated dairy products. This is in line with the low digestibility of HT-MP observed in our previous study [29]. Next to digestibility, dephosphorylation could also decrease IgE binding capacity, indicating that overall IgE binding capacity to linear IgE binding epitopes could be lower for HT-MP [43]. In contrast, the region f22–39 of β-casein showed increasing number of phosphorylated peptides in HT while progressing digestion (Supplementary Materials: Figure S2c), indicating that in some cases a slower release of peptides could also be a possible explanation for the absence of phosphorylated peptides after 10 min in the IP. Glycated peptides resulted in a higher number of peptides in some areas (Figure 3a,c). This could possibly affect immunoreactivity if present on an epitope or by binding of these peptides to AGE receptors; however, it should be noted that quantities of glycated peptides were not measured and that it is not clear which effect the glycation of peptides has for epitope recognition. In summary, dry heating of MP decreases the number of peptides released upon simulated infant in vitro digestion and results in lower sequence coverage after 10 min in the IP. The discrepancies in sequence coverage of specific regions when comparing heat treatments can also be relevant for sIgE binding and T-cell epitope presentation. At the same time, the process of digestion results in fewer differences, suggesting that digestion kinetics are important determinants for differential release of immunoreactive digestion-derived peptides when comparing heat treatments.

4.2. Hydrolysis Resistant Areas

Most regions of κ-casein and α-lactalbumin from which peptides were released after 10 min in the IP (Figure 3d,f) were also detected after 60 min in the IP (Supplementary Materials: Figure S2). However, none of them were identified as areas of interest for possible immunological consequences. For α_{s2}-casein, decreasing sequence coverage was only observed for NT-MP. This was related to the disappearance of the phosphorylated peptides (f16–34) and the peptide at f40–50, of which only low numbers were detected after 10 min in the IP. However, no potential epitopes were identified after 60 min in the IP. For α_{s1}-casein and β-casein, sequence coverage and number of peptides showed only minor decrease with prolonged digestion (Figure 2 and Supplementary Materials: Figure S1). For β-casein, the peptide pattern between samples showed only minor differences and therefore also a comparable persistence of peptides carrying an sIgE binding epitope (f96–110, f123–134, and f164–177) (Supplementary Materials: Table S1). For α_{s1}-casein, the region f202–213 was solely covered in heated samples by glycated peptides after 60 min in the IP, suggesting a higher digestion resistance of this area due to glycation. At the same time, f140–155 of α_{s1}-casein, which has previously been reported to maintain high residual immunoreactivity after simulated in vitro digestion of spray dried milk powder [7], was only partly preserved until the end of the IP (Supplementary Materials: Figure S2a). A higher number of peptides in NT-MP-digests originating from this region could possibly result in a higher immunoreactivity of this sample. Next to this, a larger number of peptides in f119–134, which contained a potential T-cell epitope (Table 3) was still found at the end of intestinal digestion in all samples (Supplementary Materials: Figure S4a). In contrast to the two caseins, β-lactoglobulin showed large decreases of sequence coverage in all samples, related to the disappearance of f17–39 and f42–55. At the same time, the regions f57–73 and f139–154 of β-lactoglobulin were highly resistant to digestion until the end of intestinal digestion (Supplementary Materials: Figure S2e), independent from the heat treatment. This was in line with the findings by Egger at al. [44] who observed a high frequency of peptides within particularly these two areas of β-lactoglobulin until 120 min in the IP of a static in vitro model. Moreover, the findings for both β-lactoglobulin and α_{s1}-casein were similar to previous findings by Picariello et al. [45], who described the sequence part f141–151 of β-lactoglobulin and f119–134 of α_{s1}-casein as highly resistant to gastrointestinal digestion after simulated adult in vitro digestion. While there is no direct evidence for the presence of an immunoreactive structure within this region of β-lactoglobulin, f119–134 of α_{s1}-casein partly covers the sequence of an sIgE binding epitope (Figure 3a) and was also identified as potential T-cell epitope (Table 3). Together with its high resistance until the end of intestinal digestion, this suggest a potential role of f119–134 in sIgE binding to the digest of MP, but independently from the heat treatment. To summarize these findings, caseins generally showed a higher resistance over large parts of their protein sequence until the end of gastrointestinal digestion, which was unaffected by the applied heat treatment. Therefore, no conclusions can be drawn from the resistance of specific areas within the protein sequence regarding differential immunoreactivity of dry heated MP compared to NT-MP.

4.3. *Effect of Heat Treatment on Identification of IgE Binding Epitopes*

Digestion-derived peptides were reported as potential sIgE binding epitopes if they covered at least 80% of the sequence of a linear sIgE binding epitope known from the literature [35]. The three proteins showing the highest numbers of digestion-derived peptides, α_{s1}-casein, β-casein, and β-lactoglobulin (Supplementary Materials: Figure S1a) also led to the highest number of sIgE binding epitopes (Table 1) after 10 min in the IP. Most sIgE binding epitopes were found in the NT-MP digest, when compared to the heated samples (Table 1), which was in line with the higher number of peptides (Figure 1a) and the higher sequence coverage of NT-MP (Figure 2a). This suggests a higher availability of linear sIgE binding epitopes in NT-MP compared to dry heated MP. The opposite trend was observed for sIgE binding epitopes carrying a glycation side (Table 2). However, this trend did not continue until 60 min in the IP (Supplementary Materials: Table S1). After 60 min in the IP, most sIgE binding epitopes were found in the digest of NT-MP, of which the majority were, however, present in all samples. This can be explained by the overall smaller differences between samples with progressing digestion, which is possibly related to differences in digestion kinetics especially in the beginning of the IP. The effect of heating and glycation on sIgE binding has been subject to previous studies on isolated milk proteins or in mixture [15–17,19]. As reviewed by Nowak-Wegrzyn et al. [18], milk proteins show reduced sIgE binding upon extensive glycation via the MR. However, these observations are based on studies of undigested milk proteins and not of linear epitopes exclusively, and thus can probably not be extrapolated for all MRPs and peptides. A study by Gasparini et al. [46] reported an approach creating the basis for studying the effect of lactosylation on linear epitopes. However, data comparing sIgE binding of lactosylated vs. non-modified peptides is not available at this time. With respect to the predicted T-cell epitopes, ~50% were specifically found in the heated samples, but most of those peptides were glycated. To our knowledge only data on T-cell epitopes from α_{s1}-casein, β-lactoglobulin, and α-lactalbumin are available in the literature. In a previous study it was shown that a peptide f118–135 of α_{s1}-casein, which is similar to the T-cell epitope identified in our study (f119–135, Table 3) is recognized by 1 out of 10 cow's milk allergic children. However, none of the major T-cell epitopes of α_{s1}-casein identified in previous studies have been observed in the digests in our study [47–49]. The T-cell epitopes identified in β-lactoglobulin partly overlapped with the peptide sequences identified previously (f41–55) [50]. All T-cell epitopes from α_{s1}-casein and β-lactoglobulin were predicted as ligands for multiple HLA alleles, indicating that their recognition could be less affected by individual differences in patients. While the majority of α_{s1}-casein derived T-cell epitopes was phosphorylated, which in a previous study did not show consistent differences in epitope recognition [47], most of the α_{s2}-casein derived T-cell epitopes were glycated (Table 3). To our knowledge there is no study directly showing the effect of glycation on T-cell epitope recognition. However, as glycation of food proteins has been shown to affect T-cell immunogenicity [21,51], it could be hypothesized that glycation of peptides matching a T-cell epitope could affect its immunogenicity. To summarize, dry heating of MP resulted in a lower number of peptides that match to known sIgE binding epitopes but a higher number of glycated sIgE binding epitopes. Moreover, T-cell epitopes were identified in the digest and equally distributed between samples, while glycated T-cell epitopes were solely found in heated samples. The consequences of glycation on sIgE epitope and T-cell epitope binding are, however, not clear.

4.4. Identification of Peptides on the Basolateral Side of the Caco-2 Cell Monolayer

Transport across the epithelial layer was assessed using a Caco-2 cell monolayer model. Peptide length distribution found on the basolateral side (Figure 4) indicated a favored transport of peptides up to 13 amino acids compared to the distribution in the digest (Supplementary Materials: Figure S3). Interestingly, peptides with a length up to 24 amino acids were also found on the basolateral side (Figure 4). The peptides in the larger size ranges (f17–19 and f20–22, and f23–25) were mainly non-glycated peptides derived from β-casein, which originated from hydrophobic patches within the sequence suggesting a passage via transcytosis [52]. Availability of larger peptides increases the possibility of recognition by the immune system. Moreover, transport of peptides carrying sIgE binding epitope sequences (e.g., f159–177) via transcytosis enables the peptide to reach the *lamina propria* intact, indicating the importance of also monitoring transport pathways when studying the availability of immunoreactive digestion-derived peptides. Most digestion-derived peptides on the basolateral side were derived from α_{s1}-casein, β-casein, and β-lactoglobulin, which is probably related to the higher number of peptides in the digest (Supplementary Materials: Figure S2a). Consequently, sIgE binding epitopes found on the basolateral side were only identified for α_{s1}-casein, β-casein, and β-lactoglobulin (Table 4). Moreover, sIgE binding epitopes were most abundant in NT-MP which was in line with the total number of digestion-derived peptides between samples (Table 1a) and presence amongst proteins (Supplementary Materials: Figure S1a). However, only 19% of the sIgE binding epitopes (non-modified and glycated) and one T-cell epitope identified in the digests were also found on the basolateral side (Tables 1, 2 and 4), while on average 49% of the number of peptides in the digest were found on the basolateral side, suggesting some sort of epitope-excluding effect of the epithelial layer. For T-cell epitopes it could be hypothesized that this was related to size, as the size ranges 8–10 as well as 11–13 were preferably transported, while T-cell epitopes normally have a length between 15–24 amino acids [36]. In contrast, most sIgE binding epitopes were identified within the smaller size ranges. Next to peptide size, the transport across the Caco-2 cell monolayer can also be determined by charge and hydrophobicity [52]. However, further studies would be necessary to determine peptide properties to find the reasons for the observed restriction of epitope transport across the Caco-2 cell monolayer. Moreover, it should be noted that in vivo a larger number of M-cells as well as specialized dendritic cells are present in the small intestine, that are able to directly sample antigens from the intestinal lumen [53]. It is thus hypothesized that the translocation of IgE and T-cell epitopes in vivo could be directed towards specialized cells rather than transport via normal enterocytes.

In contrast to this, transport of a relatively higher number of glycated peptides was observed on the basolateral side, e.g., in dry heated samples, compared to the composition in the digest. Moreover, data suggested a possible preference for the transport of lactosyl lysine and glucosyl lysine-modified peptides amongst all samples, as the percentage of these peptides showed an increasing trend on the basolateral side compared to the digest (Supplementary Materials: Figures S4 and S5). As reviewed by Moradi et al. [54], N- and O-glycosylation with different mono- and polysaccharides of therapeutic peptides has been shown to increase their transport across various biological membranes including Caco-2 cells. For example, Varamini et al. [55] observed a 700-fold increased transport across the Caco-2 cell monolayer after modification of the N-terminal amino group from endomorphin-1 with lactose and suggested that this transport took place via a lactose-selective transporter. Such transporter-mediated translocation could be a possible explanation for the facilitated migration of glucosyl lysine and lactosyl lysine-modified digestion-derived peptides across the Caco-2 cell monolayer. However, it should be noted that the position and type of linkage (N- or O-linked) can strongly affect the structure, functionality and transporter mediated uptake of the peptides [54,56]. Therefore, an extrapolation of these findings to any peptide and any kind of modification is probably not possible. With respect to the potential immunological consequences, it is suggested that glycation if present on a linear sIgE binding epitope can affect the interaction between the peptide and the antibody [46]. Moreover, AGEs themselves have also been reported to modulate inflammatory pathways by binding to receptors for AGEs [14]. For the example of peptide-bound CML, it has been shown that it is a potent ligand for the receptor for AGEs and thus possibly affects inflammatory pathways [20]. This study showed that glycated peptides larger than 7 amino acids are transported independent of the type of modification (Figure 5). The findings of this study suggested that diets with high AGE content can also result in higher uptake of AGE-modified peptides. As recently shown, the binding of AGE receptors depends on the concentration of food protein bound AGEs [22]. Therefore, quantitative data would be necessary to better judge the impact of the transport of AGE-modified peptides on the gastrointestinal immune system as well as the involved transport pathways. To summarize, results indicated that several potentially immunoreactive peptides are transported across a model epithelial barrier. In general, the presence of peptides on the basolateral side is more affected by the overall composition of the digest rather than the selective transport of specific peptides. Nevertheless, transport seemed to be favored for smaller peptides (up to 13 amino acids) as well as peptides modified to lactosyl lysine and glucosyl lysine. This should, however, be further investigated using quantitative data on selected modified vs. non-modified peptides. At the same, time transport of sIgE binding epitopes and T-cell epitopes was limited, which is possibly related to some intrinsic properties of these peptides.

This study aimed to give an overview of the composition and transport of peptides derived after simulated infant in vitro digestion of differentially dry heated MP. However, this also resulted in some limitations, as only qualitative data was presented and allergenicity as well as immunogenicity was not measured directly. Moreover, other structures that could affect immunogenicity as well as allergenicity, such as aggregated protein that might also resist in vitro digestion, have not been considered [57,58].

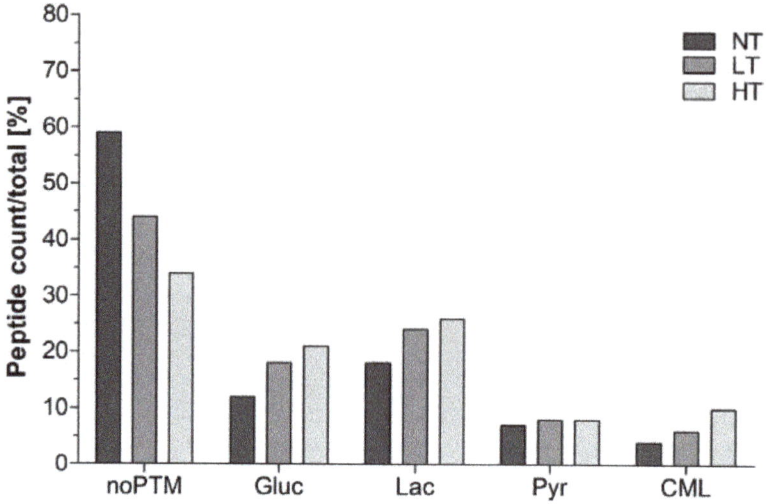

Figure 5. Digestion-derived peptides identified on the basolateral side of Caco-2 cells exposed to in vitro digests of cow's milk protein, non-treated (NT), heated in the presence of lactose at low temperature (LT), and high temperature (HT). Peptides without posttranslational modification (noPTM), as well as modification to glucosyl lysine (Gluc), lactosyl lysine (Lac), pyrroline (Pyr), and carboxymethyl lysine (CML) are shown expressed as peptide count relative to the total number (NT: 181, LT: 129, HT: 121) of peptides in one sample.

5. Conclusions

This study showed that different peptide profiles are generated during simulated infant in vitro digestion of milk that was dry heated in the presence of lactose. HT dry heating had the largest effects on peptide generation, resulting in much lower numbers of peptides and a lower sequence coverage. Moreover, a much lower number of sIgE binding epitopes but a larger proportion of glycated sIgE binding epitopes and T-cell epitopes in heated samples indicated that immunogenicity and allergenicity of these samples could be affected. However, this needs to be further tested. Transport studies showed that the transport of sIgE epitopes and T-cell epitopes across the Caco-2 cell monolayer is limited, highlighting the importance of evaluating different transport pathways. It is hypothesized that transport of lactosyl lysine and glucosyl lysine-modified peptides was favored, while CML and pyrroline-modified peptides were transported depending on their presence in the digest. This resulted in relatively more glycated peptides on the basolateral side in heated samples, indicating that if the initial level of MR is high, this will also increase the transport of glycated peptides and can thereby possibly affect immunoreactivity via interaction with AGE receptors. This pointed out the importance of studying the effect of glycation on the peptide level on immunogenicity and allergenicity.

Supplementary Materials: The following are available online at http://www.mdpi.com/2072-6643/12/8/2483/s1, Figure S1: Peptides derived from each of the five major milk proteins after infant in vitro digestion of cow's milk protein, Figure S2: Sequence alignment of digestion-derived peptides after 60 min in the intestinal phase, Figure S3: Peptide length distribution identified in the in vitro digests of cow's milk protein after 10 min in the intestinal phase, Figure S4: Non-modified vs. glycated digestion-derived peptides identified after 10 min in the intestinal phase, Figure S5: Summed intensities of peptides associated with a specific modification, within the same amino acid sequence. Table S1: sIgE binding epitopes [1] identified in digestion-derived peptides derived from cow's milk protein after 60 min in the intestinal phase, Table S2: Potential T-cell epitopes identified after 60 min in the intestinal phase.

Author Contributions: Conceptualization, H.E.Z. and K.A.H.; methodology, H.E.Z.; software, H.E.Z., S.B.; validation, H.E.Z., S.B. and M.M.M.T.; formal analysis, H.E.Z. and M.M.M.T.; investigation, H.E.Z.; resources, K.A.H., H.J.W.; data curation, H.E.Z. and S.B.; writing—original draft preparation, H.E.Z.; writing—review and editing, K.A.H., H.W.D.J. and N.W.D.J.; visualization, H.E.Z.; supervision, K.A.H. and H.W.D.J. project administration, K.A.H.; funding acquisition, K.A.H. and N.W.D.J. All authors have read and agreed to the published version of the manuscript.

Funding: This work is part of the research program iAGE/TTW with project number 14536, which is (partly) financed by the Netherlands Organization for Scientific Research (NWO).

Acknowledgments: We would like to thank Coen Covers for his support during the planning and conduction of the Caco-2 cell experiments. Furthermore, we thank Renata Ariens for her help to organize some of the materials.

Conflicts of Interest: The authors declare no conflict of interest.

References

1. Kim, J.S.; Nowak-Wgrzyn, A.; Sicherer, S.H.; Noone, S.; Moshier, E.L.; Sampson, H.A. Dietary baked milk accelerates the resolution of cow's milk allergy in children. *J. Allergy Clin. Immunol.* **2011**, *128*, 125–131. [CrossRef] [PubMed]
2. Van Boekel, M.A.J.S. Effect of heating on Maillard reactions in milk. *Food Chem.* **1998**, *62*, 403–414. [CrossRef]
3. Arena, S.; Renzone, G.; D'Ambrosio, C.; Salzano, A.M.; Scaloni, A. Dairy products and the Maillard reaction: A promising future for extensive food characterization by integrated proteomics studies. *Food Chem.* **2017**, *219*, 477–489. [CrossRef] [PubMed]
4. Birlouez-Aragon, I.; Pischetsrieder, M.; Leclère, J.; Morales, F.J.; Hasenkopf, K.; Kientsch-Engel, R.; Ducauze, C.J.; Rutledge, D. Assessment of protein glycation markers in infant formulas. *Food Chem.* **2004**, *87*, 253–259. [CrossRef]
5. Pischetsrieder, M.; Henle, T. Glycation products in infant formulas: Chemical, analytical and physiological aspects. *Amino Acids* **2012**, *42*, 1111–1118. [CrossRef]
6. Dupont, D.; Mandalari, G.; Mollé, D.; Jardin, J.; Rolet-Répécaud, O.; Duboz, G.; Léonil, J.; Mills, C.E.N.; Mackie, A.R. Food processing increases casein resistance to simulated infant digestion. *Mol. Nutr. Food Res.* **2010**, *54*, 1677–1689. [CrossRef]
7. Dupont, D.; Boutrou, R.; Menard, O.; Jardin, J.; Tanguy, G.; Schuck, P.; Haab, B.B.; Leonil, J. Heat treatment of milk during powder manufacture increases casein resistance to simulated infant digestion. *Food Dig.* **2010**, *1*, 28–39. [CrossRef]
8. Sánchez-Rivera, L.; Ménard, O.; Recio, I.; Dupont, D. Peptide mapping during dynamic gastric digestion of heated and unheated skimmed milk powder. *Food Res. Int.* **2015**, *77*, 132–139. [CrossRef]
9. Kopf-Bolanz, K.A.; Schwander, F.; Gijs, M.; Vergères, G.; Portmann, R.; Egger, L. Impact of milk processing on the generation of peptides during digestion. *Int. Dairy J.* **2014**, *35*, 130–138. [CrossRef]
10. Torcello-Gómez, A.; Dupont, D.; Jardin, J.; Briard-Bion, V.; Deglaire, A.; Risse, K.; Mechoulan, E.; Mackie, A. The pattern of peptides released from dairy and egg proteins is highly dependent on the simulated digestion scenario. *Food Funct.* **2020**. [CrossRef]
11. Wada, Y.; Phinney, B.S.; Weber, D.; Lönnerdal, B. In vivo digestomics of milk proteins in human milk and infant formula using a suckling rat pup model. *Peptides* **2017**, *88*, 18–31. [CrossRef] [PubMed]
12. Zenker, H.E.; Van Lieshout, G.A.A.; Van Gool, M.P.; Bragt, M.C.E.; Hettinga, K.A. Lysine blockage of milk proteins in infant formula impairs overall protein digestibility and peptide release. *Food Funct.* **2020**, *11*, 358–369. [CrossRef] [PubMed]
13. Zhao, D.; Li, L.; Le, T.T.; Larsen, L.B.; Xu, D.; Jiao, W.; Sheng, B.; Li, B.; Zhang, X. Digestibility of glycated milk proteins and the peptidomics of their in vitro digests. *J. Sci. Food Agric.* **2019**, *99*, 3069–3077. [CrossRef] [PubMed]
14. Teodorowicz, M.; Van Neerven, J.; Savelkoul, H. Food processing: The influence of the maillard reaction on immunogenicity and allergenicity of food proteins. *Nutrients* **2017**, *9*, 835. [CrossRef] [PubMed]
15. Xu, Q.; Shi, J.; Yao, M.; Jiang, M.; Luo, Y. Effects of heat treatment on the antigenicity of four milk proteins in milk protein concentrates. *Food Agric. Immunol.* **2016**, *27*, 401–413. [CrossRef]
16. Ehn, B.M.; Ekstrand, B.; Bengtsson, U.; Ahlstedt, S. Modification of IgE Binding during Heat Processing of the Cow's Milk Allergen β-Lactoglobulin. *J. Agric. Food Chem.* **2004**, *52*, 1398–1403. [CrossRef]

17. Taheri-Kafrani, A.; Gaudin, J.C.; Rabesona, H.; Nioi, C.; Agarwal, D.; Drouet, M.; Chobert, J.M.; Bordbar, A.K.; Haertle, T. Effects of heating and glycation of β-lactoglobulin on its recognition by ige of sera from cow milk allergy patients. *J. Agric. Food Chem.* **2009**, *57*, 4974–4982. [CrossRef]
18. Nowak-Wegrzyn, A.; Fiocchi, A. Rare, medium, or well done? The effect of heating and food matrix on food protein allergenicity. *Curr. Opin. Allergy Clin. Immunol.* **2009**, *9*, 234–237. [CrossRef]
19. Corzo-Martínez, M.; Soria, A.C.; Belloque, J.; Villamiel, M.; Moreno, F.J. Effect of glycation on the gastrointestinal digestibility and immunoreactivity of bovine β-lactoglobulin. *Int. Dairy J.* **2010**, *20*, 742–752. [CrossRef]
20. Xue, J.; Rai, V.; Singer, D.; Chabierski, S.; Xie, J.; Reverdatto, S.; Burz, D.S.; Schmidt, A.M.; Hoffmann, R.; Shekhtman, A. Advanced glycation end product recognition by the receptor for AGEs. *Structure* **2011**, *19*, 722–732. [CrossRef]
21. Heilmann, M.; Wellner, A.; Gadermaier, G.; Ilchmann, A.; Briza, P.; Krause, M.; Nagai, R.; Burgdorf, S.; Scheurer, S.; Vieths, S.; et al. Ovalbumin modified with pyrraline, a maillard reaction product, shows enhanced T-cell immunogenicity. *J. Biol. Chem.* **2014**, *289*, 7919–7928. [CrossRef]
22. Zenker, H.E.; Teodorowicz, M.; Ewaz, A.; van Neerven, R.J.J.; Savelkoul, H.F.J.; De Jong, N.W.; Wichers, H.J.; Hettinga, K.A. Binding of CML-Modified as Well as Heat-Glycated β-lactoglobulin to Receptors for AGEs is Determined by Charge and Hydrophobicity. *Int. J. Mol. Sci.* **2020**, *21*, 4567. [CrossRef] [PubMed]
23. Foerster, A.; Henle, T. Glycation in food and metabolic transit of dietary AGEs (advanced glycation end-products): Studies on the urinary excretion of pyrraline. *Biochem. Soc. Trans.* **2003**, *31*, 1383–1385. [CrossRef] [PubMed]
24. Roncero-Ramos, I.; Delgado-Andrade, C.; Tessier, F.J.; Niquet-Léridon, C.; Strauch, C.; Monnier, V.M.; Navarro, M.P. Metabolic transit of Nε-carboxymethyl-lysine after consumption of AGEs from bread crust. *Food Funct.* **2013**, *4*, 1032–1039. [CrossRef] [PubMed]
25. Hellwig, M.; Geissler, S.; Matthes, R.; Peto, A.; Silow, C.; Brandsch, M.; Henle, T. Transport of Free and Peptide-Bound Glycated Amino Acids: Synthesis, Transepithelial Flux at Caco-2 Cell Monolayers, and Interaction with Apical Membrane Transport Proteins. *ChemBioChem* **2011**, *12*, 1270–1279. [CrossRef] [PubMed]
26. O'Hagan, D.T.; Palin, K.J.; Davis, S.S. Intestinal absorption of proteins and macromolecules and the immunological response. *Crit. Rev. Drug Carr. Syst.* **1987**, *4*, 197–220.
27. Picariello, G.; De Cicco, M.; Nocerino, R.; Paparo, L.; Mamone, G.; Addeo, F.; Berni Canani, R. Excretion of dietary cow's milk derived peptides into breast milk. *Front. Nutr.* **2019**, *6*. [CrossRef]
28. Zhao, D.; Li, L.; Le, T.T.; Larsen, L.B.; Su, G.; Liang, Y.; Li, B. Digestibility of Glyoxal-Glycated β-Casein and β-Lactoglobulin and Distribution of Peptide-Bound Advanced Glycation End Products in Gastrointestinal Digests. *J. Agric. Food Chem.* **2017**, *65*, 5778–5788. [CrossRef]
29. Zenker, H.E.; Raupbach, J.; Boeren, S.; Wichers, H.J.; Hettinga, K.A. The effect of low vs. high temperature dry heating on solubility and digestibility of cow's milk protein. *Food Hydrocoll.* **2020**, *109*, 106098. [CrossRef]
30. Ménard, O.; Bourlieu, C.; De Oliveira, S.C.; Dellarosa, N.; Laghi, L.; Carrière, F.; Capozzi, F.; Dupont, D.; Deglaire, A. A first step towards a consensus static in vitro model for simulating full-term infant digestion. *Food Chem.* **2018**, *240*, 338–345. [CrossRef]
31. Dingess, K.A.; De Waard, M.; Boeren, S.; Vervoort, J.; Lambers, T.T.; Van Goudoever, J.B.; Hettinga, K. Human milk peptides differentiate between the preterm and term infant and across varying lactational stages. *Food Funct.* **2017**, *8*, 3769–3782. [CrossRef] [PubMed]
32. Liu, M.Q.; Zeng, W.F.; Fang, P.; Cao, W.Q.; Liu, C.; Yan, G.Q.; Zhang, Y.; Peng, C.; Wu, J.Q.; Zhang, X.J.; et al. PGlyco 2.0 enables precision N-glycoproteomics with comprehensive quality control and one-step mass spectrometry for intact glycopeptide identification. *Nat. Commun.* **2017**, *8*, 438. [CrossRef]
33. Cox, J.; Mann, M. MaxQuant enables high peptide identification rates, individualized p.p.b.-range mass accuracies and proteome-wide protein quantification. *Nat. Biotechnol.* **2008**, *26*, 1367–1372. [CrossRef] [PubMed]
34. Boggs, I.; Hine, B.; Smolenski, G.; Hettinga, K.; Zhang, L.; Wheeler, T.T. Changes in the repertoire of bovine milk proteins during mammary involution. *Eupa Open Proteom.* **2015**, *9*, 65–75. [CrossRef]
35. Matsuo, H.; Yokooji, T.; Taogoshi, T. Common food allergens and their IgE-binding epitopes. *Allergol. Int.* **2015**, *64*, 332–343. [CrossRef]

36. Knol, E.F.; de Jong, N.W.; Ulfman, L.H.; Tiemessen, M.M. Management of cow's milk allergy from an immunological perspective: What are the options? *Nutrients* **2019**, *11*, 2734. [CrossRef]
37. Milkovska-Stamenova, S.; Hoffmann, R. Identification and quantification of bovine protein lactosylation sites in different milk products. *J. Proteom.* **2016**, *134*, 112–126. [CrossRef]
38. Donato, L.; Guyomarc'h, F. Formation and properties of the whey protein/κ-casein complexes in heated skim milk—A review. *Dairy Sci. Technol.* **2009**, *89*, 3–29. [CrossRef]
39. Alexander, L.J.; Hayes, G.; Pearse, M.J.; Stewart, A.F.; Willis, I.M.; Mackinlay, A.G. Complete sequence of the bovine β-lactoglobulin cDNA. *Nucleic Acids Res.* **1989**, *17*, 6739. [CrossRef]
40. Vetri, V.; Militello, V. Thermal induced conformational changes involved in the aggregation pathways of beta-lactoglobulin. *Biophys. Chem.* **2005**, *113*, 83–91. [CrossRef]
41. Van Boekel, M.A.J.S. Heat-induced deamidation, dephosphorylation and breakdown of caseinate. *Int. Dairy J.* **1999**, *9*, 237–241. [CrossRef]
42. Wada, Y.; Lönnerdal, B. Effects of Industrial Heating Processes of Milk-Based Enteral Formulas on Site-Specific Protein Modifications and Their Relationship to in Vitro and in Vivo Protein Digestibility. *J. Agric. Food Chem.* **2015**, *63*, 6787–6798. [CrossRef]
43. Bernard, H.; Meisel, H.; Creminon, C.; Wal, J.M. Post-translational phosphorylation affects the IgE binding capacity of caseins. *Febs. Lett.* **2000**, *467*, 239–244. [CrossRef]
44. Egger, L.; Ménard, O.; Baumann, C.; Duerr, D.; Schlegel, P.; Stoll, P.; Vergères, G.; Dupont, D.; Portmann, R. Digestion of milk proteins: Comparing static and dynamic in vitro digestion systems with in vivo data. *Food Res. Int.* **2019**, *118*, 32–39. [CrossRef]
45. Picariello, G.; Ferranti, P.; Fierro, O.; Mamone, G.; Caira, S.; Di Luccia, A.; Monica, S.; Addeo, F. Peptides surviving the simulated gastrointestinal digestion of milk proteins: Biological and toxicological implications. *J. Chromatogr. B Anal. Technol. Biomed. Life Sci.* **2010**, *878*, 295–308. [CrossRef]
46. Gasparini, A.; Buhler, S.; Faccini, A.; Sforza, S.; Tedeschi, T. Thermally-induced lactosylation of whey proteins: Identification and synthesis of lactosylated β-lactoglobulin epitope. *Molecules* **2020**, *25*, 1294. [CrossRef]
47. Ruiter, B.; Trégoat, V.; M'Rabet, L.; Garssen, J.; Bruijnzeel-Koomen, C.A.F.M.; Knol, E.F.; Van Hoffen, E. Characterization of T cell epitopes in αs1-casein in cow's milk allergic, atopic and non-atopic children. *Clin. Exp. Allergy* **2006**, *36*, 303–310. [CrossRef]
48. Elsayed, S.; Eriksen, J.; Øysæd, L.K.; Idsøe, R.; Hill, D.J. T cell recognition pattern of bovine milk αS1-casein and its peptides. *Mol. Immunol.* **2004**, *41*, 1225–1234. [CrossRef]
49. Nakajima-Adachi, H.; Hachimura, S.; Ise, W.; Honma, K.; Nishiwaki, S.; Hirota, M.; Shimojo, N.; Katsuki, T.; Ametani, A.; Kohno, Y.; et al. Determinant analysis of IgE and IgG4 antibodies and T cells specific for bovine alpha(s)1-casein from the same patients allergic to cow's milk: Existence of alpha(s)1-casein-specific B cells and T cells characteristic in cow's-milk allergy. *J. Allergy Clin. Immunol.* **1998**, *101*, 660–671. [CrossRef]
50. Gouw, J.W.; Jo, J.; Meulenbroek, L.A.P.M.; Heijjer, T.S.; Kremer, E.; Sandalova, E.; Knulst, A.C.; Jeurink, P.V.; Garssen, J.; Rijnierse, A.; et al. Identification of peptides with tolerogenic potential in a hydrolysed whey-based infant formula. *Clin. Exp. Allergy* **2018**, *48*, 1345–1353. [CrossRef]
51. Ilchmann, A.; Burgdorf, S.; Scheurer, S.; Waibler, Z.; Nagai, R.; Wellner, A.; Yamamoto, Y.; Yamamoto, H.; Henle, T.; Kurts, C.; et al. Glycation of a food allergen by the Maillard reaction enhances its T-cell immunogenicity: Role of macrophage scavenger receptor class A type I and II. *J. Allergy Clin. Immunol.* **2010**, *125*, 175–183.e111. [CrossRef]
52. Xu, Q.; Hong, H.; Wu, J.; Yan, X. Bioavailability of bioactive peptides derived from food proteins across the intestinal epithelial membrane: A review. *Trends Food Sci. Technol.* **2019**, *86*, 399–411. [CrossRef]
53. Reitsma, M.; Westerhout, J.; Wichers, H.J.; Wortelboer, H.M.; Verhoeckx, K.C.M. Protein transport across the small intestine in food allergy. *Mol. Nutr. Food Res.* **2014**, *58*, 194–205. [CrossRef]
54. Moradi, S.V.; Hussein, W.M.; Varamini, P.; Simerska, P.; Toth, I. Glycosylation, an effective synthetic strategy to improve the bioavailability of therapeutic peptides. *Chem. Sci.* **2016**, *7*, 2492–2500. [CrossRef]
55. Varamini, P.; Mansfeld, F.M.; Blanchfield, J.T.; Wyse, B.D.; Smith, M.T.; Toth, I. Synthesis and biological evaluation of an orally active glycosylated endomorphin-1. *J. Med. Chem.* **2012**, *55*, 5859–5867. [CrossRef]
56. Polt, R.; Porreca, F.; Szabò, L.Z.; Bilsky, E.J.; Davis, P.; Abbruscato, T.J.; Davis, T.P.; Horvath, R.; Yamamura, H.I.; Hruby, V.J. Glycopeptide enkephalin analogues produce analgesia in mice: Evidence for penetration of the blood-brain barrier. *Proc. Natl. Acad. Sci. USA* **1994**, *91*, 7114–7118. [CrossRef]

57. Zenker, H.E.; Ewaz, A.; Deng, Y.; Savelkoul, H.F.J.; Van Neerven, R.J.J.; De Jong, N.; Wichers, H.J.; Hettinga, K.A.; Teodorowicz, M. Differential effects of dry vs. Wet heating of β-lactoglobulin on formation of sRAGE binding ligands and sIgE epitope recognition. *Nutrients* **2019**, *11*, 1432. [CrossRef]
58. Pinto, M.S.; Léonil, J.; Henry, G.; Cauty, C.; Carvalho, A.F.; Bouhallab, S. Heating and glycation of β-lactoglobulin and β-casein: Aggregation and in vitro digestion. *Food Res. Int.* **2014**, *55*, 70–76. [CrossRef]

nutrients

MDPI

Article

Introduction of Heated Cow's Milk Protein in Challenge-Proven Cow's Milk Allergic Children: The iAGE Study

Nicolette W. de Jong [1,2,*,†], Marloes E. van Splunter [1,†], Joyce A. M. Emons [2], Kasper A. Hettinga [3], Roy Gerth van Wijk [1], Harry J. Wichers [3], Huub F. J. Savelkoul [4], Aline B. Sprikkelman [5,6], R. J. Joost van Neerven [4,7], Liu Liu [8], Gerbrich van der Meulen [9], Irene Herpertz [10], Yvonne C. M. Duijvestijn [11], Mijke Breukels [12], Marianne I. Brouwer [13], Jaap Schilperoord [14], Olga van Doorn [14], Berber Vlieg-Boerstra [14], Jochum van den Berg [15], Linette Pellis [16], Severina Terlouw [1], Astrid I. Hendriks [2], Marco W. J. Schreurs [17], Frank E. van Boven [1] and Nicolette J. T. Arends [2,†]

1 Department of Internal Medicine, Section of Allergology & Clinical Immunology, Erasmus MC, University Medical Centre Rotterdam, 3015GD Rotterdam, The Netherlands; m.e.vansplunter@erasmusmc.nl (M.E.v.S.); r.gerthvanwijk@erasmusmc.nl (R.G.v.W.); s.terlouw@erasmusmc.nl (S.T.); f.boven@erasmusmc.nl (F.E.v.B.)
2 Depertment of Peadiatric Allergology, Sophia Children Hospital, Erasmus MC, University Medical Centre Rotterdam, 3015GD Rotterdam, The Netherlands; j.a.m.emons@erasmusmc.nl (J.A.M.E.); a.hendriks-vantoor@erasmusmc.nl (A.I.H.); n.arends@erasmusmc.nl (N.J.T.A.)
3 Food Quality & Design Group, Wageningen University & Research Centre, 6708PB Wageningen, The Netherlands; kasper.hettinga@wur.nl (K.A.H.); harry.wichers@wur.nl (H.J.W.)
4 Cell Biology & Immunology Group, Wageningen University & Research Centre, 6708PB Wageningen, The Netherlands; huub.savelkoul@wur.nl (H.F.J.S.); joost.vanneerven@wur.nl (R.J.J.v.N.)
5 Department of Peadiatric Pulmonology & Allergology, UMCG, University Medical Center Groningen, 9713GZ Groningen, The Netherlands; a.b.sprikkelman@umcg.nl
6 University Medical Center Groningen, GRIAC Research Institute, University of Groningen, 9713GZ Groningen, The Netherlands
7 FrieslandCampina, 3811LP Amersfoort, The Netherlands
8 Department of Rheumatoloy, Leiden University, 2311BD Leiden, The Netherlands; l.liu@outlook.com
9 Department of Peadiatrics, Martini Hospital, 9728NT Groningen, The Netherlands; g.vander.meulen@mzh.nl
10 Paediatric Allegology Center, VieCuri Medisch Centrum, 5912BL Venlo, The Netherlands; I.Herpertz@mzh.nl
11 Department of Peadiatrics, Noordwest Ziekenhuisgroep, 1815JD Alkmaar, The Netherlands; y.c.m.duijvestijn@nwz.nl
12 Deparment of Peadiatrics, Elkerliek, 5707HA Helmond, The Netherlands; mbreukels@elkerliek.nl
13 Department of Peadiatrics, Canisius Wilhelmina Ziekenhuis, 6532SZ Nijmegen, The Netherlands; m.l.brouwer@cwz.nl
14 Department of Peadiatrics, OLVG Hospital, 1091AC Amsterdam, The Netherlands; j.v.schilperoort@olvg.nl (J.S.); o.vandoorn@olvg.nl (O.v.D.); b.vlieg-boerstra@olvg.nl (B.V.-B.)
15 Department of Peadiatrics, Zuyderland MC, 6419PC Heerlen, The Netherlands; joc.vandenberg@zuyderland.nl
16 Ausnutria B.V., 8025BM Zwolle, The Netherlands; Linette.Pellis@ausnutria.nl
17 Department of Immunology, Erasmus MC, University Medical Centre Rotterdam, 3015GD Rotterdam, The Netherlands; m.schreurs@erasmusmc.nl
* Correspondence: n.w.dejong@erasmusmc.nl; Tel.: +31-6-21697954
† These authors contributed equally to this work.

Citation: de Jong, N.W.; van Splunter, M.E.; Emons, J.A.M.; Hettinga, K.A.; Gerth van Wijk, R.; Wichers, H.J.; Savelkoul, H.F.J.; Sprikkelman, A.B.; van Neerven, R.J.J.; Liu, L.; et al. Introduction of Heated Cow's Milk Protein in Challenge-Proven Cow's Milk Allergic Children: The iAGE Study. *Nutrients* **2022**, *14*, 629. https://doi.org/10.3390/nu14030629

Academic Editor: Elvira Verduci

Received: 27 December 2021
Accepted: 27 January 2022
Published: 31 January 2022

Abstract: The introduction of baked milk products in cow's milk (CM) allergic children has previously been shown to accelerate induction tolerance in a selected group of children. However, there is no standardized baked milk product on the market. Recently, a new standardized, heated and glycated cow's milk protein (HP) product was developed. The aim of this study was to measure safety and tolerability of a new, well characterized heated CM protein (HP) product in cow's milk allergic (CMA) children between the age of 3 and 36 months. The children were recruited from seven clinics throughout The Netherlands. The HP product was introduced in six incremental doses under clinical supervision. Symptoms were registered after introduction of the HP product. Several questionnaires were filled out by parents of the children. Skin prick tests were performed with CM and HP product, sIgE to CM and α-lactalbumin (Bos d4), β-lactoglobulin (Bos d5), serum albumin (Bos d 6), lactoferrin (Bos d7) and casein (Bos d8). Whereas 72% percent (18 out of 25) of the children tolerated the HP product, seven children experienced adverse events. Risk factors for intolerance to the HP product were higher skin prick test (SPT) histamine equivalent index (HEP) results with CM and the HP product, higher specific IgE levels against Bos d4 and Bos d8 levels and Bos d5 levels. In conclusion the HP product was tolerated by 72% of the CM allergic children. Outcomes of SPT with CM and the HP product, as well as values of sIgE against caseins, α-lactalbumin, and β-lactoglobulin may predict the tolerability of the HP product. Larger studies are needed to confirm these conclusions.

Keywords: allergy; baked milk; cow's milk; tolerance

1. Introduction

The prevalence of food allergies varies considerably, depending on self-reported food allergy (FA) sensitization to food allergens or confirmed food allergy by an open food challenge (OFC) test, or, preferably, a double-blind, placebo-controlled food challenge (DBPCFC) test. In Europe, the prevalence of challenge-proven/confirmed cow's milk allergy (CMA) in young children (< age 3 year) varies between 0.35 and 2.0% [1,2].

About 70% of CMA children reach clinical tolerance to milk proteins before the age of two years, and 60% before the age of three years, indicating transient CMA [3]. Factors that are most predictive for spontaneous resolution of CMA are: low milk-specific IgE level (<2 kU/L), small-wheal CM skin prick test (SPT) (<5 mm) and absence of (or mild) atopic dermatitis (AD) [4].

Several studies have reported high percentages (59–81%) of tolerance to baked milk in CMA children [4–7]. Children not tolerant towards baked milk products have an increased risk of developing a persistent CMA compared to children tolerant of baked milk (relative hazard ratio: 0.28 vs. 4.1) [3].

CMA children are mostly sensitized (presence of specific IgE to cow's milk proteins) to multiple cow's milk proteins. Sensitization can be found to caseins, including αs1-, αs2-, β- and kappa casein (together constituting 80% of cow's milk proteins (CMP)), and/or to whey proteins, such as α-lactalbumin (Bos d4) and β-lactoglobulin (Bos d5) [8–10]. To a lesser extent, sIgE against bovine serum albumin and other whey proteins is also found [8]. In particular, high levels of sIgE against αs1-, β-casein, Bos d4 and Bos d5 are associated with persistent CMA [11–13]. In addition to IgE-mediated cow's milk allergy, a substantial part of the children has non-IgE-mediated CMA, resulting in delayed-type reactions (>2 h) lacking the typical IgE-mediated symptoms (urticaria, angioedema, respiratory and/or gastro-intestinal symptoms or anaphylaxis) [1,13,14]. About 90% of patients with CMA also react to goat's milk, due to high protein homology (95%) and high protein identity (>84%) [15].

Consumption of baked milk products in CMA children appeared to accelerate tolerance induction in a selected group of children [6,16]. Moreover, a recent meta-analysis showed that most studies were observational, lacking an appropriate control group [17]. Processing milk proteins changes their immunogenic and allergenic properties and can lead to both allergy and the induction of tolerance [18,19]. For example, when milk sugar lactose and CMP are heated together, glycation takes place, causing sugars to be linked to the free amino groups of amino acids [20]. This can lead to the formation of advanced glycation end products (AGEs), which can mask existing epitopes but can also create new immunogenic structures [20–22].

In making baked milk products for previous studies on tolerability, both the form of the product (e.g., cake, bread, or cheese on pizza), and the precise heating process were found to be highly variable. Especially when baking products such as cakes, the internal temperature of the product may strongly differ from the surface temperature, leading to different heat-induced protein modifications throughout the product, which may even leave some of the CMP intact. For better understanding of the effect of baked milk on CMA, a standardized, heated and glycated CMP product should be produced, in which the extent of heat-induced modifications, including glycation, have been well characterized [6,16,23–25].

The aim of this study was to measure safety and tolerability of a new, well characterized heated cm protein (HP) product in CMA children between the age of 3 and 36 months.

2. Materials and Methods

2.1. Study Population

This study was performed in seven hospitals throughout The Netherlands: two university hospitals and five large referral hospitals with expertise in paediatric allergy.

Children between three months and three years of age with cow's milk allergy diagnosed by a doctor (paediatric allergist from participating centres) were approached for participation in the study.

After inclusion, a DBPCFC or OFC with cow's milk was performed. Children with a recent (<6 months) positive CM challenge or a previously severe allergic reaction after CM consumption reported by the participating clinicians did not undergo a cow's milk challenge. Sensitization to cow's milk was measured in SPT and sIgE. Parents filled out several questionnaires.

During a whole-day visit at the outpatient clinic, the HP product was introduced to the diet of the child. Either parents or legal guardians had to understand the Dutch language and signed informed consent. The medical ethical committee of each participating centre approved the study. (NL61774.078.17).

2.2. Double-Blind, Placebo Controlled Food challenge (DBPCFC) with CM

A DBPCFC with CM was performed according to the guidelines. Children consuming extensively hydrolysed formula or formula based on amino acids were challenged using standardized kits. Children consuming breastmilk or alternative milk (e.g., soy milk or rice milk) were challenged using the matrix they consumed. The dosage of CMP in the challenge test remained standard (1-3-10-30-100-300 mg, etc.; Table 1). Symptoms (both subjective and objective) were scored according PRACTALL [26,27].

Table 1. Dosages in double-blind, placebo-controlled CM challenge and HP product introduction, as well as cumulative dosage.

	CM DBPCFC	CM DBPCFC		Open Introduction	Open Introduction
Step	CM protein (mg)	Cumulative dosage (mg)	Step	HP product (mg)	Cumulative dosage (mg)
1	1	1		-	-
2	3	4		-	-
3	10	14	1	10	10
4	30	44	2	30	40
5	100	144	3	100	140
6	300	444	4	300	440
7	1000	1444	5	1000	1440
8	Age-dependent	Age-dependent	6	2000	3440

CM: cow's milk; DBPCFC: double-blind, placebo-controlled food challenge; mg: milligrams; HP product: heated cow's milk product.

2.3. Study Product: Heated Glycated CM Protein (HP)

The HP product was produced by FrieslandCampina (Amersfoort, The Netherlands) and was a powdered product that contained a mixture of whey (20%) and casein protein (80%). The HP product was treated at ultra-high temperature (UHT) (120 °C) for 20 min, spray-dried and subsequently canned. The cans were stored at 60 °C for 14 days, resulting in glycation of the CMPs. The products and procedures for making the AGE products were judged and approved by the Quality assurance/Quality Control (QA/QC) department of FrieslandCampina to be compliant with IFT guidelines. The amount of carboxymethyllysine (CML) (a measure for glycation) in regular infant formula is 28–81 ng CML/mg protein [28]. CML analysis of the new HP product showed a result of 300 ng CML/mg protein, which is comparable to evaporated milk.

2.4. Introduction of HP Product

The HP product was introduced in incremental doses in 30 min intervals (10-30-100 300 mg etc. Table 1) [27]. The HP product was dissolved in the participant's individual daily milk formula (5% of total protein intake/day). The symptoms were recorded in a database/chart according to PRACTALL and scored on severity retrospectively by two independent paediatricians [29]. Serious adverse events (SAEs) and adverse events (AEs) were reported to a data safety monitoring board (DSMB).

2.5. Skin Prick Tests (SPT)

An SPT was performed by applying a drop of skimmed CM (FrieslandCampina), goat's milk (Ausnutria B.V., Zwolle, The Netherlands), the HP study product, a positive (histamine) and a negative (PBS) control. Subsequently, the surface area was measured with an area scanner and compared with the positive control, which gives the histamine equivalent prick (HEP) index score as described [30]. No threshold values have yet been defined for the skin prick test (SPT)–HEP index values for cow's milk allergy. SPTs were considered positive at values > 3 mmØ [31].

2.6. Serum Collection and sIgE Measurements

Blood samples were collected using either a finger prick (age < 6 months) or a vena puncture (≥6 months of age). Serum was collected and stored at −20 °C. ISAC (81-1020-01, Thermo Fisher Scientific B.V, Breda, The Netherlands) was used to identify specific sIgE against CM protein. In addition, sIgE against total CMP was performed using Immunocap (f2) (14-4112-01, Thermo Fisher Scientific B.V., Breda, The Netherlands).

The following specific recombinant allergens were measured by both ISAC and ImmunoCap: α-lactalbumin (Bos d4), β-lactoglobulin (Bos d5), bovine serum albumin (Bos d6), (immunoglobulins/lactoferrin (Bos d7) and whole casein (Bos d8) [32].

2.7. Questionnaires

Validated questionnaires, as used in the "Generation R" study, were used to assess the medical history of mother, father and child [33]. Data collected included date of birth, sex, race, ethnicity, height, weight and relevant medical history. To measure eczema severity, POEM and eczema area and severity index (EASI) scores were collected [34]. Furthermore, a questionnaire specifically designed for this study was used to gather information from the parents about the child's current situation in relation to, e.g., atopy, dairy consumption, type of formula, feeding or breastfeeding, introduction of other (solid) food and type of symptoms. The Food allergy quality of life questionnaire (FAQLQ) was implemented according to Velde et al. [35].

2.8. Open Clinica Database

All patient-related information of this study (Case report forms (CRFs)) is kept blinded at Erasmus MC, and data were digitalized in an Open Clinica (OpenClinica, LCC, Waltham, MA, USA) study database (version 3.12.2).

2.9. Statistical Analysis

This study was originally part of a large long-term clinical trial to test tolerance-induction of the HP product. As the trial was hampered by a low inclusion rate, we decided to perform a small exploratory study with the available patients, focusing on measuring safety and tolerability of the product. For further analyses focused on this aim, the Bayesian approach is the recommended method for evaluating small samples [36]. The Bayes factor (BF) produces the likelihood ratio of the alternative hypothesis (H1) (difference between groups) and the null hypothesis (H0) (no difference between groups) [37]. Evidence for the alternative hypothesis (H1) was set as BF > 3 (moderate), BF > 10 (strong), BF > 30 (very strong) and BF > 100 (extreme), and evidence for the null hypothesis (H0) was set as BF < 1 [37]. BF was calculated for proportions of positive sIgE, contingency tables and two-sample designs of SPT-HEP indexes by the Bayes factor package in R, version 4.0.4/ (https://CRAN.R-project.org/package=BayesFactor, accessed on 15 February 2021). Priors in the proportions were set to low, mediocre, high or unknown probability. In the two-sample designs, the prior distribution was set to a Cauchy with rscale = 0.707 [38].

3. Results

In total, 25 CMA children participated in this study: 9 girls and 16 boys, with a mean age of 14.5 months (range: 6–37 months). A total of 18 children were tolerant to the HP product (HPt group), and seven children developed an allergic reaction to the HP product (HPr group). No differences were found in baseline characteristics (e.g., eczema, rhinitis and asthma) between both groups, as shown in Table 2.

Table 2. Baseline characteristics of the HP-tolerant (HPt) and HP-reactive (HPr) children.

		HPt Children (*n* = 18)				HPr Children (*n* = 7)			
		Mean	**Range**	**N Pos (T)**	**%**	**Mean**	**Range**	**N Pos (T)**	**%**
	Age (months)	14.6	(6.5–22.5)	18		13.3	(6.1–37)	7	
	Gender: F(tot)			6 (18)	33.3			3 (7)	42.9
Atopy *	Eczema			11 (18)	61.1			3 (6)	50.0
	EASI			5 (18)	27.8			2 (6)	33.3
	POEM			10 (18)	55.6			3 (6)	50.0
	Rhinitis			4 (18)	22.2			0/6	0
	Asthma-like symptoms			3 (18)	16.7			1 (6)	16.6
	Asthma + medication			2 (18)	11.1			0 (6)	0
Exclusively breastfed	Period, (months)	5.2	(1–9)	9 (18)	50%	3.2	(2–7)	7 (7)	100%
Formula use at inclusion visit	eHF			13 (18)	72%			5 (7)	71%
	AA			5 (18)	27%			2 (7)	29%
Multiple food allergy	Egg, peanut and/or nuts			4 (18)	22%			0 (7)	0%

HPt: HP-product-tolerant; HPr: HP-product-reactive; *: parent reported; EASI: eczema area and severity index, T: totals; eHF: extra hydrolysed formula; AA: amino acid formula.

Baseline measurements comparing differences in sensitization (SPT, sIgE) to CM and CM components are shown in Table 3. In the HPt group, 10/17 (59%) children, and in the HPr group 4/5 (80%) children, had a positive SPT (>3 mmØ) for CM (BF 0.6). Specific serum IgE to CM was positive in 10/15 (67%) children in the HPt group and in 4/5 (80%) children in the HPr group (BF 2.4). Most children were sensitized both in SPT and sIgE to whole CM, but in some cases, only one was positive. Results show that the group does not contain non-IgE-mediated CMA children, although in some individual cases, symptoms in DBPCFC occurred > 2 h after the challenge. The HPt group showed lower sIgE levels to Bos d4 (BF 6,2) and Bos d8 (BF 17,8) in comparison with the HPr group. On the contrary, the HPr group showed lower Bos d5 sIgE levels (BF 6,2). The SPT with HP product was found in only half of all children who tested positive (11/22): six (35%) in the HPt group and four (80%) in the HPr group (BF 6.5). SIgE values measured with ISAC against house dust mite, grass pollen and birch pollen were negative in all children.

Table 3. Baseline sensitization profiles of the HP-product-tolerant (HPt) and HP-reactive (HPr) children.

		HPt Children (n = 18)				HPr Children (n = 7)				BF
		Mean	Range	N Pos (T)	%	Mean	Range	N Pos (T)	%	
SPT *	CM	0.72	(0–2.98)	10 (17)	58.8	1.17	(0–1.72)	4 (5)	80	0.6
	Goat's milk	0.54	(0–4.23)	6 (15)	40	0.86	(0.22–1.16)	4 (4)	100	0.5
	HP product	0.23	(0–1.58)	6 (17)	35.3	1.06	(0–2.33)	4 (5)	80	6.5
sIgE	CM (kU/L)	3.08	(0–17.2)	10 (15)	66.7	14.93	(0.01–49.6)	4 (5)	80	2.4
	α-lactalbumin Bos d4 (ISU)	0.41	(0–3.36)	5 (15)	33.3	1.48	(0–5.45)	4 (5)	80	6.2
	β-lactoglobulin Bos d5 (ISU)	1.47	(0–10.7)	5 (15)	33.3	0.42	(0–0.8)	4 (5)	80	6.2
	Bovine serum albumin Bos d6 (ISU)	0.16	(0–0.79)	4 (15)	26.7	0.5	(0–2.5)	1 (5)	20	1.3
	Casein Bos d8 (ISU)	0.09	(0–1)	1 (15)	6.7	0.74	(0–2.3)	3 (5)	60	17.8
	Lactoferrin (ISU) Bos d7 (ISU)	-	-	0/15	0	-	-	0 (5)	0	NA

HPt: HP-product-tolerant; HPr: HP-product-reactive; *: HEP index; CM: cow's milk; BF: Bayes factor theorem; H0 two values/means are equal; H1 two values/means are not equal. BF < 1 = H0 most likely; BF ≥ 3 = H1 most likely.

No differences were observed for characteristics of the parents, households and the background of children, e.g., familial atopic diseases, between the HPt group and HPr group. (Table 4) The use of antibiotics in the children was higher in the HPt group versus the HPr group (BF 9.4). The percentage of children going to a day-care facility was higher in the HPt group versus the HPr group (BF 3.1), and breastfeeding (ever) was higher in the HPt group versus the HPr group (BF 29.1). More children in the HPt group received CM formula feeding in the first week of life compared to the HPr group (61% vs. 28.6%, respectively).

Table 4. Characteristics of the parents, households and the background of children.

		HPt Children (n = 18)		HPr Children (n = 7)		BF	Prior Chance
		N Pos (T)	%	N Pos (T)	%		
Atopy (parents)	Mother	13 (18)	72.2	5 (7)	71.4	1.1	mediocre
	Father	11 (18)	61.1	4 (7)	57.1	1.3	mediocre
	Both parents	7 (18)	38.9	2 (7)	28.5	2.6	low
	Both parents not	1 (18)	5.5	0 (7)	0	NA	NA
Background (child)	Breastfeeding (ever)	9 (18)	50	7 (7)	100	29.1	high
	Antibiotics use (child)	12 (18)	66.7	2 (7)	28.5	9.4	low
Pregnancy Mother	Antibiotics use	12 (18)	17.6	0(7)	0	1.3	mediocre
	Folic acid	16 (17)	94.1	6 (7)	85.7	0.4	unknown
	Vitamin D suppl.	2 (16)	12.5	1 (7)	14.3	1.8	low
	Ω-3 fatty acid suppl.	2 (16)	12.5	1 (7)	14.3	1.8	low
	fish oil capsules	0 (15)	0	1 (7)	14.3	0.6	unknown
	multivitamin suppl.	13/17	76.5	4 (7)	57.1	0.7	unknown
Exposure to smoke	During pregnancy *	4 (17)	23.5	2 (7)	28.6	2.3	low
	Current smoking	1 (17)	5.9	0 (7)	0	0.7	unknown
Pet keeping	Currently	1 (17)	58.8	4 (7)	57.1	1.3	mediocre
Other	Day care	11 (18)	61.1	2 (7)	28.6	3.1	mediocre

HPt: HP-product-tolerant; HPr: HP-product-reactive; *: passive smoking + current smoking; n: number; BF = Bayes factor: H0 two values/means are equal; H1 two values/means are not equal. BF < 1 = H0 most likely; BF ≥ 3 H1 most likely.

Table 5 shows the adverse events in the HPr group ($n = 7$) during introduction of the HP product. Ref. [31] One patient experienced an SAE (ID: 111002), which started 15 min after the second step (3 mg CM protein). Six patients experienced AEs. An overview of the category and type of symptoms developed at a certain step of HP product is given in Supplementary Table S1.

Table 5. Serious adverse events and adverse events (SAE/AE).

Patient ID	Age Months/ M/F	Allergic Reactions at Step	Minutes after Intake	Medication, According to National Anaphylaxis Protocol	Stopped at Dose/Outcome *	SAE/AE Sampson Scale
111002	13/M	Step 2: Eczema lips (15′ diminished) Step 2 repeated: Stridor, cough, crying	Step 2: 2 min Step 2 repeated: 15 min	Step 2: Adrenaline auto-injector (0.15 mg), Xyzal suspension (2.5 mg), dexamethason (4 mg)	Dose: 2/ After 5 h released from hospital	SAE 4
555004	6/F	Step 6: Eczema feet, back, belly; vomiting, itch, eczema face, diarrhea	Step 6: 15 min; 6–9 h	None	Full challenge/ <24 h	AE/2
888004	37/M	Dry cough, stridor	Step 4: 5–10 min	Step 4: Aerius suspension (2.5 mL)	Dose: 4/ After 2 h released from hospital	AE/4
888005	8/M	Step 1: Sneezing, erythema chin 5: Cough, redness face, nausea	Step 1 repeated: 5 min 5: 10–15 min	Step 5: Aerius-suspension (2x 2.2 mL)	Dose: 5/ After 2 h released from hospital	AE/3
999001	11/M	Step: 2: increased eczema 3: Sneezing, cough, runny nose, increased eczema, wheezing	Step 3: 25 min	Step 3: Aerius suspension (2.5 mL) and prednisone	Dose: 3/ 15 min no more wheezing	AE/3
999002	10/F	Step: 4: Sneezing, cough, starting urticaria 5: Runny nose, redness face, urticaria 6: (after pause): urticaria	Step: 4: 15–20 min 5: directly 6: after 10 min	none	Dose: 6/ Not specified	AE/3
999004	10/F	Step:	Step:	Step 2: Aerius suspension (2.5 mL) and 1.6 mL prednisone	Dose: 2/ After 4 h released from hospital	AE/2

Abbreviations: Step 1 (10 mg), step 2 (30 mg), step 3 (100 mg), step 4 (300 mg), step 5 (1000 mg), step 6 (3000 mg). * Hours until symptoms disappeared; SAE: serious adverse events; AE: adverse events; M/F: male/female; mL: millilitres; min: minutes.

No differences were found in baseline CM challenge (DBPCFC or OFC) between HPt and HPr groups. In total, 17/25 challenges were double blind. In one child, no challenge was performed due to two anaphylactic reactions to CM in recent history. In some individual cases, symptoms in DBPCFC occurred > 2 h after the challenge. The symptoms that occurred in these individuals were skin disorders and vomiting. Epinephrine auto-injectors had to be used three times in the HPt group, and no epinephrine was administered in the HPr group (Supplementary Table S2).

In the FAQLQ, no differences were found between groups, except one: we found a lower parental perception (<2 = very small chance) on appropriate response by others to allergic reactions in their child in the HPr group (HPt group mean: 2.28; range 0–6; HPr group mean: 1.5; range 1–2) [6] (Supplementary Table S3).

4. Discussion

This study aimed to investigate the safety and tolerability of a new heated and glycated cow's milk protein product, the HP product. This HP product aimed to mimic and to have tolerance-inducing capacities similar to those of "baked milk" products, with well-defined production methods, e.g., exact heating time and temperature during glycation process, in order to achieve a predetermined glycation level.

Due to low inclusion numbers, the primary aim to study tolerance-inducing capacity of the HP product could not be reached, and planned statistical analyses could not be performed. Nevertheless, in alternative analyses (e.g., Bayes factor) differences between HPt and HPr children could be analysed and were, in some cases, significantly high and should be considered exploratory.

Low inclusion numbers in intervention studies in children with cow's milk allergy have been described previously. Many studies on specific food allergy in children are underpowered, according to a recent Cochrane database systematic review on effects of eHF use in CMA children [38]. Reasons for the low inclusion numbers in the current study were less motivated parents due to the many planned hospital visits with the child during the study, following a strictly cow's-milk-free diet and low numbers of positive cow's milk DBPCFC in children with suspected CMA. The latter is not surprising, as recent research by Vlieg et al. in The Netherlands reported a considerable percentage of overdiagnoses of CMA in children (66%) when children did not react in a DBPCFC [39].

Eighteen out of the 25 included CMA children (72%) were tolerant of the HP product (HPt group). Sensitization patterns differed between the two groups. Children in the HPr group showed a higher sensitization profile. This might be one of the reasons for allergic reactions to the HP product [40]. It is known from literature that, in contrast to whey proteins, caseins do not denature and aggregate upon heat treatment but can be glycated [41]. These lesser changes due to heat treatment of the caseins may explain the intolerance of the HP product in this group. Sensitization to caseins was much higher in the HPr group (60%) in comparison with the HPt group (only one child; 7%).

At the same time, the HPr group had clearly lower sIgE against β-lactoglobulin (Bos d5). β-lactoglobulin can be denatured and aggregated after heating, but this HPr group was borderline sIgE-positive (0.42 ISU) for β-lactoglobulin, so that could not affect the tolerability to the heated cow's milk protein in this group. Sensitization level in SPT to goat's milk and CM was comparable between groups (BF 0.6 and 0.5, respectively). Apparently, the HP product cannot be tolerated in a substantial number of children with goat's milk sensitization.

Regarding baseline characteristics, in the HPr group, the use of antibiotics (BF: 9.4) and attendance at day care was lower (BF 3.1) Furthermore, all children in the HPr group received breast feeding (ever), whereas only half of the children in the HPt group did (BF 29.1). When children attend day care to a lesser extent, a lower use of antibiotics seems logical, as infections occur less often in "no-day-care" children [42], and longer breastfeeding is easier in practical terms. Although breastfeeding has many benefits, it does not reduce the risk of CMA [43].

Combining all results of this study, we hypothesize that a possible cause of intolerance of the HP product (HPr group) lies in an overall lower general exposure to allergens (less day care and infections) in the first months of life, as previously described by McGowan et al. [44], and to milk allergens in specific.

Seventy-two per cent of the patients could tolerate the HP product. This is comparable with data from the "baked milk studies" [4–7,16]. In a more recent study by Agyemang et al. among 84 children, 72% were tolerant to muffins containing CM [40]. Furthermore, the HPt group showed > 90% negative sIgE to caseins, most likely representing a group with transient CMA more likely to tolerate baked milk products [8].

In the current study, the safety profile of the HP product was found to be comparable with larger studies with baked milk challenges [45].

The tolerance-inducing effects of baked milk products are described in many studies, but the effect of heating and glycation on tolerance-inducing effects is only sparsely investigated. With this new HP product, we tried to mimic "baked milk" products, while standardizing its characteristics and production. This is, as far as we could find in the literature, the first well described "baked milk" product. The powder can be easily added to the daily formula of very young CMA children, who, in some cases, might not yet be able to consume baked products, e.g., cake. However, the introduction can cause mild to severe allergic reactions and should therefore be supervised by a clinician. Sensitization profiles to CM can be useful to pre-select children who will most likely tolerate the HP product. Further trials with this promising new HP product should be performed in larger groups of children to the measure the tolerance-inducing capacity.

In summary, a new HP product was found to be safe and was tolerated by 72% of challenge-proven CMA children. Outcomes of SPT with CM and the HP product, as well as values of sIgE against caseins, α-lactalbumin and β-lactoglobulin, may predict the tolerability of the HP product. Larger studies are needed to confirm these conclusions.

Supplementary Materials: The following are available online at https://www.mdpi.com/article/10.3390/nu14030629/s1, Table S1: Characteristics of patients with a reaction during introduction of the heated cow's milk protein study product. Table S2: Baseline CM food challenge; comparison between HP-product-tolerant children (HPt) and HP-product-reactive (HPr) in developed symptoms and emergency medication. Table S3: Results FAQ-LQ (D-Q1) parents of children in HPt group and HPr group.

Author Contributions: Conceptualization, N.W.d.J., A.B.S., K.A.H., H.J.W., J.A.M.E., R.J.J.v.N., H.F.J.S., M.E.v.S., N.J.T.A. and R.G.v.W.; methodology, N.W.d.J., J.A.M.E., M.E.v.S. and N.J.T.A.; formal analysis, F.E.v.B.; investigation, N.W.d.J., M.E.v.S., J.A.M.E., K.A.H., R.G.v.W., A.B.S., L.L., G.v.d.M., I.H., Y.C.M.D., M.B., M.I.B., J.S., O.v.D., B.V.-B., J.v.d.B., L.P., S.T., A.I.H., M.W.J.S., F.E.v.B. and N.J.T.A.; writing—original draft preparation, N.W.d.J., M.E.v.S., J.A.M.E., N.J.T.A., K.A.H., R.G.v.W., H.J.W., H.F.J.S., A.B.S., F.E.v.B. and N.J.T.A.; writing—review and editing, N.W.d.J., M.E.v.S., J.A.M.E., N.J.T.A., K.A.H., R.G.v.W., H.J.W., H.F.J.S., A.B.S. and F.E.v.B.; funding acquisition, N.W.d.J., K.A.H., H.J.W., R.J.J.v.N., H.F.J.S. and A.B.S. All authors have read and agreed to the published version of the manuscript.

Funding: This work is part of the iAGE/TTW research programme, project number 14536, which is (partly) financed by The Netherlands Organisation for Scientific Research (NWO).

Informed Consent Statement: The medical ethical committee of each participating centre approved the study. (NL61774.078.17). Written informed consent has been obtained from the patient(s) to publish this paper.

Acknowledgments: We would like to thank all the children and their parents for participating in the iAGE study.

Conflicts of Interest: R. J. Joost van Neerven is an employee of FrieslandCampina. Linette Pellis is an employee of Ausnutria. All authors declare no conflict of interest.

References

1. Schoemaker, A.A.; Sprikkelman, A.B.; Grimshaw, K.E.; Roberts, G.; Grabenhenrich, L.; Rosenfeld, L.; Siegert, S.; Dubakiene, R. Rudzeviciene, O.; Reche, M.; et al. Incidence and natural history of challenge-proven cow's milk allergy in European children—EuroPrevall birth cohort. *Allergy* **2015**, *70*, 963–972. [CrossRef] [PubMed]
2. Nwaru, B.I.; Hickstein, L.; Panesar, S.S.; Roberts, G.; Muraro, A.; Sheikh, A.; Halkene, S.; Hoffmann-Sommergruber, K. Bindslev-Jensen, C.; Worm, M.; et al. Prevalence of common food allergies in Europe: A systematic review and meta-analysis *Allergy* **2014**, *69*, 992–1007. [CrossRef]
3. Wood, R.A.; Sicherer, S.H.; Vickery, B.P.; Jones, S.M.; Liu, A.H.; Fleischer, D.M.; Henning, A.K.; Mayer, L.; Wesley Burks, A. Grishin, A.; et al. The natural history of milk allergy in an observational cohort. *J. Allergy Clin. Immunol.* **2013**, *131*, 805–812 [CrossRef] [PubMed]
4. Miceli Sopo, S.; Greco, M.; Monaco, S.; Bianchi, A.; Cuomo, B.; Liotti, L.; Iacono, D.I. Matrix effect on baked milk tolerance in children with IgE cow milk allergy. *Allergol Immunopathol.* **2016**, *44*, 517–523. [CrossRef] [PubMed]
5. Barbosa, C.P.G.; Castro, A.P.M.; Yonamine, G.H.; Gushken, A.K.F.; Beck, C.M.L.; Macedo, P.R.C.; Dorna, M.D.; Santos, C.J.N. Pastorino, A.C.; Jacob, C.M.A. Baked milk tolerant patient: Is there any special feature? *Allergol. Immunopathol.* **2017**, *45*, 283–289 [CrossRef]
6. Kim, J.S.; Nowak-Węgrzyn, A.; Sicherer, S.H.; Noone, S.; Moshier, E.L.; Sampson, H.A. Dietary baked milk accelerates the resolution of cow's milk allergy in children. *J. Allergy Clin. Immunol.* **2011**, *128*, 125–131. [CrossRef] [PubMed]
7. Nowak-Wegrzyn, A.; Bloom, K.A.; Sicherer, S.H.; Shreffler, W.G.; Noone, S.; Wanich, N.; Sampson, H.A. Tolerance to extensively heated milk in children with cow's milk allergy. *J. Allergy Clin. Immunol.* **2008**, *122*, 342–347. [CrossRef]
8. Bartuzi, Z.; Cocco, R.R.; Muraro, A.; Nowak-Węgrzyn, A. Contribution of Molecular Allergen Analysis in Diagnosis of Milk Allergy. *Curr. Allergy Asthma Rep.* **2017**, *17*, 46. [CrossRef]
9. Kuitunen, M.; Englund, H.; Remes, S.; Movérare, R.; Pelkonen, A.; Borres, M.P.; Mäkelä, M.J. High IgE levels to α-lactalbumin, β-lactoglobulin and casein predict less successful cow's milk oral immunotherapy. *Allergy* **2015**, *70*, 955–962. [CrossRef]
10. Wal, J.M. Bovine milk allergenicity. *Ann. Allergy Asthma Immunol.* **2004**, *93*, 2–11. [CrossRef]
11. Caubet, J.C.; Lin, J.; Ahrens, B.; Gimenez, G.; Bardina, L.; Niggemann, B.; Sampson, H.A.; Beyer, K. Natural tolerance development in cow's milk allergic children: IgE and IgG4 epitope binding. *Allergy* **2017**, *72*, 1677–1685. [CrossRef] [PubMed]
12. Sánchez-Valverde, F.; Etayo, V.; Gil, F.; Aznal, E.; Martínez, D.; Amézqueta, A.; Mendizábal, M.; Galbete, A.; Pastor, N.; Vanderhoof, J. Factors Associated with the Development of Immune Tolerance in Children with Cow's Milk Allergy. *Int. Arch. Allergy Immunol.* **2019**, *179*, 290–296. [CrossRef] [PubMed]
13. Uncuoglu, A.; Cogurlu, M.T.; Eser Simsek, I.; Ergul, N.; Baydemir, C.; Aydogan, M. Predicting outgrowth of IgE-mediated cow's milk allergy: Diagnostic tests in children under two years of age. *Allergol. Immunopathol.* **2019**, *47*, 449–456. [CrossRef] [PubMed]
14. Haselden, B.M.; Kay, A.B.; Larché, M. Immunoglobulin E-independent major histocompatibility complex-restricted T cell peptide epitope-induced late asthmatic reactions. *J. Exp. Med.* **1999**, *189*, 1885–1894. [CrossRef]
15. Nachshon, L.; Goldberg, M.R.; Levy, M.B.; Epstein-Rigbi, N.; Appel, M.Y.; Katz, Y.; Elizur, A. Cross-desensitization to goat and sheep milk following cow's milk oral immunotherapy. *J. Allergy Clin. Immunol. Pract.* **2020**, *8*, 358–360. [CrossRef]
16. Uncuoglu, A.; Yologlu, N.; Simsek, I.E.; Uyan, Z.S.; Aydogan, M. Tolerance to baked and fermented cow's milk in children with IgE-mediated and non-IgE-mediated cow's milk allergy in patients under two years of age. *Allergol. Immunopathol.* **2017**, *45*, 560–566. [CrossRef]
17. Lambert, R.; Grimshaw, K.E.C.; Ellis, B.; Jaitly, J.; Roberts, G. Evidence that eating baked egg or milk influences egg or milk allergy resolution: A systematic review. *Clin. Exp. Allergy* **2017**, *47*, 829–837. [CrossRef]
18. Hilmenyuk, T.; Bellinghausen, I.; Heydenreich, B.; Ilchmann, A.; Toda, M.; Grabbe, S.; Saloga, J. Effects of glycation of the model food allergen ovalbumin on antigen uptake and presentation by human dendritic cells. *Immunology* **2010**, *129*, 437–445. [CrossRef]
19. Stojadinovic, M.; Pietersm, R.; Smit, J.; Velickovic, T.C. Cross-linking of β-lactoglobulin enhances allergic sensitization through changes in cellular uptake and processing. *Toxicol. Sci.* **2014**, *140*, 224–235. [CrossRef]
20. Iwan, M.; Vissers, Y.M.; Fiedorowicz, E.; Kostyra, H.; Kostyra, E.; Savelkoul, H.F.; Wichers, H.J. Impact of Maillard reaction on immunoreactivity and allergenicity of the hazelnut allergen Cor a 11. *J. Agric. Food Chem.* **2011**, *59*, 7163–7171. [CrossRef]
21. Lehmann, K.; Schweimer, K.; Reese, G.; Randow, S.; Suhr, M.; Becker, W.M.; Vieths, S.; Rösch, P. Structure and stability of 2S albumin-type peanut allergens: Implications for the severity of peanut allergic reactions. *Biochem J.* **2006**, *395*, 463–472. [CrossRef] [PubMed]
22. Suhr, M.; Wicklein, D.; Lepp, U.; Becker, W.M. Isolation and characterization of natural Ara h 6: Evidence for a further peanut allergen with putative clinical relevance based on resistance to pepsin digestion and heat. *Mol. Nutr. Food Res.* **2004**, *48*, 390–399. [CrossRef] [PubMed]
23. Bavaro, S.L.; De Angelis, E.; Barni, S.; Pilolli, R.; Mori, F.; Novembre, E.M.; Monaci, L. Modulation of Milk Allergenicity by Baking Milk in Foods: A Proteomic Investigation. *Nutrients.* **2019**, *11*, 1536. [CrossRef] [PubMed]
24. Kiykim, A.; Karakoc-Aydiner, E.; Gunes, E.; Nain, E.; Ogulur, I.; Yazici, D.; Aktac, S.; Bicer, A.H.; Sackesen, C.; Baris, S.; et al. Evaluation of a Standardized Bakery Product (SUTMEK) as a Potential Tool for Baked-Milk Tolerance and Immunotherapy Research Studies. *Int. Arch. Allergy Immunol.* **2019**, *178*, 1–9. [CrossRef] [PubMed]
25. Robinson, M.L.; Lanser, B.J. The Role of Baked Egg and Milk in the Diets of Allergic Children. *Immunol. Allergy Clin. North Am.* **2018**, *38*, 65–76. [CrossRef]

26. Boyce, J.A.; Assa'ad, A.; Burks, A.W.; Jones, S.M.; Sampson, H.A.; Wood, R.A.; Plaut, M.; Cooper, S.F.; Fenton, M.J.; Arshad, S.H.; et al. Guidelines for the diagnosis and management of food allergy in the United States: Summary of the NIAID-sponsored expert panel report. *Nutr. Res.* **2011**, *31*, 61–75. [CrossRef]
27. Sampson, H.A.; Gerth van Wijk, R.; Bindslev-Jensen, C.; Sicherer, S.; Teuber, S.S.; Burks, A.W.; Dubois, A.E.J.; Beyer, K.; Eigenmann, P.A.; Spergel, M.J.; et al. Standardizing double-blind, placebo-controlled oral food challenges: American Academy of Allergy, Asthma & Immunology-European Academy of Allergy and Clinical Immunology PRACTALL consensus report. *J. Allergy Clin. Immunol.* **2012**, *130*, 1260–1274.
28. Chen, Z.; Kondrashina, A.; Greco, I.; Gamon, L.F.; Lund, M.N.; Giblin, L.; Davies, M.J. Effects of Protein-Derived Amino Acid Modification Products Present in Infant Formula on Metabolic Function, Oxidative Stress, and Intestinal Permeability in Cell Models. *J. Agric. Food Chem.* **2019**, *67*, 5634–5646. [CrossRef]
29. Sampson, H.A. Anaphylaxis and emergency treatment. *Pediatrics* **2003**, *111*, 1601–1608. [CrossRef]
30. van der Valk, J.P.; Gerth van Wijk, R.; Hoorn, E.; Groenendijk, L.; Groenendijk, I.M.; de Jong, N.W. Measurement and interpretation of skin prick test results. *Clin. Transl. Allergy* **2015**, *6*, 8. [CrossRef]
31. Sampson, H.A.; Albergo, R. Comparison of results of skin tests, RAST, and double-blind, placebo-controlled food challenges in children with atopic dermatitis. *J. Allergy Clin. Immunol.* **1984**, *74*, 26–33. [CrossRef]
32. Hochwallner, H.; Schulmeister, U.; Swoboda, I.; Spitzauer, S.; Valenta, R. Cow's milk allergy: From allergens to new forms of diagnosis, therapy and prevention. *Methods* **2014**, *66*, 22–33. [CrossRef] [PubMed]
33. de Jong, N.W.; Elbert, N.J.; Mensink-Bout, S.M.; van der Valk, J.P.M.; Pasmans, S.; Jaddoe, V.W.V.; de Jongste, J.C.; van Wijk, R.G.; Duijts, L. Parental and child factors associated with inhalant and food allergy in a population-based prospective cohort study: The Generation R Study. *Eur. J. Pediatr.* **2019**, *178*, 1507–1517. [CrossRef]
34. Chalmers, J.R.; Haines, R.H.; Bradshaw, L.E.; Montgomery, A.A.; Thomas, K.S.; Brown, S.J.; Ridd, M.J.; Lawton, S.; Simpson, E.L.; Cork, M.J.; et al. Daily emollient during infancy for prevention of eczema: The BEEP randomised controlled trial. *Lancet* **2020**, *395*, 962–972. [CrossRef]
35. van der Velde, J.L.; Flokstra-de Blok, B.M.; Dunngalvin, A.; Hourihane, J.O.; Duiverman, E.J.; Dubois, A.E. Parents report better health-related quality of life for their food-allergic children than children themselves. *Clin. Exp. Allergy* **2011**, *41*, 1431–1439. [CrossRef] [PubMed]
36. Billingham, L.; Malottki, K.; Steven, N. Small sample sizes in clinical trials: A statistician's perspective. *Clin. Investig.* **2012**, *2*, 655–657. [CrossRef]
37. Makowski, D.; Ben-Shachar, M.S.; Chen, S.H.A.; Lüdecke, D. Indices of Effect Existence and Significance in the Bayesian Framework. *Front. Psychol.* **2019**, *10*, 2767. [CrossRef]
38. Osborn, D.A.; Sinn, J.K.; Jones, L.J. Infant formulas containing hydrolysed protein for prevention of allergic disease and food allergy. *Cochrane Database Syst. Rev.* **2017**, *3*. [CrossRef]
39. Vlieg-Boerstra, B.J.; Tissen, I.; Wensing, C.; Meijer, Y.; de Vries, E.; Hendriks, T.; Agache, I. Overdiagnosis of cow's milk allergy with home reintroduction. *Pediatr. Allergy Immunol.* **2020**, *31*, 70–706. [CrossRef]
40. Agyemang, A.; Feuille, E.; Tang, J.; Steinwandtner, I.; Sampson, H.; Nowak-Węgrzyn, A. Outcomes of 84 consecutive open food challenges to extensively heated (baked) milk in the allergy office. *J. Allergy Clin. Immunol. Pract.* **2018**, *6*, 653–655. [CrossRef]
41. Moeckel, U.; Duerasch, A.; Weiz, A.; Ruck, M.; Henle, T. Glycation Reactions of Casein Micelles. *J. Agric. Food Chem.* **2016**, *64*, 2953–2961. [CrossRef] [PubMed]
42. Nordlie, A.L.; Andersen, B.M. Children in day care centers, infections and use of antibiotics. *Tidsskr. Nor. Laegeforen.* **2002**, *122*, 2707–2710. [PubMed]
43. de Silva, D.; Halken, S.; Singhm, C.; Muraro, A.; Angier, E.; Arasi, S.; Arshad, H.; Beyer, K.; Boyle, R.; du Toit, G.; et al. Preventing food allergy in infancy and childhood: Systematic review of randomised controlled trials. *Pediatr. Allergy Immunol.* **2020**, *31*, 813–826. [CrossRef] [PubMed]
44. McGowan, E.C.; Bloomberg, G.R.; Gergen, P.J.; Visness, C.M.; Jaffee, K.F.; Sandel, M.; O'Connor, G.; Kattan, K.; Gern, J.; Wood, R.A. Influence of early-life exposures on food sensitization and food allergy in an inner-city birth cohort. *J. Allergy Clin. Immunol.* **2015**, *135*, 171–178. [CrossRef] [PubMed]
45. Al Enezi, M.; Lackm, G.; Foxm, A.T.; Anagnostou, K. Safety and allergic reaction profiles of children undergoing baked milk and egg challenges: A 6-year experience from a pediatric tertiary referral center. *J. Allergy Clin. Immunol. Pract.* **2018**, *6*, 674–676. [CrossRef] [PubMed]

nutrients

MDPI

Article

Methylglyoxal Decoration of Glutenin during Heat Processing Could Alleviate the Resulting Allergic Reaction in Mice

Yaya Wang [1], Xiang Li [1], Sihao Wu [1], Lu Dong [1], Yaozhong Hu [1], Junping Wang [2], Yan Zhang [1] and Shuo Wang [1,*]

1 Tianjin Key Laboratory of Food Science and Health, School of Medicine, Nankai University, Tianjin 300071, China; 18829349424@163.com (Y.W.); lixiang0015@mail.nankai.edu.cn (X.L.); 2120181359@mail.nankai.edu.cn (S.W.); 819063@nankai.edu.cn (L.D.); yzhu@nankai.edu.cn (Y.H.); yzhang@tust.edu.cn (Y.Z.)

2 College of Food Engineering and Biotechnology, Tianjin University of Science & Technology, Tianjin 300457, China; wangjp@tust.edu.cn

* Correspondence: wangshuo@nankai.edu.cn; Tel.: +86-22-8535-8445

Received: 23 August 2020; Accepted: 15 September 2020; Published: 17 September 2020

Abstract: Background: It is widely believed that Maillard reactions could affect the sensitization of allergens. However, the mechanism of action of methylglyoxal (MGO) production in Maillard reactions in the sensitization variation of glutenin (a predominant allergen in wheat) during heat processing is still unclear. Methods: This research evaluated the effect of MGO on the immune response against glutenin in a mouse model. The resulting variations in conformation and corresponding digestibility of glutenin were determined. The immune response and gut microflora variation in mice were analyzed following administering of glutenin and MGO-glutenin. Results: The results of the study showed that MGO-glutenin induced a lower immune response than native glutenin. Cytokine analysis showed that MGO-glutenin regulated mouse immune response by inducing Treg differentiation. MGO decoration changed the structure and digestibility of glutenin. In addition, MGO-glutenin contributes to the maintenance of the beneficial gut microflora. Conclusion: MGO decoration of glutenin during heat processing could alleviate the resulting allergic reaction in mice. Decoration with MGO appears to contribute to the aggregation of glutenin, potentially masking surface epitopes and abating sensitization. Furthermore, Bacteroides induced regulatory T-cell (Treg) differentiation, which may contribute to inhibition of the Th2 immune response and stimulation of immune tolerance.

Keywords: glutenin; methylglyoxal; allergic reaction; gut microflora; heat-processing

1. Introduction

As an important food resource, wheat has been processed into diverse foodstuffs to meet different tastes. While high consumption of wheat can ensure adequate energy and nutrition supply it can also increase the risk of developing allergic disease, including mild and acute reactions induced by wheat allergens. This has prompted research into studying the food allergies of wheat (FAW) toward determining appropriate interventions or for alleviation of allergic reactions. As the predominant allergen, gluten is involved in this immune disorder, especially in the form of glutenin, which is the main protein in terms of FAW [1].

Over the past few decades, the consumption of ultraprocessed foods derived from wheat has increased dramatically. On the one hand, ultraprocessing has been reported to increase the level of side products that are harmful to the host either by direct interplay or via accumulation as intermediates. On the other hand, the Maillard reaction that occurs during this process has the potential to conformationally modify allergen proteins and, as a result, alter the allergenicity of related allergens. Recently, pyrrolidine produced from heat-processed ovalbumin has been shown to have increased immunogenicity, enhancing dendritic cell uptake and IgE production [2], whereas sensitization toward vicilin was decreased after its interaction with glucose during Maillard reaction [3]. Herein, it was concluded that the complicated effects of protein glycation on sensitization are not consistent for different conditions (some increase but some decrease). Allergen processing in the digestive tract plays an important role in determining the allergenicity of food proteins. While a previous report suggested a correlation between digestive stability and allergenicity [4], further research indicated that this correlation was not rigorous [5]. In addition, the changes in allergen structure caused by heat processing are closely related to alterations in the allergenicity of food allergens [6]. An important reason for the increased stability of food allergens during heat processing is the formation of new disulfide bonds or the maintenance of inherent disulfide bonds [7]. However, the role of glycosylation of food allergens on the regulation of immunological properties remains obscure and needs to be elucidated. Gut-associated lymphoid tissue (GALT), as part of the mucosal immune system, is the main tissue responsible for the allergic reactions occurring in the digestive system, especially through oral intake. In GALT, various immune cells and cytokines are involved in the eventual immune response including the allergen-related allergic symptoms or immune tolerance. However, the mechanism by which the individuals can modulate the immune response against allergens remains unclear.

The hygiene hypothesis provides the basis for the correlation between allergic reaction and microbes and addresses environmental changes as a major factor for the development of allergies [8]. Therefore, gut microflora have been demonstrated to play a crucial role in food allergy. The evidence suggests that specific bacterial species from healthy gut microflora play an important role in regulating immune tolerance, as well as their metabolites, such as short-chain fatty acids [9]. It has been demonstrated that the imbalance of gut microflora, characterized by changes in the composition and functional imbalance of intestinal microorganisms, contributes to the development of food allergy (FA) [10]. However, the published conclusions on the features of gut microflora associated with FA still seem preliminary given the generally small number of observations [11].

The Maillard reaction is the reaction of reducing sugars and sugar degradation products with proteins. α-Dicarbonyl compounds like methylglyoxal (MGO) and glyoxal are important intermediates in the Maillard reaction. Of these, MGO has the highest reactivity. In the past, we have studied the effect of MGO on the digestibility of glutenin and its mechanism. Herein, the current project aimed to determine the effect of MGO on the allergenicity of glutenin based on the BALB/c mouse model pre-sensitized to native glutenin, heated glutenin, and MGO-glutenin in order. The changes of structure and digestibility of glutenin and gut microflora in mice were analyzed to elucidate the detailed mechanism by which the potential for allergic reaction is reduced as a result of MGO decoration.

2. Materials and Methods

2.1. Materials

Wheat was purchased from the local commercial market, and the variety was Jimai 22. The water used in this study was manufactured by Milli-Q Ultrapure Water Systems (Shanghai, China). Pepsin from porcine gastric mucosa (>2500 U/mg), trypsin from porcine pancreas (1655 U/mg), and chymotrypsin (>40 U/mg) were purchased from Sigma-Aldrich Chemical Corporation. Unless otherwise specified, all used chemicals were higher than analytical grade. Unless otherwise stated, all chemicals were obtained from Sigma-Aldrich (Shanghai, China).

2.2. Protein Sample Preparation

The glutenin used in the experiment was prepared according to the method in our previous paper [12]. Briefly, n-hexane (1:20, n-hexane: gluten; *w/v*) was added to gluten and then stirred for 1 h to remove fat, followed by placing the gluten suspension in a fuming cupboard overnight to remove the n-hexane. A 0.4 mol/L NaCl solution (1:20, NaCl: gluten; *w*/*v*) was added to the gluten, followed by stirring again for 1 h. The suspension was then centrifuged to collect the precipitate. Ultrapure water (1:20, *w*/*v*) was added to the precipitate and then stirred for 1 h to remove NaCl, followed by centrifugation to collect precipitate. The precipitate was again dissolved in 70% alcohol (1:40, *w*/*v*) and centrifuged at 10,000× *g* for 20 min to remove the prolamin. Each extraction step was repeated three times.

2.3. MGO-Glutenin Preparation

A given mass of glutenin powder was placed in a mortar and fully ground. After grinding to an ultrafine powder, glutenin powder was added to ultrapure water (the mass ratio of glutenin to water was 1:1), which was then homogenized at 10,000 rpm using a high-speed blender (Ika T18 Basic, Staufen, Germany) for several rounds until the glutenin powder was stably suspended in the ultrapure water. In order to make the MGO and glutenin fully react, the mass ratio of MGO and glutenin was selected to be 1:8 (slightly higher than that of MGO and glutenin in actual food). MGO was added to the suspension to provide a mass ratio of glutenin to MGO of 1:8. The suspension was then heated to 100 °C for 15 min to simulate heat processing. The glutenin was mixed with water and underwent the same protocol as MGO-glutenin and served as a control (heated glutenin). Samples were then freeze-dried after ultrafiltration to obtain MGOglutenin.

2.4. Structural Characterization of Glutenin

The secondary structure of glutenin and its related reaction products was determined by Fourier transform infrared spectroscopy (FT-IR). A scanning band of 4000–400 cm^{-1} was then used for FT-IR spectroscopy using 32 scanning frames. The corresponding resolution of the spectra was 4 cm^{-1} [13]. The disulfide bond (SS) contents of glutenin were determined using a previously described method [14]. The surface hydrophobicity index (H_0) of glutenin and its corresponding reaction product was determined by the method published in a former study [15]. The extent of proteolytic hydrolysis (DH%) was determined using methods previously described by Wenjun Wen et al. [16]. All measurements were performed in triplicate.

2.5. Determination of Digestibility

In this study, digestibility was determined according to the method in our previous paper [12]. Simulated gastric and intestinal fluids were prepared based on US Pharmacopoeia formulae. The enzyme used in the gastric digestion phase was pepsin (182 U/mg proteins), and the small intestinal digestion phase included trypsin (40 U/mg proteins) and chymotrypsin (0.5 U/mg proteins). The extent of proteolytic hydrolysis (*DH*) was calculated using the following equation:

$$DH(\%) = \frac{h_s}{h_{total}} \times 100\% \quad (1)$$

where h_s is the concentration (mmol) of free amine groups per gram of protein in the sample and h_{total} is the concentration (mmol) of free amino groups per gram of protein, assuming complete hydrolysis of the protein (8.83 mmol/g protein). All measurements were made in triplicate.

2.6. Mice

Female BALB/c mice aged 6 to 8 weeks were purchased from SiBeiFu Experimental Animal Breeding Co. Ltd. (Beijing, China). All mice used in this study were treated according to the guidelines for the care and use of Laboratory Animals published by the US National Institutes of Health, and all experimental procedures were approved by the Animal Care Review Committee of Tianjin University of Science and Technology. Animals were housed in an air-conditioned room (23 ± 2 °C) with a

12 h light/12 h dark cycle. All mice were allowed free access to food and purified water. All animal experiments began one week after feeding.

2.7. Experimental Design

The mice were divided into four groups: three groups of mice sensitized to either native glutenin, heated glutenin, or MGO-glutenin in addition to unsensitized (control group). Mice ($n = 8$) were intraperitoneally sensitized with 10 μg of glutenin mixed on aluminum hydroxide (Sigma-Aldrich, Saint Quentin Fallavier, France) on days 0, 7, 14, 21, and 28. Then, mice were intragastrically administered with 20 mg of glutenin on day 35 [17,18]. The schematic diagram of the experimental design for glutenin-sensitizing of mice is shown in Figure S1.

2.8. Allergy Evaluation

First, anaphylaxis symptoms were scored by visually monitoring mice for 1 h after challenge. Anaphylactic symptoms were rated as 0 = no symptoms; 1 = hair up, scratching head and ear; 2 = reduced activity; 3 = swelling around the eyes and mouth; 4 = loss of consciousness, no activity upon prodding; and 5 = convulsion, death.

Blood was then taken from the retro-orbital plexus on day 36 (24 h after intragastric stimulation) and then centrifuged at 3000× g for 20 min at 4 °C. The serum was then stored at −80 °C until use. Levels of serum total IgE, histamine, mast cell tryptase (MCT), and serum mouse mast cell protease 1 (mMCP-1) were measured using a commercial ELISA kit according to the manufacturer's recommendations (NanJingJianCheng Co. Ltd., Nanjing, China).

2.9. Cell Separation

GALT was prepared according to the method of Resendiz-Albor A. et al. [19]. The mice were sacrificed by dislocation of cervical vertebra. All small intestines were taken and soaked in pre-cooled D-Hank's solution, the mesentery was carefully removed, and Peyer's patches were then carefully cut out and collected. The intestine was repeatedly washed with D-Hank's solution containing 5% fetal calf serum, and the washed small intestine was cut along the longitudinal axis, placed in a centrifuge tube containing EDTA–DTT (Ethylene Diamine Tetraacetic Acid-dithiothreitol) digest, and then oscillated for 40 min at 180 r/min at 37 °C. The cell suspension was passed through a mesh filter and then collected by centrifugation. The cells were resuspended in 5 mL of 40% Percoll solution, then carefully added to 4 mL of 70% Percoll solution and centrifuged at 1000 r/min for 30 min. The cells collected at the interface were intestinal intraepithelial lymphocytes (IELs). The mesentery and the collected Peyer's patches (PPs) were ground, filtered with mesh filter, and centrifuged. The cells were resuspended in 5 mL of 40% Percoll solution, carefully added to 4 mL of 70% Percoll solution, and centrifuged at 1000 r/min for 30 min, and the cells at the interface were collected to obtain mesenteric lymph nodes (MLNs) and PPs. Spleens was collected upon sacrifice under sterile conditions. Single-cell suspensions were prepared from spleen by pressing through a cell strainer using a piston, and the collected cells were washed with PBS. To isolate splenocytes, red blood cells were removed by treatment with RBC lysis buffer (Beyotime, Jiangsu, China). The lymphocytes and splenocytes were then used as material for further cytokine assays.

2.10. Detection of Cytokines by ELISA

Suspensions of GALT cells (5×10^6 cells/well) were prepared in RPMI-1640 medium containing 100 μg mL^{-1} of either glutenin, heated glutenin, or MGO-glutenin and incubated at 37 °C. The RPMI-1640 medium contained 1% penicillin–streptomycin, 10% fetal bovine serum, 25 mM Hepes buffer, and 5×10^5 M 2-mercaptoethanol. After 24 h, the supernatants of the cultures from each mouse were collected and pooled. Cytokines (IFN-γ, IL-4, IL-10, and TGF-β) in the culture supernatants were determined by commercial ELISA kits (NanJingJianCheng Co. Ltd., Nanjing, China) following the manufacturer's recommendations.

2.11. High-Throughput Sequencing and Bioinformatic Analysis

Isolated fecal DNA was treated as previously reported. Sequencing of 16S rRNA gene amplicons was performed on the Illumina MiSeq platform (Nuohe Zhiyuan Bio-informatics Technology Co. Ltd., Tianjin, China), according to a previous report, to determine sequences of primers targeting the V4 hypervariable region of the bacterial 16S rRNA genes. Bioinformatic analysis of sequencing data was conducted using the Quantitative Insights Into Microbial Ecology (QIIME) software. Briefly, raw data from all samples were filtered and spliced to obtain high-quality clean reads. Taxonomic ranks were assigned to OTU (operational taxonomic unit) representative sequences using Ribosomal Database Project (RDP) Classifier v 2.2. Finally, an OTU table and a phylogenetic tree were produced according to diversity (within sample) and β diversity (between samples) analysis.

2.12. Statistical Analysis

Data are presented as mean ± standard deviation (SD) or standard error of mean (SEM) and were analyzed using the SPSS 19.0 software (International Business Machines Corporation, New York, NY, USA). Data were tested by Student's *t*-test. Differences were considered significant when $p < 0.05$.

3. Results

3.1. MGO Induced Conformational Changes of Glutenin

To investigate whether MGO modification resulted in changes to the secondary and spatial structure of glutenin, native glutenin, and its related reaction products were analyzed using FT-IR, and the secondary structure was determined using OMICN software based on Fourier transform infrared spectroscopy results. The results show that the unwinding of the α-helix is accompanied by a parallel interaction with the β-sheet, indicating direct transformation of glutenin into a regular structure induced by MGO decoration (Figure 1A). The heat treatment of glutenin without MGO resulted in decreased α-helix structure. In addition, the significant difference in the structural changes of heated glutenin and MGO-glutenin compared with native glutenin was demonstrated by SS and H_0 analysis, with the heated glutenin showing significantly higher SS content and lower H_0 (Figure 1C,E). The digestibility of MGO-glutenin was characterized by determining its degree of hydrolysis using the OPA method, and the results show lower DH% of MGO-glutenin compared with heated glutenin (Figure 1D).

3.2. MGO-Glutenin Induced a Lower Immune Response than Native Glutenin

Our study evaluated the allergic responses induced by native glutenin, heated glutenin, and MGO-glutenin in a mouse model (Figures 2 and 3). Hypersensitivity symptoms were indicated by scoring from 0 to 5 within 1–1.5 h after each challenge of the mice. All mice in the control group exhibited no hypersensitivity symptoms. Heated glutenin elicited significant hypersensitivity with symptoms of systemic anaphylaxis while mice sensitization with MGO-glutenin showed a lower hypersensitivity reaction (all scores ≤ 1), and three out of 10 were negatively symptomatic. The anaphylactic response score for MGO-glutenin group were relatively discrete. This is because the individual difference is great under this experimental condition, and some mice are more sensitive to MGO-glutenin. Although MGO decoration of glutenin alleviated the resulted allergic reaction, some mice showed reduced activity after the MGO-glutenin challenge, leading to slightly higher allergy scores. (Figure 2). The levels of total IgE and histamine in the serum of MGO-glutenin-stimulated mice were significantly lower than those of the glutenin group (Figure 3A,B). Since the high levels of total IgE and histamine represent a typical allergic response, these results indicate that the sensitivity to MGO-glutenin, in comparison with glutenin, was significantly decreased. The MCT content in mice serum in response to MGO-glutenin was significantly lower than that of the glutenin group, but higher compared with the control group (Figure 3C). No significant difference in the levels of mMCP-1 was observed between the glutenin group and MGO-glutenin group (Figure 3D). The ability of glutenin to induce mast cell degranulation by binding to a specific IgE was significantly reduced by MGO modification but not completely abolished.

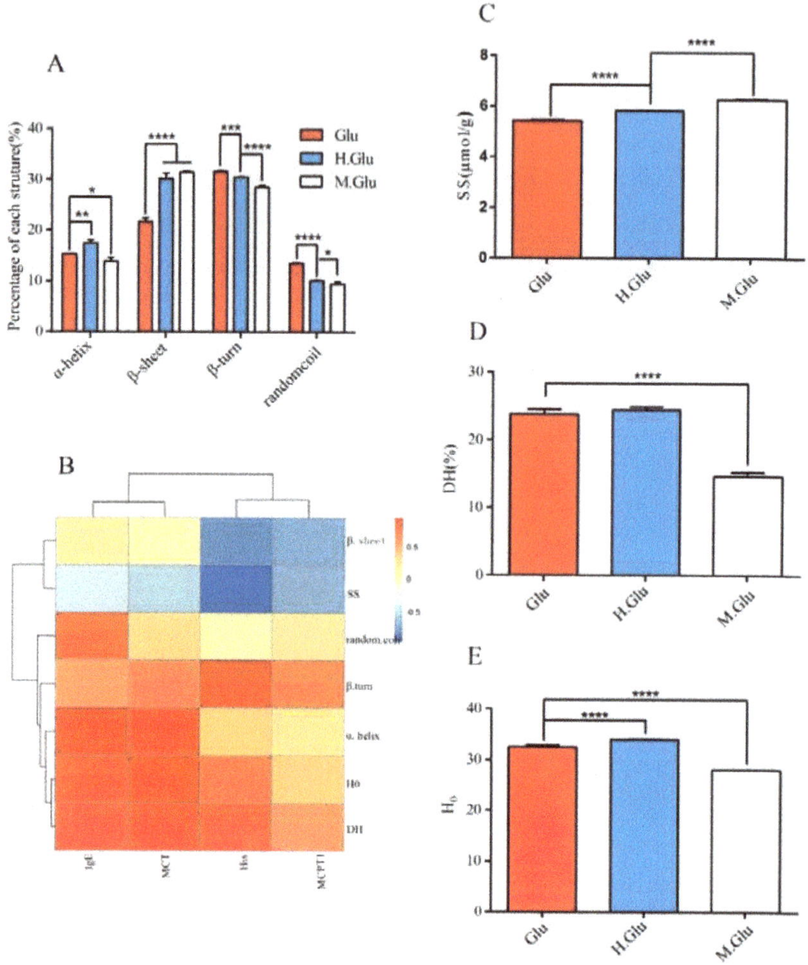

Figure 1. Methylglyoxal (MGO) changes the structure of glutenin. (**A**) Secondary structure changes. Significant differences among α-helix, β-sheet, β-turn, and random coil were indicated as * $p < 0.05$, ** $p < 0.01$, *** $p < 0.001$, **** $p < 0.0001$. (**B**) Correlation analysis between structural changes and sensitivity of glutenin. (**C**) Disulfide bond (SS) levels of glutenin and its corresponding reaction products. (**D**) Proteolytic hydrolysis (DH%) levels of glutenin and its corresponding reaction products. (**E**) H_0 levels of glutenin and its corresponding reaction products. Glu, H.Glu, and M.Glu represent native glutenin, heated glutenin, and MGO-glutenin, respectively.

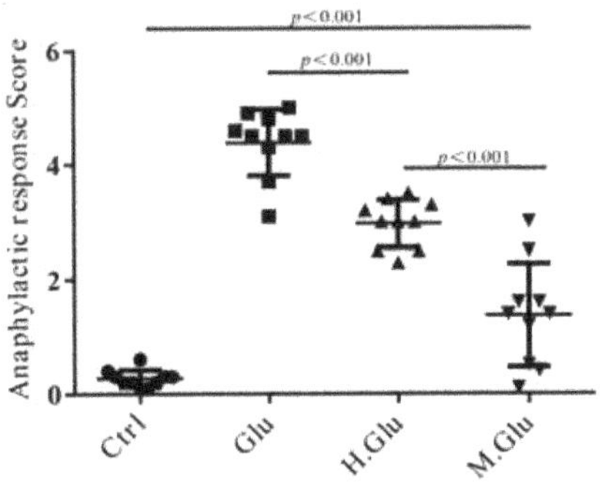

Figure 2. Hypersensitivity symptom scores of mice. Mice sensitization with MGO-glutenin showed a low hypersensitivity reaction (all scores ≤ 1). BALB/c mice were treated with normal saline (control group), glutenin, heated glutenin, and heated MGO-glutenin. Ctrl represents the control group and Glu, H.Glu, and M.Glu represent native glutenin, heated glutenin, and MGO-glutenin, respectively. The triangle represents the control group, the square represents native glutenin, the upward triangle represents heated glutenin, and the downward triangle represents MGO-glutenin.

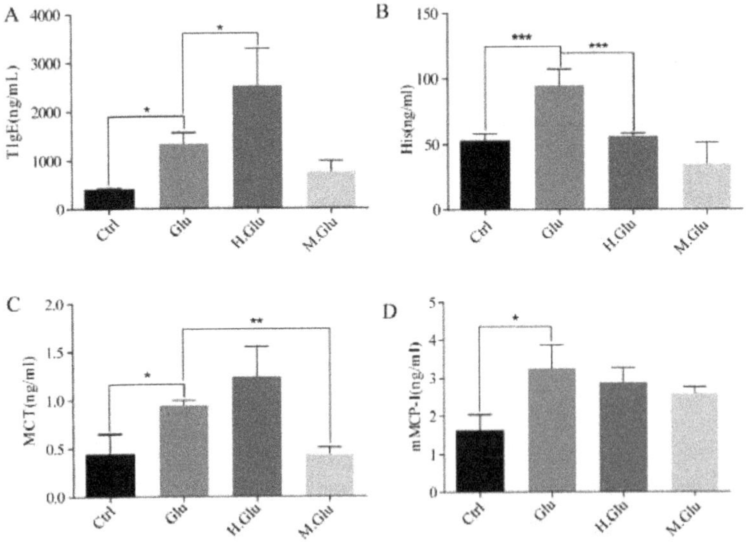

Figure 3. MGO-glutenin induced a lower immune response than glutenin. Ctrl represents the control group and Glu, H.Glu, and M.Glu represent native glutenin, heated glutenin, and MGO-glutenin, respectively. The bars indicate the serum levels of total IgE (**A**), His (**B**), MCT (**C**), and mMCP-1 (**D**). Data are presented as mean ± SEM. Significant differences of Glu versus H.Glu and M.Glu groups are indicated as * $p < 0.05$, ** $p < 0.01$, *** $p < 0.001$.

3.3. MGO-Glutenin Regulated Mouse Immune Response by Inducing Treg Differentiation

To further depict the mechanisms of allergy, the level of cytokines secreted by immune cells was determined based on the supernatants of spleen, thymus, and Peyer's patch cells. After specific activation by glutenin, the secretion of INF-γ into the cell supernatants of spleen and GALT from glutenin-treated mice was reduced compared with control mice. However, after specific activation of MGO-glutenin, the secretion of IFN-γ in the cell supernatants of GALT from MGO-glutenin-treated mice increased compared with that of glutenin-treated mice (Figure 4A). No such difference was observed in the cell supernatant of spleen and GALT except for MLN cells in heated-glutenin-treated mice. Th2 polarization was indicated by IL-4 secretion, which was found to be increased in the in spleen and GALT of glutenin-treated mice compared to control. IL-4 secretion in spleen and GALT was decreased in heated-glutenin-treated and MGO-glutenin-treated compared with glutenin-treated mice (Figure 4B). As an indicator of Treg activation, the levels of IL-10 and TGF-β in the cell supernatant of spleen and gut-associated tissues were determined. Low levels of TGF-β and IL-10 were found in the cell supernatants of spleen and GALT from the glutenin group, whereas high levels of both cytokines were quantified in the control group. In addition, secretion of TGF-β but not IL-10 was increased in the spleen and GALT from heated-glutenin-treated mice compared to glutenin mice. Stimulated secretion of IL-10 was also observed in GALT from heated-glutenin-treated mice, but no significant difference was observed in spleen (Figure 4C,D).

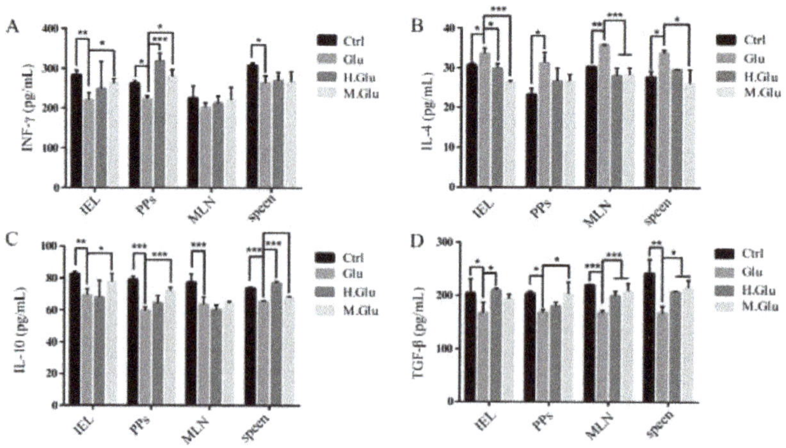

Figure 4. MGO-glutenin regulated mouse immune response by inducing Treg differentiation. The production of IFN-γ (**A**), IL-4 (**B**), IL-10 (**C**), and TGF-β (**D**) cytokines measured by ELISA in the supernatant from lymphoid cells. Ctrl represents the control group and Glu, H.Glu, and M.Glu represent native glutenin, heated glutenin, and MGO-glutenin, respectively. Significant differences of Glu versus H.Glu and M.Glu group are indicated as * $p < 0.05$, ** $p < 0.01$, *** $p < 0.001$. IEL: intestinal intraepithelial lymphocytes, PPs:, MLN: mesenteric lymph nodes.

3.4. MGO-Glutenin but Not Glutenin Contributes to the Maintenance of the Beneficial Gut Microflora

The composition of gut microflora was determined by analyzing the abundance of bacteria in feces using 16S genome sequencing to explore the influence of MGO-glutenin on gut microflora. Alpha diversity was used to describe the variation of microbiologic species diversity in the experimental groups. As shown in Figure 5A, the results of ecological indicators showed that the gut microflora richness of glutenin, heated-glutenin, and MGO-glutenin treated mice all increased compared with the control group. Principal coordinate analysis (PCoA) was used to examine the changes in fecal microbiota across groups (Figure 5B). Compared with the glutenin group, a similar PCoA distance between the MGO-glutenin group and control group was observed. To more precisely determine the effect of glutenin and MGO-glutenin on the distribution disparity in gut microflora composition, LEfSe (LDA effect size) analysis was performed to visualize gut microflora abundance at various taxonomic levels. Fifteen taxa with significant differences were verified in four groups (Figure 5C), among which only four taxa with significant differences were observed at the genus level. At the family level, Bacteroidaceae abundance is a significant distinguishing feature of MGO-glutenin challenge, and this is also reflected at the genus level. Another discriminative feature of MGO-glutenin-stimulated mice are *Bacteroides acidifaciens* profiles. In the glutenin group, distinguishing bacteria at the phylum, class, family, and order levels belong to Firmicutes, Clostridia, Ruminococcaceae, and Clostridiales, respectively. In the glutenin group, significant discriminative genera include bacterium_str_77003, proteobacterium UMB8H, and *Actinobacillus*. Thus, *proteobacterium* UMB8H and *Actinobacillus* may be closely related to the development of glutenin-related food allergy.

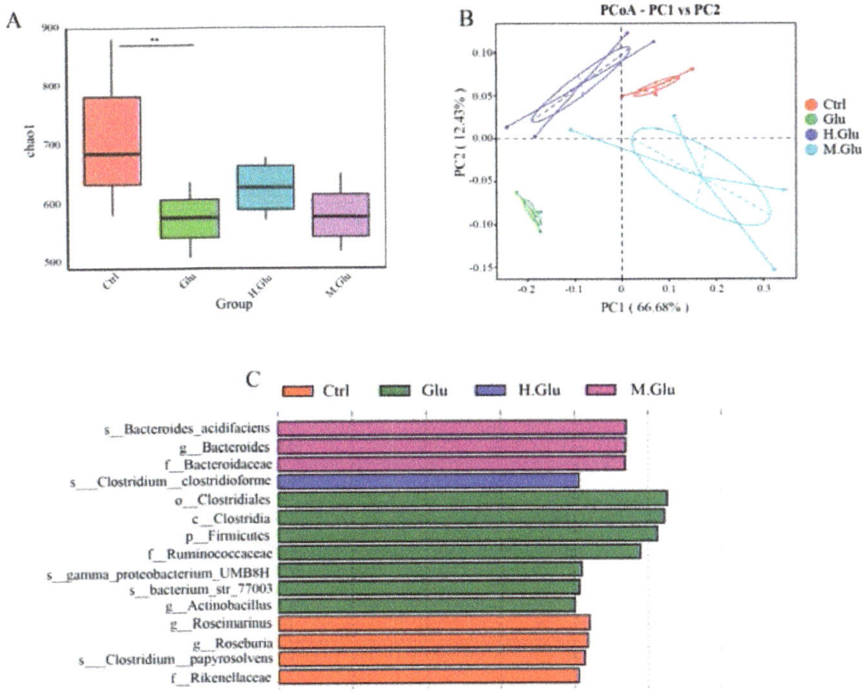

Figure 5. MGO regulates immune response by regulating the microbial composition of mice feces. Ctrl represents the control group and Glu, H.Glu, and M.Glu represent native glutenin, heated glutenin, and MGO-glutenin, respectively. Genomic DNA was extracted from the fecal samples taken from mice just before sacrifice. (**A**) Microbiota diversity of each group. (**B**) Principal coordinate analysis (PCoA) of each group. (**C**) LEfSe analysis microbiota diversity of each group. ** $p < 0.01$.

4. Discussion

Protein glycation based on Maillard reaction may lead to structural changes of allergens, thus changing the allergenicity. However, the role of glycation of allergens in regulating immunological properties remains unclear. In this study, the BALB/c mouse model was used to study the effect of MGO decoration of glutenin on its allergic reaction during heat processing. The mechanism of glutenin allergic reaction changes was elucidated by analyzing the structural and digestive changes and the characteristics of gut microflora in mice.

The effect of food processing methods on the allergenicity of allergens is usually dependent on induced structural variations and epitope redistribution as well as alterations to intrinsic biophysical and chemical properties. Moreover, food allergens have the potential to interact with other food matrixes and subsequently play an important role in allergenic regulation [6]. Analysis of the correlation between glycosylation and allergenicity variation indicated an uncertain effect of either increased or decreased allergenicity induced by Maillard reaction during food processing [20]. Nevertheless, reducing the allergenicity of food allergens using glycosylation or the Maillard reaction is a promising strategy that has been studied in related research to determine the correlation between the secondary structure of proteins and the resulting allergic reactions [21]. Furthermore, the loosing of conformational structure during the process of Maillard reaction would increase the possibility to form polymerization and aggregation [22]. Han et al. reported on the hypoallergenic products of the Maillard reaction in which the underlying mechanism of altered allergenicity was induced structural changes [23]. The commonly employed heat-processing strategy of wheat could affect epitope accessibility through induction of aggregation and may irreversibly destroy conformational epitopes. MGO mainly modifies the side chains of Lysine, arginine, histidine, cysteine, tryptophan, and methionine in proteins during heat processing. The allergen epitopes containing these amino acid residues in glutenin might be destroyed and their sensitivity reduced. The identification of the effect of MGO on the epitope of glutenin allergens is of great significance to our experiment. Unfortunately, due to the complex structure of glutenin, it is difficult to identify the change of its epitope in our experiment. FT-IR was used to analyze the secondary structure, and it was found that unwinding of the α-helix in MGO-glutenin may contribute to the decreased allergenicity compared with glutenin. In addition, the increased level of disulfide bonds and reduced surface hydrophobicity could be observed in MGO-glutenin compared with glutenin and heated glutenin. The reduction of hydrophobicity was likely due to the induction of intermolecular crosslinks by MGO that led to the burying of hydrophobic groups within glutenin. It is well known that heating above 55 °C can promote the glutenin disulfide bond/sulfhydryl exchange reaction, thereby promoting the formation of new disulfide bonds [1]. Therefore, the supramolecular structure formed by polymerization and aggregation may appear during the heating process of glutenin [24]. As shown in the Figure S2, the solubility of heated glutenin decreased slightly and the heated MGO-glutenin increased significantly. Martin et al. have shown that disulfide bond production during heat treatment drives the insolubilize of glutenin [25]. High molecular weight glutenin subunits are related to the solubility of glutenin [26]. Early studies proved that fatty acids may bind to cysteine or lysine residues in high molecular weight glutenin subunit, thereby destroying the glutenin structure and changing the solubility of subunits [27–29]. In this study, MGO may also destroy the lysine and arginine residues of glutenin, thereby reducing its solubility. However, other factors such as oxidation reaction are not excluded. In general, it is meaningful to further study the mechanism of the solubilization of glutenin by MGO. As a result, a general conclusion could be proposed that MGO alters the structure of glutenin and leads to aggregation. Toheder Rahaman et al. showed that heating to 100 °C induced gliadin aggregation that resulted in decreased digestibility and less availability of antigenic components and therefore minimum antigenicity. This study is similar to our results. Therefore, we speculate that the polymerization and aggregation of glutenin results in the decrease of its digestibility and the decrease of antigen utilization, thus leading to the decrease of antigenicity [30]. In addition, the formation of aggregates may destroy or mask allergen epitopes, thereby preventing IgE binding and cross-linking and subsequent mediator release, which ultimately leads to reduced allergenicity.

It is generally believed that there is a close relationship between protein digestibility and sensitization. In order to trigger an allergic immune response, food proteins (or peptides) must be retained in the gastrointestinal tract to allow sufficient time to induce sensitization. From this point of view, the sensitivity of food proteins to intestinal digestion seems to be an important factor for determining their allergenicity [31]. As shown in Figure 1C, the digestibility of MGO-glutenin was significantly lower than that of glutenin. This is because glycosylated lysine and arginine residues are less susceptible to pepsin/trypsin proteolysis by masking the sites of cleavage. Research shows MGO decoration of protein leads to the destruction of amino acid residues with affinity side chains (such as lysine, arginine, cysteine, and histidine), resulting in reducing its digestibility [32]. It is generally accepted that the resistance of protein to gastrointestinal digestion is an indicator of potential allergic reactions. For instance, comparison of digested and non-digested β-lactoglobulin (β-Lg) in rat models found that non-digested β-Lg induced more IgE and more severe allergic reactions, directly linking β-Lg digestion to allergenicity [33]. In general, high levels of glycosylation will reduce the digestibility of the protein, resulting in increased IgE reactivity of the hydrolysate. However, it has also been demonstrated that aggregate formation leads to increased resistance to protein digestion and reduced sensitization [34]. As in current research, MGO modification resulted in reduced glutenin digestibility and sensitization. This is most probably because the poor absorption of digestive products and the delay in their sensitivity to the immune system lead to a weakened immune response in the mouse model [34].

Next, we studied the effect of MGO modified glutenin on allergic reactions. It will be more meaningful if the serum pool of human allergic to glutenin can be used in the experiment to study the sensitization of glutenin by MGO modification. Taking into account the ethical issues and the convenience of the experiment, we use mice model for this study. The experimental design in this study can also support our conclusions. As concluded, our study shows that MGO decoration of glutenin results in reduced sensitization toward glutenin as demonstrated using a mouse model. First, MGO-glutenin-sensitized mice were observed with relatively minor hypersensitive symptoms and a lower level of total IgE compared with glutenin-sensitized and control mice. The consistent variation in allergic symptom score and IgE levels reflects the difference in the degree of sensitization for MGO-glutenin. The degranulation of mast cells occurs during the effector phase of the food allergy, which can reflect the ability of antigens to stimulate allergic reactions. In the case of the re-interaction with the antigen, mast cells degranulate and release effectors such as histamine directly related to the pathological clinical symptoms of food allergy [35]. In this study, it was also observed that the level of histamine in glutenin-sensitized mice was elevated compared with control mice, was restored in MGO-glutenin allergic mice. This result further suggests the lower immune response induced by MGO-glutenin. The development of allergic reactions is accompanied by the release of trypsin by mast cells. The increased activity of mast cell trypsin in mice is due to the release of mast cell trypsin in secretory granules following the activation of mast cells. Mast cell trypsin can destroy the integrity of the relevant tissue membrane, promoting tissue remodeling and the progression of inflammation as induced allergic symptoms. Therefore, MCT is a more selective marker of mast cell-mediated inflammatory response [36]. However, a low level of MCT was observed in MGO-glutenin-sensitized mice, which suggested the reduced sensitization potential resulting from MGO decoration in BALB/c mice. No significant difference in mMCP-1 contents was observed between the glutenin group and MGO-glutenin group. As one of the bioactive substances released after mast cell degranulation, mMCP-1 can be used as the indicator of this process [37]. Though no significant difference in mMCP-1 levels was verified between the glutenin group and MGO-glutenin group to reflect the difference of the effect stage after activation, mMCP-1 levels in the MGO-glutenin group still showed a downward trend compared with the glutenin group. In general, glycosylation of food allergens may alter their immunological behavior.

Correlation analysis was performed to unveil the relationship between structure change and sensitization of glutenin. As shown in Figure 1B, α-helix structure and H_0 are positively correlated with IgE and MCT, while β-sheet formation is negatively correlated with mMCP-1. The change in in protein secondary structure caused by glycation is an effective strategy to mask epitopes related to the allergenicity of food allergens. Several studies have demonstrated the suppressed allergic response resulting from the alteration of secondary structure [38,39]. Gupta R.K. et al. demonstrated that reduced α-helix structure induced by glycation may shield the epitopes of protein and lead to a reduction in allergenicity compared to native chickpea albumin [8], which is consistent with the results of this study. Otherwise, changes in disulfide bonds are also an important factor related to the epitope variation. Mameri, H. et al. demonstrated that the binding ability of gliadin to IgE was reduced due to disulfide bond changes under heating conditions [40]. Similarly, in this study, the presence of disulfide bonds was negatively correlated with MCT, mMCP-1, and His levels. We have recently demonstrated that glutenin can form aggregates through hydrophobic interactions as a result of MGO modification [12]. The formation of aggregates may prevent binding between antigen and antibody epitopes. Moreover, cross-linking of proteins seems to reduce epithelial uptake, which has been demonstrated using crossed-linked β-Lg [39]. In addition, the larger agglomerates can be further metabolized by intestinal microbes, resulting in the formation of new bioactive compounds with the function of modulating the gut microflora composition [41]. Epitope modification induced by the Maillard reaction may affect the allergenicity of wheat protein. Lysine, arginine, histidine, cysteine, tryptophan, and methionine may be involved in the Maillard reaction between MGO and glutenin [42], which facilitates the conclusion that glutenin epitopes containing the above amino acids might be destroyed to reduce their affinity for IgE binding, resulting in reduced sensitization toward glutenin. The above results indicate that MGO can alter the secondary and tertiary structures of glutenin under heating conditions, thereby affecting epitope accessibility by inducing aggregation and resulting in irreversible conformational epitope destruction. It seems to be of special relevance to consider that impaired enzymatic protein digestion is associated with enhanced allergenicity of food proteins. The reason is that reduced digestive capacity results in larger protein fragments that are recognized by the cells of the immune system [43]. However, in this study, DH% is positively correlated with IgE. This may be because of the poor absorption of digestive products and a delay in their sensitivity to the immune system. In addition, longer peptides could alter the composition of the microbiome [44]. In other words, the gut microflora could degrade MGO-glutenin and produce hypoallergenic glutenin peptides.

The interaction between GALT cells and non-immune cells is important for the maintenance of immune tolerance. A previous study has shown that oral tolerance to ovalbumin cannot be induced in PP-deficient mice, which suggests the essential role of PPs in the mucosal immune response [45]. There is clear evidence that MLNs play a key role in inducing mucosal immunity or tolerance. Within hours of the protein entering the intestine, antigen recognition occurred in MLNs, and natural T-cell activation and division took place primarily in MLNs. Previous studies have shown that Tregs can reduce the occurrence of allergies against food allergens, and Treg induction may be the basis for the protective effect of certain dietary interventions in the food allergy model [46]. Transforming growth factor-β (TGF-β) and IL-10 are Treg-related cytokines that are essential for maintaining immune tolerance and reducing allergic reactions [47]. MGO-glutenin leads to increased TGF-β and IL-10, which suggests that MGO-glutenin may induce Treg differentiation. In addition, there is increasing evidence that oral tolerance is mediated through immunosuppressive activation in the gut. The main cells involved in this process are Tregs, which are derived from T-cells after exposure to allergens in the presence of TGF-β [48]. IFN-γ and IL-4 are cytokines released by Th1 and Th2 cells, respectively. Under normal conditions, Th1/Th2 cells are in a relatively balanced state in mice. When dysfunctions such as allergic reactions occur, the balance shifts to Th2 cells, leading to a range of symptoms, including ear swelling, IgE increase, and mast cell threshing [45]. The increased IL-4 levels and the decreased IFN-γ levels in the supernatant of all cells in the MGO-glutenin group indicate that MGO-glutenin inhibited allergic reactions by inhibiting Th2 cell differentiation. These effects coincide with the lower production of Th2-related cytokines, which might dampen Th2 response. This means that MGO-glutenin inhibits Th2 cell differentiation and participates in the induction of beneficial effects by inducing regulatory T-cells and has a profound effect on tolerance maintenance.

The changes of gut microflora in mice were studied by 16S sequencing, which could only reflect the relative changes of flora abundance. To study the effect of MGO-glutenin on the composition of the gut microflora, we used PCoA to compare the glutenin group, heated-glutenin group, and MGO-glutenin group. The PCoA showed marked difference in gut microflora between the glutenin group and MGO-glutenin group. Thus, it was suggested that MGO-glutenin had a significant effect on the intestinal microbes in mice. LEfSe analysis was used to detect the key taxa that differ between the four groups, so as to compare the relative contributions of the discrepant taxa. A total of 15 taxa of different levels were identified to have significant abundance differences across the four groups. In our study, the most significant difference between the MGO-glutenin and other groups in terms of gut microflora composition was associated with the phylum Bacteroidetes, especially for the family Bacteroidaceae and the genus *Bacteroides*, whose presence is an important characteristic of the

MGO-glutenin group. *Bacteroides* has been identified to be conducive to promoting the development of Treg, and to promote tolerance to dietary antigens by inducing the expression of transcription factor RORgt in nascent Treg cells through the upstream myd88-dependent mechanism. Our results provide some interesting insights into the relationship between changes of gut microflora caused by protein glycation and immune response regulation. In future, we plan to conduct studies to verify that *Bacteroides*, which is characteristic of the MGO-glutenin group, participates in immune regulation. Bacteroidetes are thought to be involved in metabolic transformation, usually associated with protein degradation, which is essential for the host. Unlike other allergen proteins, gluten is rich in proline and glutamine residues, which are exceptionally resistant to enzyme degradation in mammalian digestive tracts [49]. This incomplete digestion facilitates the production of longer oligopeptides for the interaction with antigen-presenting cells, which can activate the T-cell response associated with wheat protein allergy [50]. Bacteroidetes can metabolize the digested glutenin into small molecules without associated immunogenicity. Recent studies have shown that symbiotic microorganisms that colonize the intestinal tract have a strong regulatory effect on Th2 immune responses [51]. An obvious example was the observation that in the absence of microorganisms, mice that had not been treated with antibiotics were susceptible to allergy and had elevated levels of IL-4, basophils, and serum IgE, thereby enhancing the Th2 immune response, which suggested that microorganisms are important players for the modulation of Th2 immune response [52]. The potential pathway may depend on the mediation of gut microbes to regulate the differentiation of induced Tregs, thereby suppressing the Th2 immune response. Ohnmacht's research showed that the gut microbes stimulated the expression of RORγt in Tregs and inhibited the Th2 cells, so as to avoid the formation of IL-4 and IgE. This suggested that in the MGO-glutenin group, Bacteroidetes, promote the differentiation of Tregs, which may function in suppressing the Th2 response. Considering the effects of Tregs and Th2 cells on allergic diseases, mucosal immunity, and intestinal flora regulation, a correlation could be established between the immune response and the gut microflora through the determined increased level of Tregs and decreased Th2 cells in the MGO-glutenin group, with further evidence that Bacteroidetes alleviated the immune response in the MGO-glutenin group. Similar to our results, Caminero A. et al. reported a reduction in specific bacterial populations such as *Lactobacilli* and *Bacteroides* that metabolize gluten in celiac disease patients [53]. The most significant differences between the glutenin group and other groups at phylum, family, and genus level were the phylum Bacteroidetes, the family Ruminococcaceae, and the genus *Actinobacillus*. Ruminococcaceae is positively correlated with inflammation-related diseases [54]. In this study, *Actinobacillus* was the only genus of the biological marker in the glutenin group, which has not been previously reported. Therefore, the relative abundance of *Actinobacillus* may be closely related to the development of glutenin-related food allergy. Pearson correlation analysis showed that the abundance of actinomycetes was positively correlated with Th2 cytokines and negatively correlated with Th1 cytokines, which indicates that actinomycetes may be involved in the promoted Th2 cell differentiation, but the detailed mechanism remains to be elucidated. The current research demonstrates the reduced immune response of MGO-glutenin compared with glutenin, which mainly depends on the protective effect of *Bacteroides* in promoting Treg differentiation and inhibiting Th2 differentiation.

In summary, our study proved the previous hypothesis that MGO decoration of glutenin would alleviate allergic reactions in mice. MGO decoration may contribute to the aggregation of glutenin caused by conformational changes in the secondary and tertiary structures, which has the potential to mask or even destroy surface epitopes and mitigate sensitization. In addition, MGO-glutenin alters the composition of gut microflora. *Bacteroides*, which may be a marker microorganism in the feces of MGO-glutenin-sensitized mice, functions in inducing the polarization of Tregs to facilitate the stimulation of immune tolerance and inhibition of the Th2 immune response as part of a general effect that dampens the immune response.

5. Conclusions

This study is the first to investigate the effect of MGO decoration of glutenin on the resulting allergic reaction in mice during heat processing. The current research results show that MGO decoration of glutenin could alleviate the resulting allergic reaction in mice. This remission is achieved by changing the structure and digestibility of glutenin and the intestinal flora of mice. This study provides a theoretical basis for alleviating glutenin allergic reactions through processing. However, the mechanism by which *Bacteroides* challenged mice with MGO-gluten induces Treg cell polarization and suppresses Th2 immune response needs to be further elucidated.

Supplementary Materials: The following are available online at http://www.mdpi.com/2072-6643/12/9/2844/s1, Figure S1: Schematic diagram of mouse sensitization model. Figure S2. Solubility of native glutenin, heated glutenin, and heated MGO-glutenin samples.

Author Contributions: Conceptualization, Y.W. and S.W. (Shuo Wang); methodology, Y.W., X.L., and S.W. (Sihao Wu); software, Y.W., S.W. (Sihao Wu), and L.D.; formal analysis, Y.W. and J.W.; resources, S.W. (Shuo Wang), Y.W., and J.W.; data curation, Y.W. and L.D.; writing—original draft preparation, Y.W. and Y.H.; writing—review and editing, J.W., Y.H., Y.Z., and S.W. (Shuo Wang); visualization, X.L., S.W. (Sihao Wu), and Y.Z.; supervision, Y.W.; project administration, Y.W.; and funding acquisition, S.W. (Shuo Wang). All authors have read and agreed to the published version of the manuscript.

Funding: This work was supported by the grants from the National Natural Science Foundation of China (No. 31772095) and the key research and development plan of Shanxi Province (No. 201903D211006).

Conflicts of Interest: The authors declare no conflict of interest.

References

1. Lupi, R.; Denery-Papini, S.; Claude, M.; Tranquet, O.; Drouet, M.; Masci, S.; Larré, C. Thermal treatment reduces gliadin recognition by IgE, but a subsequent digestion and epithelial crossing permits recovery. *Food Res. Int.* **2019**, *118*, 22–31. [CrossRef] [PubMed]
2. Heilmann, M.; Wellner, A.; Gadermaier, G.; Ilchmann, A.; Briza, P.; Krause, M.; Nagai, R.; Burgdorf, S.; Scheurer, S.; Vieths, S.; et al. Ovalbumin Modified with Pyrraline, a Maillard Reaction Product, shows Enhanced T-cell Immunogenicity. *J. Boil. Chem.* **2014**, *289*, 7919–7928. [CrossRef] [PubMed]
3. Bielikowicz, K.; Kostyra, H.; Kostyra, E.; Teodorowicz, M.; Rigby, N.; Wojtacha, P. The influence of non-enzymatic glycosylation on physicochemical and biological properties of pea globulin 7S. *Food Res. Int.* **2012**, *48*, 831–838. [CrossRef]
4. Astwood, J.D.; Leach, J.N.; Fuchs, R.L. Stability of food allergens to digestion in vitro. *Nat. Biotechnol.* **1996**, *14*, 1269–1273. [CrossRef]
5. Fu, T.-J.; Abbott, U.R.; Hatzos, C. Digestibility of Food Allergens and Nonallergenic Proteins in Simulated Gastric Fluid and Simulated Intestinal FluidA Comparative Study. *J. Agric. Food Chem.* **2002**, *50*, 7154–7160. [CrossRef]
6. Gupta, R.K.; Raghav, A.; Sharma, A.; Gupta, K.; Neelabh; Mandal, P.; Tripathi, A.; Ansari, I.A.; Das, M.; Dwivedi, P.D.; et al. Glycation of clinically relevant chickpea allergen attenuates its allergic immune response in Balb/c mice. *Food Chem.* **2017**, *235*, 244–256. [CrossRef]
7. Mills, E.N.C.; Sancho, A.I.; Rigby, N.M.; Jenkins, J.A.; Mackie, A. Impact of food processing on the structural and allergenic properties of food allergens. *Mol. Nutr. Food Res.* **2009**, *53*, 963–969. [CrossRef]
8. Neau, E.; Delannoy, J.; Marion, C.; Cottart, C.H.; Waligora-Dupriet, A.J. Three Novel Candidate Probiotic Strains with Prophylactic Properties in a Murine Model of Cow's Milk Allergy. *Appl. Environ. Microbiol.* **2016**, *82*, 1722. [CrossRef]
9. Feehley, T.; Aitoro, R.; Stefka, A.; Nagler, C. Bacterial metabolites reduce intestinal permeability to protect against food allergen sensitization (MUC8P.724). *J. Immunol.* **2015**, *194*, 2.
10. Lynch, S.V.; Pedersen, O. The Human Intestinal Microbiome in Health and Disease. *N. Engl. J. Med.* **2016**, *375*, 2369–2379. [CrossRef]
11. Canani, R.B.; Gilbert, J.A.; Nagler, C.R. The role of the commensal microbiota in the regulation of tolerance to dietary allergens. *Curr. Opin. Allergy Clin. Immunol.* **2015**, *15*, 243–249. [CrossRef] [PubMed]

12. Wang, Y.; Wang, J.; Wang, S.; Guo, J.; Wang, S.; Junping, W. Modification of Glutenin and Associated Changes in Digestibility Due to Methylglyoxal during Heat Processing. *J. Agric. Food Chem.* **2019**, *67*, 10734–10743. [CrossRef] [PubMed]
13. Hong, G.-P.; Avramenko, N.; Stone, A.; Abbott, D.; Classen, H.; Nickerson, M. Extractability and Molecular Modifications of Gliadin and Glutenin Proteins Withdrawn from Different Stages of a Commercial Ethanol Fuel/Distillers Dried Grains with Solubles Process Using a Wheat Feedstock. *Cereal Chem. J.* **2012**, *89*, 276–283. [CrossRef]
14. Mir, A.R.; Uddin, M.; Alam, K.; Ali, A. Methylglyoxal mediated conformational changes in histone H2A—generation of carboxyethylated advanced glycation end products. *Int. J. Boil. Macromol.* **2014**, *69*, 260–266. [CrossRef]
15. Alizadeh-Pasdar, N.; Li-Chan, E.C.Y. Comparison of protein surface hydrophobicity measured at various pH values using three different fluorescent probes. *J. Agric. Food Chem.* **2000**, *48*, 328–334. [CrossRef] [PubMed]
16. Wen, W.; Li, S.; Gu, Y.; Wang, S.; Wang, J.; Junping, W. Effects of Starch on the Digestibility of Gluten under Different Thermal Processing Conditions. *J. Agric. Food Chem.* **2019**, *67*, 7120–7127. [CrossRef]
17. Castan, L.; Cheminant, M.-A.; Colas, L.; Brouard, S.; Magnan, A.; Bouchaud, G. Food allergen-sensitized CCR9+ lymphocytes enhance airways allergic inflammation in mice. *Allergy* **2018**, *73*, 1505–1514. [CrossRef]
18. Denery-Papini, S.; Bodinier, M.; Pineau, F.; Triballeau, S.; Tranquet, O.; Adel-Patient, K.; Moneret-Vautrin, D.; Bakan, B.; Marion, D.; Mothes, T.; et al. Immunoglobulin-E-binding epitopes of wheat allergens in patients with food allergy to wheat and in mice experimentally sensitized to wheat proteins. *Clin. Exp. Allergy* **2011**, *41*, 1478–1492. [CrossRef]
19. Reséndiz-Albor, A.A.; Esquivel, R.; López-Revilla, R.; Verdin, L.; Moreno-Fierros, L. Striking phenotypic and functional differences in lamina propria lymphocytes from the large and small intestine of mice. *Life Sci.* **2005**, *76*, 2783–2803. [CrossRef]
20. Gupta, R.K.; Gupta, K.; Sharma, A.; Das, M.; Ansari, I.A.; Dwivedi, P.D. Maillard reaction in food allergy: Pros and cons. *Crit. Rev. Food Sci. Nutr.* **2017**, *58*, 208–226. [CrossRef]
21. Rupa, P.; Nakamura, S.; Katayama, S.; Mine, Y. Effects of ovalbumin glycoconjugates on alleviation of orally induced egg allergy in mice via dendritic-cell maturation and T-cell activation. *Mol. Nutr. Food Res.* **2013**, *58*, 405–417. [CrossRef] [PubMed]
22. Van Nguyen, C.; Van Chuyen, N. Toxicity of the AGEs generated from the Maillard reaction: On the relationship of food-AGEs and biological-AGEs. *Mol. Nutr. Food Res.* **2006**, *50*, 1140–1149. [CrossRef] [PubMed]
23. Han, X.-Y.; Yang, H.; Rao, S.; Liu, G.-Y.; Hu, M.-J.; Zeng, B.-C.; Cao, M.-J. The Maillard Reaction Reduced the Sensitization of Tropomyosin and Arginine Kinase from Scylla paramamosain, Simultaneously. *J. Agric. Food Chem.* **2018**, *66*, 2934–2943. [CrossRef]
24. Nicolai, T.; Durand, M. Controlled food protein aggregation for new functionality. *Curr. Opin. Colloid Interface Sci.* **2013**, *18*, 249–256. [CrossRef]
25. Martin, C.; Morel, M.-H.; Reau, A.; Cuq, B. Kinetics of gluten protein-insolubilisation during pasta processing: Decoupling between time- and temperature-dependent effects. *J. Cereal Sci.* **2019**, *88*, 103–109. [CrossRef]
26. Huang, D.Y.; Khan, K. Quantitative Determination of High Molecular Weight Glutenin Subunits of Hard Red Spring Wheat by SDS-PAGE. I. Quantitative Effects of Total Amounts on Breadmaking Quality Characteristics. *Cereal Chem. J.* **1997**, *74*, 781–785. [CrossRef]
27. Rousselphilippe, C.; Pina, M.; Graille, J. Chemical lipophilization of soy protein isolates and wheat gluten. *Eur. J. Lipid Sci. Technol.* **2000**, *102*, 97–101. [CrossRef]
28. Pommet, M.; Redl, A.; Morel, M.-H.; Guilbert, S. Study of wheat gluten plasticization with fatty acids. *Polymers* **2003**, *44*, 115–122. [CrossRef]
29. Wilkes, M.; Copeland, L. Storage of Wheat Grains at Elevated Temperatures Increases Solubilization of Glutenin Subunits. *Cereal Chem. J.* **2008**, *85*, 335–338. [CrossRef]
30. Rahaman, T.; Vasiljevic, T.; Ramchandran, L. Effect of heat, pH and shear on digestibility and antigenic characteristics of wheat gluten. *Eur. Food Res. Technol.* **2016**, *242*, 1829–1836. [CrossRef]
31. Peram, M.R.; Loveday, S.M.; Ye, A.; Singh, H. In vitro gastric digestion of heat-induced aggregates of β-lactoglobulin. *J. Dairy Sci.* **2013**, *96*, 63–74. [CrossRef] [PubMed]
32. Milkovska-Stamenova, S.; Mnatsakanyan, R.; Hoffmann, R. Protein carbonylation sites in bovine raw milk and processed milk products. *Food Chem.* **2017**, *229*, 417–424. [CrossRef] [PubMed]

33. Bøgh, K.L.; Barkholt, V.; Madsen, C.B. The Sensitising Capacity of Intact β-Lactoglobulin Is Reduced by Co-Administration with Digested β-Lactoglobulin. *Int. Arch. Allergy Immunol.* **2013**, *161*, 21–36. [CrossRef]
34. Watanabe, H.; Toda, M.; Sekido, H.; Wellner, A.; Fujii, T.; Henle, T.; Hachimura, S.; Nakajima-Adachi, H. Heat treatment of egg white controls allergic symptoms and induces oral tolerance to ovalbumin in a murine model of food allergy. *Mol. Nutr. Food Res.* **2013**, *58*, 394–404. [CrossRef] [PubMed]
35. Schöll, I.; Kalkura, N.; Shedziankova, Y.; Bergmann, A.; Verdino, P.; Knittelfelder, R.; Kopp, T.; Hantusch, B.; Betzel, C.; Dierks, K.; et al. Dimerization of the Major Birch Pollen Allergen Bet v 1 Is Important for its In Vivo IgE-Cross-Linking Potential in Mice. *J. Immunol.* **2005**, *175*, 6645–6650. [CrossRef]
36. Caughey, G.H. Mast cell tryptases and chymases in inflammation and host defense. *Immunol. Rev.* **2007**, *217*, 141–154. [CrossRef]
37. Bredehorst, R.; David, K. What establishes a protein as an allergen? *J. Chromatogr. B Biomed. Sci. Appl.* **2001**, *756*, 33–40. [CrossRef]
38. Zhang, Z.; Xiao, H.; Zhang, X.; Zhou, P. Conformation, allergenicity and human cell allergy sensitization of tropomyosin from Exopalaemon modestus: Effects of deglycosylation and Maillard reaction. *Food Chem.* **2019**, *276*, 520–527. [CrossRef] [PubMed]
39. Rahaman, T.; Vasiljevic, T.; Ramchandran, L. Effect of processing on conformational changes of food proteins related to allergenicity. *Trends Food Sci. Technol.* **2016**, *49*, 24–34. [CrossRef]
40. Mameri, H.; Brossard, C.; Gaudin, J.; Gohon, Y.; Paty, E.; Beaudouin, E.; Moneret-Vautrin, D.-A.; Drouet, M.; Solé, V.; Wien, F.; et al. Structural Basis of IgE Binding to α- and γ-Gliadins: Contribution of Disulfide Bonds and Repetitive and Nonrepetitive Domains. *J. Agric. Food Chem.* **2015**, *63*, 6546–6554. [CrossRef]
41. Teodorowicz, M.; Świątecka, D.; Savelkoul, H.; Wichers, H.; Kostyra, E. Hydrolysates of glycated and heat-treated peanut 7S globulin (Ara h 1) modulate human gut microbial proliferation, survival and adhesion. *J. Appl. Microbiol.* **2013**, *116*, 424–434. [CrossRef] [PubMed]
42. Matsuo, H.; Yokooji, T.; Taogoshi, T. Common food allergens and their IgE-binding epitopes. *Allergol. Int.* **2015**, *64*, 332–343. [CrossRef] [PubMed]
43. Pali-Schöll, I.; Untersmayr, E.; Klems, M.; Jensen-Jarolim, E. The Effect of Digestion and Digestibility on Allergenicity of Food. *Nutrients* **2018**, *10*, 1129. [CrossRef]
44. Diesner, S.C.; Bergmayr, C.; Pfitzner, B.; Assmann, V.; Krishnamurthy, D.; Starkl, P.; Endesfelder, D.; Rothballer, M.; Welzl, G.; Rattei, T.; et al. A distinct microbiota composition is associated with protection from food allergy in an oral mouse immunization model. *Clin. Immunol.* **2016**, *173*, 10–18. [CrossRef] [PubMed]
45. Castan, L.; Villemin, C.; Claude, M.; Aubert, P.; Durand, T.; Neunlist, M.; Brossard, C.; Magnan, A.; Bodinier, M.; Bouchaud, G. Acid-Hydrolyzed Gliadins Worsen Food Allergies through Early Sensitization. *Mol. Nutr. Food Res.* **2018**, *62*, 1800159. [CrossRef]
46. Bouchaud, G.; Castan, L.; Chesné, J.; Braza, F.; Aubert, P.; Neunlist, M.; Magnan, A.; Bodinier, M. Maternal exposure to GOS/inulin mixture prevents food allergies and promotes tolerance in offspring in mice. *Allergy* **2015**, *71*, 68–76. [CrossRef]
47. Huang, C.-H.; Liu, D.-Z.; Jan, T.-R. Diosgenin, a Plant-Derived Sapogenin, Enhances Regulatory T-Cell Immunity in the Intestine of Mice with Food Allergy. *J. Nat. Prod.* **2010**, *73*, 1033–1037. [CrossRef]
48. Smith, K.M.; Davidson, J.M.; Garside, P. T-cell activation occurs simultaneously in local and peripheral lymphoid tissue following oral administration of a range of doses of immunogenic or tolerogenic antigen although tolerized T cells display a defect in cell division. *Immunology* **2002**, *106*, 144–158. [CrossRef]
49. Shan, L.; Molberg, Ø.; Parrot, I.; Hausch, F.; Filiz, F.; Gray, G.M.; Sollid, L.M.; Khosla, C. Structural Basis for Gluten Intolerance in Celiac Sprue. *Science* **2002**, *297*, 2275–2279. [CrossRef]
50. Sollid, L.M.; Jabri, B. Triggers and drivers of autoimmunity: Lessons from coeliac disease. *Nat. Rev. Immunol.* **2013**, *13*, 294–302. [CrossRef]
51. Canani, R.B. Gut Microbiota as a Target for food allergy.UT MICROBIOTA AS A TARGET FOR FOOD ALLERGY. *J. Pediatr. Gastroenterol. Nutr.* **2016**, *63*, S48. [CrossRef]
52. O'Brien, J.; Morrissey, P.A.; Ames, J.M. Nutritional and toxicological aspects of the Maillard browning reaction in foods. *Crit. Rev. Food Sci. Nutr.* **1989**, *28*, 211–248. [CrossRef] [PubMed]

53. Caminero, A.; Nistal, E.; Herran, A.R.; Pérez-Andrés, J.; Ferrero, M.A.; Ayala, L.V.; Vivas, S.; De Morales, J.M.G.R.; Albillos, S.M.; Casqueiro, J. Differences in gluten metabolism among healthy volunteers, coeliac disease patients and first-degree relatives. *Br. J. Nutr.* **2015**, *114*, 1157–1167. [CrossRef]
54. Meehan, C.J.; Beiko, R.G. A Phylogenomic View of Ecological Specialization in the Lachnospiraceae, a Family of Digestive Tract-Associated Bacteria. *Genome Boil. Evol.* **2014**, *6*, 703–713. [CrossRef]

nutrients MDPI

Review

Brown Seaweed Food Supplementation: Effects on Allergy and Inflammation and Its Consequences

Simone E. M. Olsthoorn [1,2], Xi Wang [1,3], Berend Tillema [4], Tim Vanmierlo [5,6], Stefan Kraan [7,8], Pieter J. M. Leenen [2,†] and Monique T. Mulder [1,*,†]

1 Laboratory of Vascular Medicine, Department of Internal Medicine, Erasmus MC, University Medical Center, 3015 GD Rotterdam, The Netherlands; olsthoorn98@gmail.com (S.E.M.O.); wangxihnzyu@gmail.com (X.W.)
2 Department of Immunology, Erasmus MC, University Medical Center, 3015 GD Rotterdam, The Netherlands; p.leenen@erasmusmc.nl
3 The Basic Medical Laboratory, Hunan University of Chinese Medicine, Changsha 410208, China
4 Seaweed Food Solutions BV, 8444 DC Heerenveen, The Netherlands; tillema@boxnv.nl
5 Neuroimmune Connections & Repair (NIC&R), Biomedical Research Institute (Biomed), Hasselt University, B-3590 Diepenbeek, Belgium; tim.vanmierlo@uhasselt.be
6 Department of Psychiatry & Neuropsychology, School for Mental Health and Neuroscience, Division Translational Neuroscience, Maastricht University, 6200 MD Maastricht, The Netherlands
7 R&D Unit, Carnmore, Oranmore, H91 E09X Galway, Ireland; stefan.kraan@theseaweedcompany.com
8 The Seaweed Company Blue Turtle, Co., Ltd., H91 E09X Galway, Ireland
* Correspondence: m.t.mulder@erasmusmc.nl
† These authors contributed equally to this work.

Citation: Olsthoorn, S.E.M.; Wang, X.; Tillema, B.; Vanmierlo, T.; Kraan, S.; Leenen, P.J.M.; Mulder, M.T. Brown Seaweed Food Supplementation: Effects on Allergy and Inflammation and Its Consequences. *Nutrients* **2021**, *13*, 2613. https://doi.org/10.3390/nu13082613

Academic Editor: Iain A. Brownlee

Received: 1 July 2021
Accepted: 27 July 2021
Published: 29 July 2021

Publisher's Note: MDPI stays neutral with regard to jurisdictional claims in published maps and institutional affiliations.

Abstract: Multiple health benefits have been ascribed to brown seaweeds that are used traditionally as dietary component mostly in Asia. This systematic review summarizes information on the impact of brown seaweeds or components on inflammation, and inflammation-related pathologies, such as allergies, diabetes mellitus and obesity. We focus on oral supplementation thus intending the use of brown seaweeds as food additives. Despite the great diversity of experimental systems in which distinct species and compounds were tested for their effects on inflammation and immunity, a remarkably homogeneous picture arises. The predominant effects of consumption of brown seaweeds or compounds can be classified into three categories: (1) inhibition of reactive oxygen species, known to be important drivers of inflammation; (2) regulation, i.e., in most cases inhibition of proinflammatory NF-κB signaling; (3) modulation of adaptive immune responses, in particular by interfering with T-helper cell polarization. Over the last decades, several inflammation-related diseases have increased substantially. These include allergies and autoimmune diseases as well as morbidities associated with lifestyle and aging. In this light, further development of brown seaweeds and seaweed compounds as functional foods and nutriceuticals might contribute to combat these challenges.

Keywords: seaweed; allergy; inflammation; oral

1. Introduction

Brown algae are one of three types of algae, i.e., brown algae (Phaeophyta), red algae (Rhodophyta) and green algae (Chlorophyta) classified based on their color and major photosynthetic pigments. Brown algae contain chlorophyll a, chlorophyll c and fucoxanthin, red algae contain chlorophyll a, chlorophyll d and phycoerythrin, while green algae contain chlorophyll a, chlorophyll b and xanthophylls. Due to the different abiotic and biotic factors in the marine environment and their distinct evolutionary origin seaweeds are a rich source of unique compounds of which several have demonstrated health benefits. Bioactive compounds of interest found in brown seaweed include polysaccharides (e.g., alginate, fucoidan), proteins (e.g., phycobiliproteins), polyphenols (e.g., phlorotannins), carotenoids (e.g., fucoxanthin), phytosterols (fucosterol) and *n*-3 long-chain polyunsaturated fatty acids (e.g., eicosapentaenoic acid) [1]. They have been reported to have beneficial effects in various diseases, including metabolic diseases, diabetes mellitus, cardiovascular disease, cancer and neurodegenerative diseases.

We structure this review according to the distinct compounds in brown seaweed that have been applied, and therefore distinguish effects of whole seaweeds or crude extracts (described in Section 3), polysaccharides such as fucoidan and alginate (Section 4), compounds with ring-shaped structures such as phytosterols (e.g., fucosterol) and (poly)phenols (e.g., phloroglucinol and phlorotannins) (Section 5) and carotenoids, in particular fucoxanthin, fucoxanthinol and meroterpenoids (Section 6).

For each of these compounds we discuss the demonstrated effects on different phases of the inflammatory response (schematically depicted in Figure 1). We describe (i) studies that investigate the brown seaweed (compound) effects on steady state immune parameters that influence subsequent responses to inflammatory challenges. Then, in view of the focus of this special issue on allergic disease, we separately discuss (ii) studies aimed at identifying the effects of intake of brown seaweed or -components on allergies and models thereof. As next steps (iii) and (iv), we distinguish brown seaweed effects on different phases in other acute inflammatory responses, with their characteristic read-outs as indicated in Table 1. Briefly, here we distinguish the first, immediate response (iii) typified by reactive oxygen radical production and release of early mediators such as IL-1, TNF and arachidonic acid metabolites. Then, we consider (iv) the seaweed effects on the second inflammation phase characterized by enhanced production of above-mentioned cytokines as well as secondary cytokines, such as IL-10, chemokines and other mediators, and infiltration of leukocytes in tissue. Finally, we discuss (v) the effects of seaweed intake on chronic inflammation, which is mostly low-grade, and their sequels. These include insulin-, leptin- or glucocorticoid resistance, which are related to the induction of diabetes, obesity and hampered down-regulation of inflammation, respectively. Furthermore, malignant transformation may be another consequence of chronic exposure to inflammatory conditions.

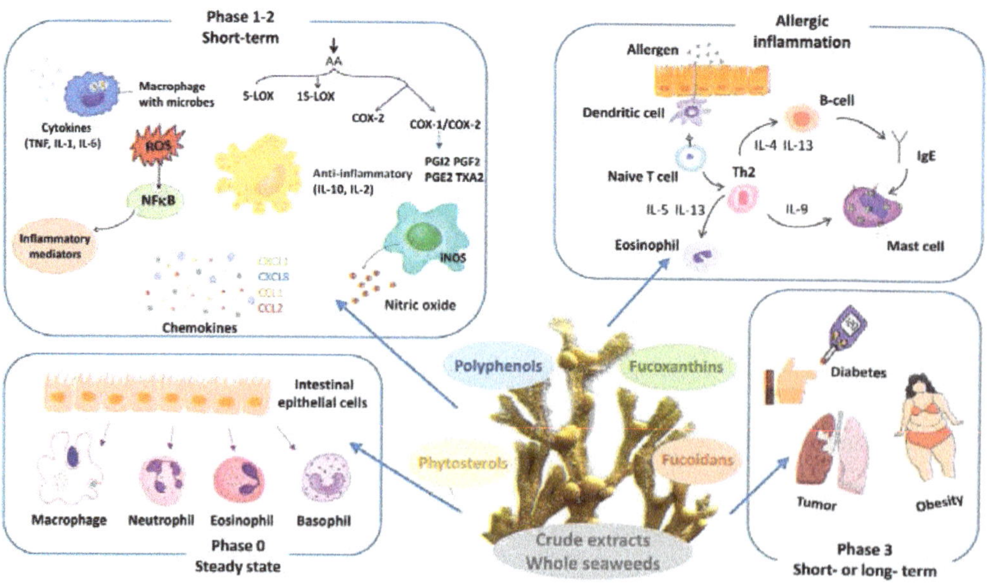

Figure 1. Schematic representation of brown seaweed and it constituents and of the phases in the immune response (Phase 0: steady state, allergy, Phase 1–2: acute inflammatory response and Phase 3: chronic inflammatory response) addressed in this review.

Table 1. Phases of the inflammatory response with characteristic features [(1)].

Phase	Description	Characteristic Read-Outs
0 steady state	homeostatic condition	- growth, organ weight, etc. - leukocyte numbers and subset composition - steady state activities, e.g., phagocytosis
1 short-term	initiation of inflammation as response to triggering by damage- or infection-related molecules	- mast cell and basophil degranulation - neuronal activation (e.g., scratching) - reactive oxygen species (ROS) - pro-inflammatory arachidonic acid (AA) metabolites (e.g., PGE_2) - phospholipase (PLA_2), cyclo-oxygenase (COX-2/PTGS2) and lipoxygenase (LOX/ALOX) activity - early pro-inflammatory cytokines (IL-1, TNF, IL-6 [(2)])
2 short-term	amplification and regulation of inflammation; initiation of adaptive immunity	- additional pro- and anti-inflammatory cytokines (e.g., IL-12, IL-10, IL-2) - soluble forms of cellular R (e.g., sCD25, sCD163) - chemokines (e.g., CXCL8, CCL2) - induction of iNOS and nitric oxide (NO) production - anti-inflammatory AA metabolites (e.g., LXA4) - endothelial activation - edema - leukocyte mobilization and tissue infiltration - acute phase proteins (e.g., CRP) - HPA-axis activation (cortisol or corticosterone) - lymphocyte proliferation - microbial infection parameters - clearance and repair in short-term
3 short- or long-term	consequences of severe or chronic inflammation	- clearance and repair in long-term - glucose, insulin resistance, diabetes - adipokines, leptin resistance, obesity - glucocorticoid resistance, stress, hampered down-regulation of inflammation - inflammation-induced malignancy

[(1)] See Box 1 for a brief introduction on inflammation. [(2)] In the inflammatory response, IL-6 is produced in a second wave, as it has to be expressed de novo, while initial TNF and IL-1β release only requires processing (i.e., membrane cut or enzymatic cleavage and secretion, resp.). Yet, IL-6 is frequently measured together with IL-1β and TNF. Therefore, IL-6 is classified in the first phase. In itself, it is an inducer of mediators of the second phase, in particular acute phase proteins.

Challenges to generating a comprehensible overview are the multifaceted aspects of allergy and inflammation, besides the large variety of brown seaweed species and preparations, as well as the different host species in which seaweed administration has been tested. Furthermore, this divergency bears a risk of overstretching conclusions based on limited findings. With these caveats in mind, we attempt to describe the commonalities between different studies, or discuss specific findings that deserve attention for future development. Detailed findings from the individual publications have been summarized in tables, linked to the compound categories mentioned above.

Box 1. Brief introduction on the inflammatory response.

Inflammation is the response of tissues to any trigger significantly disturbing homeostatic conditions. These triggers can be manifold and vary from infection or trauma, to metabolic challenges. The response is initiated by recognition of damage- or danger-related molecules (so-called DAMPs) that become exposed and are sensed by parenchymal tissue cells and resident innate immune cells, in particular macrophages and mast cells. Activation of these cells (Phase 1) stimulates a cascade of events, within the first seconds to minutes, involving local release of ready-made mediators such as histamine, inflammatory arachidonic acid products such as PGE_2, and stored cytokines. Oxygen radicals (reactive oxygen species; ROS), produced by NADPH-oxidase complex assembled upon cellular activation and by mitochondria, further amplify activation of resident cells. In addition to their essential role in redox signaling, ROS are important as microbicidal molecules and as inducers of oxidative damage. Secreted first wave mediators alarm neighboring cells and stimulate local vascular cell activation, causing upregulation of adhesion molecules on the endothelial cell surface, and vasodilation. This enables plasma fluid, including bioactive proteins, to penetrate the tissue. Subsequently, activated resident cells produce inflammatory cytokines and chemokines by de novo gene transcription (Phase 2). Inflammatory signaling pathways involving NF-κB and AP-1, triggered by the receptor-DAMP interaction, are essential in this process. The released mediators amplify the inflammatory response and enable the recruitment of leukocytes from the circulation. Based on the profile of released chemokines, neutrophils are the most numerous attracted cells in the initial response. In addition, expression of inducible nitric oxide synthase (iNOS) is stimulated by bacterial products and pro-inflammatory cytokines, in particular IFN-γ. iNOS generates nitric oxide (NO), which is an important inflammatory mediator as microbicidal product of especially activated macrophages, and as an autocrine and paracrine signaling molecule. In this second phase of inflammation, also anti-inflammatory mediators are produced, which include arachidonic products such as lipoxins and resolvins, and cytokines such as IL-10. Dependent on the strength of the inflammatory trigger, systemic responses may occur, including activation of the hypothalamus-pituitary-adrenal (HPA) axis, leading to cortisol or corticosterone release. The first systemic cytokine wave initiates the production of so-called acute phase proteins such as C-reactive protein (CRP) from the liver, which occurs from approximately 24 h after the initial trigger. An adaptive immune response is initiated as DAMPs and -related antigens are transported to draining lymph nodes via lymph and migrating antigen-presenting cells. This leads to proliferation of antigen-specific B- and T-lymphocytes that eventually leave the lymph nodes and become effective in the periphery from approximately 4 days after initiation. Dependent on the nature and severity of the initial trigger, and ability to annihilate it, repair processes are initiated after a few days. If the trigger remains, such as in case of adverse metabolic conditions or persistent infection, inflammation may become chronic. Then a new equilibrium is sought (Phase 3), which affects tissue performance and the response to further environmental triggers. Long-term exposure to adverse conditions thus may lead for instance to resistance to regulating hormones, such as insulin or cortisol. Allergic responses are inflammatory responses triggered by pre-existing antibodies or primed T-lymphocytes specific for harmless molecules from the environment. Best known are the responses caused by activation of mast cells evoked by IgE antibodies specific for innocuous antigens such as house dust mite or pollen. Other antibody subclasses with adverse specificities, for instance against certain drugs, also can cause clinical responses. Finally, T-lymphocytes with unfavorable specificities may cause so-called delayed-type hypersensitivity responses. The delay is explained by the necessity of T-lymphocytes to migrate to the site of antigen exposure and presentation, and to become activated locally. In essence, allergic responses follow the phases as outlined for the inflammatory response in general.

2. Approach to Systematic Search

The aim of this systematic review is to investigate the effect of brown seaweed components as a dietary supplement on inflammation with extra focus on allergies.

2.1. Databases and Search Strategy

We performed two searches, in May and in November 2020. Several electronic databases were used to include studies: Embase via Embase.com (1971–Present), Medline ALL via Ovid (1946–Present), Web of Science Core Collection via Web of Knowledge (1975–Present), Cochrane Central Register of Controlled Trials via Wiley (1992–Present) and Google Scholar. References were filtered for duplicates in Endnote. The systematic search was executed by Elise Krabbendam, Biomedical Information Specialist at the Medical Library of Erasmus MC. The exact search terms are shown in the Supplementary Materials, where also a detailed flow chart of paper selection is shown. First, at least two review authors independently assessed title and abstract of the articles based on inclusion and exclusion criteria mentioned below. This was carried out using Endnote X9 software, based on reviewing methods described by Bramer et al. 2017 [2]. Secondly, after independent assessment of the articles, references selected by both review authors were included in a preliminary database. For final full text inclusion all references were combined and assessed by all review authors to assure eligibility and to extract important information to be included in the tables. References were divided by seaweed component among all review authors and based on full text selected to be included or excluded from the systematic review. Third, some relevant articles were included that were not retrieved in the systematic search, primarily because searchable parts of the articles did not contain reference to terms related to inflammation.

2.2. Eligibility Criteria

For this systematic review we only included articles written in English exclusively. Review articles were excluded from the search. The aim of the first search was to include papers addressing the effect of brown seaweed as nutritional supplement on leucocytes and how this contributes to the inflammatory process. This resulted in a total of 906 references after deduplication. For extra focus on allergy and atopy the search was expanded, resulting in a final total of 1035 references after deduplication. Papers on components of brown seaweed or whole extracts were included, but only when orally administered to human or animals. Exclusion criteria for this systematic review were seaweeds other than brown species or no oral administered. Furthermore, articles mentioning fucoidan as control inhibitor/blocker for selectins or scavenger receptor exclusively, were also excluded.

3. Whole Seaweed or Crude Extract Supplementation

3.1. Whole Seaweed or Crude Extract: Effects in Steady State

Table A1 summarizes the main findings of oral supplementation of whole seaweeds or crude extracts on inflammation-related parameters. Safety aspects of seaweed consumption are important to consider. Potential overload with iodine or heavy metals are major long-term risks when unprocessed edible seaweeds are ingested [1]. Short-term monitoring after administration of single doses of crude extracts in rodents showed no effects of toxicity up to 5000 mg/kg in mice (*Sargassum micracanthum* [3]; *Cystoseira compressa* (Esper) [4]), or showed LD50 values of 1000–2000 mg/kg (*Fucus vesiculosis* [5]). Considering a dosage of 200 mg/kg/day is routinely applied for regular use, it may be argued that the safe dosage range might be limited. However, specific toxicity may be highly batch-dependent and related to toxic contaminants rather than seaweed content.

A widely studied facet of seaweed supplementation is its high anti-oxidant activity, and related to this, anti-inflammatory activity. Human studies in this direction are scarce, however. Consumption of 4.8 g dried *Sargassum muticum* per day for a period of 4 weeks by healthy volunteers stimulated an increased total anti-oxidant status in serum, correlated to decreased concentrations of oxidized LDL [6]. In contrast to general assumptions, Baldrick et al. observed no significant changes in oxidative or inflammatory parameters after oral consumption of *Ascophyllum nodosum* extract by individuals with overweight or obesity (100 mg/day, 8 weeks) [7]. An interesting aspect of the latter study is that individuals differed notably (up to >4000×) in the total amount of seaweed polyphenol metabolites present in urine. It is tempting to speculate this might be attributed to differences in microbiota composition between individuals. In a study in goat, where *Ascophyllum nodosum* extract was added (2%) to feed, an increased anti-oxidant status was shown [8]. Similarly, 4-week treatment of rats with *Fucus vesiculosus* extract stimulated increased serum paraoxonase and superoxide dismutase activities, thus leading to an increased anti-oxidant status [5].

Effects of seaweed supplementation on blood cell counts are variable; studies in human or other mammals showed limited effects [6,8], but increased counts were observed when chicken and fish were fed *Laminaria japonica* or *Sargassum oligocystum*, respectively [9,10]. Such addition to animal feed is not only associated with increased growth and feed conversion ratios in chicken and fish, but also enhanced status of innate and adaptive immune defenses and immune responsiveness and survival after infectious challenge [11–13].

3.2. Whole Seaweed or Crude Extract: Effects on Allergy

Importantly, oral brown seaweed supplementation shows consistent beneficial effects in different models of allergy (Table A1). *Eisenia* (=*Ecklonia*) *arborea* powder intake by Brown Norway rats, orally sensitized to ovalbumin, leads to decreased serum IgE and histamine levels and decreased IL-4 and IL-10 production in lymphoid organs, while IFN-γ synthesis is increased. This indicates a favorable change in Th1/Th2 balance towards the former [14].

Mouse models using 2,4-dinitrophenol (DNP) or 2,4-dinitrochlorobenzene (DNCB) sensibilization and challenge are much used in this field, and similarly show positive effects of brown seaweed intake. *Sargassum horneri* extract has anti-allergic activity by suppressing degranulation of mast cells and basophils. This reduces nasal rubbing [15] or clinical signs of atopic dermatitis, as well as inflammatory cytokine levels and leukocyte skin infiltrates [16] in these allergy models. Similarly, application of extracts from *Costaria costata* [17] and *Laminaria japonica* [18] reduce severity of allergic dermatitis and stimulate healing, possibly by decreasing inflammatory pathways in keratinocytes.

3.3. Whole Seaweed or Crude Extract: Effects in Acute and Chronic Inflammation

Without prior sensitization, irritant application on murine skin also causes signs of acute inflammation associated with local mast cell degranulation and increased vascular permeability. Feeding rats with *Laminaria japonica* extract decreases paw swelling and leukocyte infiltration induced by carrageenan application, likely by inhibiting NF-κB activation causing decreased inflammatory mediator production [19] (Table A1). Similarly, oral or topical administration of *Sargassum fusiforme* extract significantly reduces ear swelling by inhibiting mast cell degranulation and enzymes involved in production of inflammatory arachidonic acid mediators [20].

The challenge of experimental animals or isolated cells with lipopolysaccharide (LPS) is a common model of acute inflammation, mimicking the response to bacterial infection. A 5-week treatment of rats with *Eisenia (=Ecklonia) bicyclis* extract mediated a reduced inflammatory activation of peritoneal macrophages upon in vitro LPS stimulation through inhibition of NF-κB activity [21]. This was associated with reduced iNOS expression and nitric oxide (NO) production. In marked contrast, a similar study in mice showed that 3-week oral treatment with *Sargassum fusiforme* (also called *Hizikia*) extract slightly potentiated the production of IL-1β, IL-6 and TNF-α by isolated peritoneal macrophages stimulated in vitro with LPS [22]. In accordance with an inflammation-regulating effect, 4-day oral application of *Sargassum serratifolium* extract in mice caused significantly reduced production of TNF-α, IL-1β and IL-6 upon in vivo LPS challenge [23]. This confirmed the in vitro findings of direct inhibition of NF-κB activation and nuclear translocation by seaweed components.

Investigation of seaweed treatment on LPS responses in other than murine species generally corroborated an inflammation-inhibiting effect. Challenge of zebrafish embryos with LPS or H_2O_2 showed reduced reactive oxygen species (ROS) production and associated cell death when treated simultaneously with *Sargassum polycystum* or *Chnoospora minima* extract [24].

Adding *Sargassum latifolium* to sheep feed caused a reduced inflammatory response to LPS challenge and increased blood anti-oxidant defense capacity [25]. This seaweed treatment also mediated a reduced inflammatory response to heat stress challenge in these sheep [26]. The latter study confirms earlier work in lamb, showing reduction of heat stress-related effects on leukocyte oxidative and phagocytic function by *Ascophyllum nodosum* extract administration [27].

Furthermore, in mouse models of inflammatory disease, seaweed supplementation has shown beneficial effects. In dextran sulfate sodium-induced chronic colitis, application of *Turbinaria ornata* extract causes decreased disease activity as indicated by colon length, histomorphological index and myeloperoxidase activity [28]. This was accompanied by increased expression of regulatory T-cell-associated FoxP3 and anti-inflammatory IL-10. Similarly, *Laminaria japonica* extract caused a significant diminution of colitis signs in this model, and simultaneous application of bacterial probiotics showed synergistic beneficial effects on histological score and decreased levels of some proinflammatory cytokines [29]. In a mouse model of arthritis induced by bovine type II collagen immunization, oral supplementation with *Sargassum muticum* extract significantly decreased the arthritis and edema scores as well as TNF, IL-6 and IFN-γ levels [30].

In one of the scarce human studies, Cooper et al. found that individuals with active Herpes infection showed increased healing rates with *Undaria pinnatifida* consumption, while latent Herpes carriers did not experience viral reactivation [31]. Investigating the mechanisms, the authors found the extract strongly inhibited Herpes virus infectivity in vitro, and stimulated human T cell mitogenesis, thus potentiating adaptive immune responses.

A study using a phylogenetically more distinct organism, kuruma shrimp, indicated that oral supplementation with *Laminaria japonica* significantly increased survival upon White Spot Syndrome virus infection [32]. This was accompanied by enhancement of chemotaxis as well as other defense mechanisms by hemolymph leukocytes (hemocytes), including superoxide production and antioxidative phenoloxidase activity upon appropriate stimulation.

3.4. Whole Seaweed or Crude Extract: Late Consequences of Inflammation and Sequels

Acute or chronic inflammatory conditions influence local and systemic tissue responses, and thus seaweed supplementation also affects peripheral tissue function in inflammation (Table A1). In a rat model of ligature-induced periodontitis, *Sargassum fusiforme (Hizikia)* application reduced alveolar bone loss, related to decreased osteoclast and increased osteoblast gene expression in vitro [33]. Furthermore, in a model of autoimmune thyroiditis, a traditional Chinese medicine combination of 10 different herbs, including *Sargassum fusiforme*, mediated a decrease in autoimmune thyroiditis and anti-thyroid autoantibody formation [34]. Omission of *Sargassum fusiforme* in this model significantly diminished the protective effect.

In recent years, the link between adipose tissue metabolic dysregulation and inflammation has been recognized increasingly [35]. Several studies investigated metabolic effects of seaweed application, in particular *Undaria pinnatifida*, in murine models of obesity and type 2 diabetes induced by a high-fat diet [36–38]. Seaweed was in some studies combined with other nutraceuticals. In general, the obesity phenotype did not change, while improvement of glucose regulation was only observed by Oh et al. (2016) [36], but not in the other two studies. Other aspects, however, showed beneficial effects of seaweed supplementation, such as microbiome composition, MCP-1 induction [38], systolic blood pressure and non-esterified fatty acid levels [37] or presence of clusters of necrotic adipocytes surrounded by macrophages in adipose tissue (so-called crown-like structures) [36].

A pathological condition strongly related to obesity is the polycystic ovary syndrome (PCOS). In a rat model of PCOS, application of *Ecklonia cava* extract mediated a decrease in vaginal leukocyte infiltration, and restored hormonal levels and irregular ovarian cycles [39]. However, it did not inhibit the weight gain associated with PCOS induction.

4. Brown Seaweed Polysaccharide Supplementation

Among polysaccharides present in brown seaweed fucoidan has received most attention as a constituent with diverse bioactive effects. Furthermore, bioactivity has been demonstrated of the polysaccharides laminarin, a beta-glucan polysaccharide and alginate, a linear acidic soluble dietary polysaccharide.

Fucoidans are a group of polysaccharides (fucans) primarily composed of sulphated L-fucose with less than 10% of other monosaccharides. They are widely found in the cell walls of brown seaweeds, but not in other algae or higher terrestrial plants [40]. The major function of fucoidans in cell walls is mechanical support and protection against desiccation during air-exposure of the seaweed at low tide. The amount of fucoidan in brown seaweeds is variable; 8–20% of dry weight with the highest content of about 20% being detected in *Fucus vesiculosus* [41,42].

A number of health-improving effects have been ascribed to fucoidans [40,41,43]. Biological activities of fucoidans, such as antioxidant and anti-coagulant capacity, are affected by their molecular weight and sulphated ester content, both the number of sulphate groups determining the negative charge of the molecule and the position of the sulphate groups on the sugar residues [40,44]. The biological activity of fucoidan is also affected by the glucuronic acid and fucose content. The molecular weight of fucoidan ranges from for example from 50 to 80 KDa in *Undaria pinnatifida* and *Fucus vesiculosus*, respectively, to 1920 kDa in *Cladosiphon* species [45], with multiple sizes being present in certain species. Low molecular weight (LMW) fucoidan is produced by enzymatic digestion or acid hydrolysis of naturally occurring high molecular weight (HWM) fucoidan. Application of different molecular species of fucoidan obtained by different methods of purification and treatments such as hydrolysis complicates interpretation of results.

4.1. *Brown Seaweed Polysaccharide Effects in Steady State*

Fucoidan is absorbed in limited amounts in the gastrointestinal tract after oral intake [46,47]. In Japanese populations where brown seaweed is part of daily diet, systemic fucoidan uptake was shown by its presence in serum and urine [48]. Protective effects of fucoidan on the intestinal epithelial barrier function were observed in vitro. Fucoidan protected the tight junctions from oxidative injury and upregulated the expression of claudin-1 [49]. Table A2 summarizes the effects of fucoidan on different aspects of inflammation.

Fucoidan is not toxic, but high dosages can induce increased bleeding time. In rats no toxicity as observed after oral administration of a single dose of *Ascophyllum nodosum* fucoidan of 2000 mg/kg [50] or 300 mg/kg/day *Laminaria japonica* fucoidan for 6 months [51]. However, application of a daily dose of 2500 mg/kg for 6 months resulted in increased bleeding time. The application of fucoidan in food has been approved for human consumption up to 250 mg/day by the European Food Safety Authority, EFSA [1].

4.2. *Effects of Polysaccharides from Brown Seaweed in Allergy, Acute Inflammation and in the Modulation of Immune Responses*

Both pro- and anti-inflammatory effects of fucoidan have been reported. In macrophages fucoidan treatment induced NF-κB nuclear translocation, followed by iNOS and COX-2 transcription, inducing the secretion of the pro-inflammatory cytokines IFN-γ, TNF-α and IL-1β and of inflammatory mediators NO and PGE2 [52]. However, pre-treatment of macrophages and lymphocytes with fucoidan prior to stimulation with LPS or other pro-inflammatory stimuli was found to blunt the pro-inflammatory reaction or induces an anti-inflammatory effect, resulting in inhibition of NF-κB translocation and in lower levels of pro-inflammatory mediator production [53–56].

Below an elaboration on anti-allergy effects and enhanced immune effects in production animals and innate and adaptive immune system modulation studies in mice is described.

4.2.1. Anti-Allergic Effects of Brown Seaweed-Derived Polysaccharides

Overall, oral supplementation of brown seaweed polysaccharides was reported to inhibit allergic responses via multiple mechanisms. The polysaccharides were shown to be an effective agent antagonizing IgE production as examined in different ovalbumin (OVA)-sensitized mouse models [57,58], but also an allergy-specific mechanism of oral fucoidan supplementation has been found in its capacity to induce galectin-9 production from intestinal cells [59,60]. Galectin-9, belonging to a soluble lectin family, recognizes β-galactoside and prevents IgE binding to mast cells, consequently inhibiting mast cell degranulation. Accordingly, fucoidan from *Saccharina japonica* (400 μg for 4 days) was found to increase circulating galectin-9 [59]. After OVA-immunization the allergic symptoms in sensitized mice were reduced by fucoidan (60 μg/mouse/d for 17 days) via inducing galectin-9 production from colonic epithelial cells [60].

In several OVA-immunized mouse models, oral administration of fucoidan or a polysaccharide fraction was shown to have anti-allergy activity. Application of a polysaccharide fraction from *Laminaria japonica* (50 mg/kg/day for 2 weeks) in a mouse model of asthma significantly decreased the numbers of eosinophils in the bronchoalveolar fluid and alleviated lung inflammation compared to the non-treated control mice [58]. It also reduced serum IgE concentrations and decreased the concentrations of IL-13 and TGF-β1 in bronchoalveolar fluid and expression in lung, while increasing expression of IL-12. Similarly, *Laminaria japonica* fucoidan ingestion (200, 600, 1000 mg/kg for 6 weeks) decreased OVA-specific IgE in mice [61]. Fucoidan from *Undaria pinnatifida* (400 mg/kg for 7 days) inhibited particulate matter-induced exacerbation of allergic asthma [57]. Specifically, fucoidan treatment significantly attenuated lipid peroxidation, infiltration of inflammatory cells and Th2-related IL-4 concentrations. Furthermore, fucoidan suppressed mast cell activation, degranulation and IgE synthesis as well as mucus hypersecretion and goblet cell hyperplasia. This also is reflected in immunoglobulin isotypes produced as *Cladosiphon*-fucoidan dose-dependently (up to 1025 mg/kg body weight for 8 weeks) increased systemic IgM, IgG and IgA levels, while decreasing IgE and IL-4 significantly [62].

The observed changes are suggestive of a shift from Th2 to Th1 induced by orally ingested fucoidan. Enhanced IL-12 and IFN-γ production by ingestion of *Tetragenococcus halophilus* KK221, a probiotic known for its anti-allergic properties, was even further increased by combined ingestion of the probiotic and LMW fucoidan isolated from *Undaria pinnatifida* in OVA-immunized mice. The results indicated an extra shift towards Th1.

Furthermore, alginate was found to improve (food) allergy outcomes in an OVA-sensitized mouse model [63]. Ingestion of alginate (2 mg) extracted from *Laminaria japonica* one day before oral application of ovalbumin improved integrity of intestinal epithelial villi and inhibited mast cell degranulation in the jejunum. Serum levels of IgE, histamine and IL-4 were significantly lower, while IFN-γ was markedly increased. Furthermore, Tregs in spleen were increased, while OVA-induced differentiation of Th0 cells into Th2 cells was inhibited [63].

Overall, brown seaweed-derived polysaccharides generally appear to modulate the Th1/Th2 balance and mast cell degranulation in favor of an anti-allergic effect. This shows that fucoidan is potentially an effective therapeutic agent for type I allergic diseases.

4.2.2. Effects of Brown Seaweed Polysaccharides on Innate and Adaptive Immune System (Production Animals)

In search for alternatives for antibiotics in production animals, brown seaweed polysaccharides and especially fucoidan, have appeared as promising functional feed additives. Fucoidan and laminarin were found to improve the immune response of pregnant sows and piglets prior to or while suckling [64], and after weaning [65].

Dietary supplementation of sows in the final part of gestation with *Laminaria* spp. extract increased IgG and IgA in sow colostrum by 19% to 25% [64]. Consequently, also a 10% increase in piglet serum IgG was observed. This suggests an important effect of maternal diet on the immune status of piglets. Dietary supplementation with an extract of *Ascophyllum nodosum* and *Fucus* in sows (30 g/day from the 85th day of gestation until weaning) resulted in an increased population of CD4+CD8+ T cells in the thymus, spleen, mesenteric node, liver and in peripheral blood as compared to the control group [66]. Piglets from laminarin-fed sows (1.0 g/d from day 107 of gestation until weaning) showed down-regulation of IL-6 mRNA expression in the colon at weaning and of IL-8 in the ileum on day 8 post weaning compared with those from the non-laminarin-fed sows [67].

Weaning of piglets is a stressful event for piglets and is often associated with pro-inflammatory immune effects in the piglets' gastro-intestinal tract. Addition of laminarin to weaning piglets' diets resulted in lower expression of pro-inflammatory cytokines IL-1β, IL-6 and IL-17 in colonic mucosa [65]. Even though these positive effects were observed in piglets, laminarin did not result in any detectable benefits in Friesian bull calves [68].

In Salmonella-challenged broiler chickens, addition of 0.2% alginate oligosaccharides to the regular diet inhibited *Salmonella enteritis* colonization, possibly by increasing colonic anti-Salmonella IgA levels [69]. In unchallenged broiler chickens, supplementation of 0.2% alginate oligosaccharides showed dramatic immunostimulatory activity by inducing interferon-γ, IL-10 and IL-1β mRNA expression in cecal tonsils. Interestingly, the robust mucosal immune response in the absence of a challenge was related to a decline in body weight, as compared to the control group.

In aquaculture, several studies pointed at improved innate immune markers upon fucoidan and laminarin supplementation in shrimp and fish [70–76]. In addition, higher survival rates during bacterial challenges were observed in the supplemented animals, as compared with those fed a regular diet.

Taken together, enhancing the innate and adaptive immune system by oral ingestion of seaweed-derived polysaccharides and oligosaccharides is a promising solution for improving animal health, reducing infection incidence and reducing the need for antibiotics use.

4.2.3. Effects of Brown Seaweed Polysaccharides on Innate and Adaptive Immune System (Mouse Models)

Polysaccharides obtained from brown seaweed may support various aspects of the immune system in both immunocompetent and immunosuppressed states. For instance, oral ingestion of a polysaccharide extract from *Kjellmaniella crassifolia* (2 weeks) by C57BL/6 mice, resulted in enhanced IFN-γ, IL-12, IL-6 and IgA secretions by spleen cell cultures upon concanavalin-A stimulation [77]. Orally administered LMW fucoidan (200–1000 mg/kg for 6 weeks) from *Laminaria japonica* to BALB/c mice stimulated the innate immune system by increasing natural killer (NK) cell activity and peritoneal macrophage phagocytic activity [61]. LMW fucoidan also increased IL-2, IL-4 and IFN-γ secretion by splenocytes and IgG and IgA concentrations in serum, while it decreased OVA-specific IgE. In bacterial antigen-stimulated immune responses, the IgM and IgG concentrations in serum were significantly higher in the LMW fucoidan group than in the control group. In addition, an LMW fucoidan-enriched extract from *Okinawa mozuku* orally administered (up to 1025.0 mg/kg for 6 weeks) to BALB/c mice resulted in enhanced splenocyte proliferation and secreted IL-2 levels, as well as in increased macrophage phagocytic activity, and serum IgM, IgG and IgA, while splenocyte-secreted IL-4 and IL-5 were decreased, and also serum IgE was decreased significantly [62]. Interestingly, HMW fucoidan (50 g/kg) but not LMW and IMW fucoidan, increased the relative number of cytotoxic T-cells in spleens of Balb/c mice [78]. These immune-potentiating effects appear to be effective in infection as complete elimination of liver and spleen parasite burden was achieved by fucoidan (200 mg/kg, 3 times weekly, for 6-weeks) in a mouse model of *Leishmania donovanii* infection [79]. This curative effect was associated with switching of T cell differentiation from Th2 to Th1 mode.

In addition to its capacity to enhance the innate and adaptive immune system, oral fucoidan is an interesting candidate for antiviral therapies related to its intrinsic capacity as a competitive binding agent for envelope viruses, thus preventing cellular entrance [80]. Oral ingestion of fucoidan improves the outcome in virus-infection mouse models with respect to viral load [81], serum antibody levels and overall survival [82,83] in immunocompetent and immune-suppressed mice. Furthermore, fucoidan extracted from *Undaria pinnatifida* protected both immunocompetent and immunosuppressed mice from infection with HSV-1 as indicated by the improved survival rate and lesion scores [82]. In immunocompetent mice fucoidan enhanced activity of CTL and increased circulating anti-HSV antibodies in HSV-1-infected mice.

In an immunosuppressed state, selective augmentation of NK activity was observed upon oral treatment with *Undaria* fucoidan, but this induced no significant change in NK activity in immunocompetent mice where a normal level of NK activity was maintained. Fucoidan extracted from *Undaria pinnatifida* showed also beneficial effects during influenza virus infection in immunocompetent an immunosuppressed mice [83]. Fucoidan administration (7 days prior to virus inoculation until 7 days after inoculation (2 × 5 mg/day)) resulted in significant increase in neutralizing antibody titers in bronchoalveolar lavage fluids in both healthy mice and mice with suppressed immunity as compared with placebo groups.

Furthermore, in the defense against tumors fucoidans enhance innate and adaptive immune responses. Different fucoidans were found to increase immune reactions in various tumor models, leading to comparable or even better results than standard chemotherapy exclusively [84–87]. Consumption of fucoidan isolated from *Undaria pinnatifida* (1% of the diet for a period of 10 days) showed tumor inhibition in an A20 leukemia mouse model, related to enhanced Th1 and NK cell activity [84]. Oral intake of polysaccharides from *Sargassum fusiforme* (100 and 200 mg/kg for 28 days) significantly inhibited the growth of A549 lung adenocarcinoma in mice, but also remarkably promoted IL-1 and TNF-α production from peritoneal macrophages, increased serum TNF-α levels and splenocyte proliferation [87]. Oral administration of fucoidan extracted from *Cladosiphon okamuranus* (5 g/kg/day for 28 days) also inhibited tumor growth and increased survival time in a colon 26 tumor-bearing mouse model. In the spleens of these mice, an increased population of NK cells was observed. Furthermore, in an experiment applying the same fucoidan to MyD88 knockout mice, a model for investigating TLR4 signaling pathways, it was found that the observed anti-tumor effects are related to gut immunity [85]. Furthermore, polysaccharide extract from *Sargassum fusiforme* (400 mg/kg for 28 days) exerted anti-tumor and immunomodulatory activities in nasopharyngeal carcinoma [86] and hepatic carcinoma tumor-bearing mice [88]. In a xenograft tumor model orally administered fucoidan from *Fucus vesiculosus* (150 mg/kg body weight daily for 2 weeks) increased cytolytic activity of NK cells and significantly delayed tumor growth [89]. Furthermore, in a rat model for experimental mammary carcinogenesis administration of fucoidan (400 mg/kg/day for 4 months) showed protective and immunomodulatory effects [90]. Tumor growth in Sarcoma 180 (S-180)-bearing mice was delayed by ingestion of fucoidan from *Cladosiphon okamuranus*, which stimulated NO production by macrophages via NF-κB-dependent signaling pathways [52]. Oral intake of ascophyllan, a sulphated polysaccharide obtained from *Ascophyllum nodosum*, (50 and 500 mg/kg), also delayed tumor growth. Interestingly, oral ingestion significantly increased serum IL-12 and TNF-α levels and mediated better overall outcome compared to intraperitoneal application in S-180 mice, where immune markers did not change [91].

Seaweed polysaccharides can also function as immune-stimulating adjuvant in immunosuppressed states during chemotherapy. Oral intake of polysaccharide extract from *Sargassum fusiforme* (200 mg/kg for 6 days) was identified as a potent immune-enhancing agent in immunosuppressed mice [92]. Oral administration of fucoidan (150 mg/kg for 14 days) resulted in enhanced recovery of all T cell populations (CD3+, CD4+, CD8+) and of the proliferative capacity of splenocytes in immunosuppressed mice [93]. Furthermore, laminarin administration (500–1000 mg/kg/day for 10 days) induced IL-12 and IFN-γ in immunosuppressed mice [94]. Taken together, oral intake of brown seaweed polysaccharides is shown to be an effective immune enhancer in a wide variety of mouse models.

4.2.4. Anti-Inflammatory Effects of Fucoidan in Animal Models and Clinical Trials

Fucoidans extracted from different seaweed species and molecular sizes showed anti-inflammatory effect in a wide range of acute and chronic inflammation models in mice.

In an arachidonic acid-induced ear inflammation model sulphated polysaccharide extracted from *Sargassum hemiphyllum* decreased ear swelling and erythema [95]. The polysaccharides decreased the local levels of myeloperoxidase, nitric oxide, IL-1β, IL-6 and TNF-α in a dose-dependent manner (20–80 mg/kg body weight for 5 consecutive days). Histological examination revealed that the polysaccharides reduced the area of neutrophilic infiltration in inflamed ears. Similarly, oral ingestion of fucoidan extracted from *Undaria pinnatifida* (0.5 mg for 20 days) inhibited the inflammatory reaction in a mouse model where LPS was injected buccally [55]. In the same set up, but now using bacterial infection, fucoidan reduced inflammation but did not lead to clearance of the bacterial infection, nor to prevention of infection-related bone loss. In a carrageenan-induced air pouch inflammation model, a preparation of fucoidan (54 mg/kg for 7 days) inhibited inflammatory markers and showed reduced attraction of inflammatory cells as demonstrated by histology [96]. Pretreatment with orally administered fucoidan (20 mg/kg for 2 weeks reduced mucosal lining inflammation and prevented elevation of serum IL-6 levels, while levels of serum IL-10 increased in an aspirin-induced mucosal ulcer model in mice [97]. Accordingly, *Cladosiphon* fucoidan (chow containing 0.05% *w/w*) ingestion beneficially affected murine dextran sulphate sodium (DSS)-induced colitis [98]. The lamina propria of inflamed colon showed reduced amounts of IL-6 and IFN-γ, and an increase in IL-10 and TGF-β upon fucoidan treatment. Murine DSS-induced colitis significantly improved upon treatment with fucoidan (10 mg/day for 1 week) [99]. Treatment with a fucoidan polyphenol complex showed even better results as it reduced IL-12, TNF-α and IL-6 in colitis tissue and ameliorated colitis-related visible body markers, such as weight loss and blood in stool. In contrast, when this complex was injected intraperitoneally, it was unable to reduce disease severity and even deteriorated some colitis markers. In mice, colonic inflammation and microbiota dysbiosis induced by antibiotics was alleviated by administration of fucoidan extracted from *Ascophyllum nodosum* (400 mg/kg for 28 days) [100] Fucoidan prevented colon shortening and colon tissue damage, and it improved abundance of beneficial microbes while decreasing harmful microbes. In a model with chemically induced colorectal cancer, ingestion of *Fucus vesiculosus* fucoidan was shown to protect against tumorigenesis and to reduce colorectal inflammation and dysbiosis [101].

The bioactivity of fucoidan is related to its molecular weight. LMW and HMW fucoidan from *Undaria pinnatifada* were tested in a murine model of collagen-induced arthritis [102]. LMW fucoidan protected against tissue degeneration, while the same dose of HMW fucoidan worsened it. In accordance, LMWF reduced the severity of arthritis and the levels of Th1-dependent collagen-specific IgG2a, while HMWF enhanced the severity arthritis and the levels of collagen-specific antibodies.

Furthermore, in different acute inflammation models in zebrafish embryos, strikingly similar anti-inflammatory effects were noticed [53,54,56,103]. Administration of fucoidan (25–100 μg/mL) one hour prior to LPS treatment improved survival of zebrafish embryos and diminished inflammatory markers.

In support of an inflammation-regulating effect of fucoidan, oral administration of *Laminaria japonica* fucoidan (50–200 mg/kg) protected against myocardial ischemia–reperfusion injury in rats in a dose-dependent manner [104]. Furthermore, oral ingestion of enzymatically hydrolyzed fucoidan extracted from *Sargassum hemiphyllum* (200 mg/kg/day for 14 days) decreased radiation-induced pneumonitis and lung fibrosis by reducing inflammatory cytokine expression in lung tissues [105]. In both models decreased accumulation of neutrophils and macrophages was observed.

In a mouse model of chronic infection with *Schistosoma japonicum* oral ingestion of *Fucus vesiculosus* fucoidan (500 mg/kg per 2 days for 40 days) significantly reduced the hepatic granuloma size and fibrosis response [106]. Lower levels of pro-inflammatory cytokines were observed in the livers of fucoidan-treated infected mice. Infiltration of Treg cells and levels of IL-10 and TGF-β were significantly enhanced in both the livers and spleens from fucoidan-treated mice. Another study aimed to explore the effects of fucoidan from *Fucus vesiculosus* on concanavalin A (ConA)-induced acute liver injury in mice. Pretreatment with fucoidan (10–50 mg/kg for 2 weeks) protected liver function indicated by ALT, AST and histopathological changes by suppressing inflammatory cytokines, TNF-α and IFN-γ [107]. The results demonstrated that fucoidan alleviated ConA-induced acute liver injury via the inhibition of intrinsic and extrinsic apoptosis mediated by the TRADD/TRAF2 and JAK2/STAT1 pathways which were activated by TNF-α and IFN-γ.

Fucoidan from *Fucus vesiculosis* (300–600 mg/kg) was shown to be able to delay the onset and incidence of autoimmune diabetes in non-obese diabetic mice via regulating DC/Treg-induced immune tolerance via induction of IL-10 and TGF-β, while reducing the levels of IL-6 and IFN-γ [108]. In the pancreas TLR4 expression and the downstream molecules were downregulated while pancreatic internal environment was maintained in the fucoidan-treated groups.

In a clinical trial in patients with chronic hepatitis B infection, oral ingestion of a commercial oligo-fucoidan preparation (550 mg twice a day for 48 weeks) was shown to have hepatoprotective effects related to serum concentrations of vitamin D, which is known to have immunoregulatory activity [109]. A clinical trial in healthy volunteers showed anti-inflammatory effects of a blend containing fucoidan from 3 different seaweeds. Daily oral ingestion of 1000 mg for 4 weeks was found to decrease serum IL-6 levels [110]. In advanced cancer patients a mixture of enzymatically digested and undigested fucoidan from *Cladosiphon novae caledoniae* (4 weeks of 4000 mg/day) was found to reduce several major proinflammatory cytokines, including IL-1β, IL-6 and TNF-α [111]. The analyses revealed that the responsiveness of IL-1β was inversely correlated with overall survival and was suggested as a possible prognostic factor for disease outcome in advanced cancer patients receiving fucoidan.

Taken together, brown seaweed polysaccharide ingestion is shown to be effective in antagonizing the effects of acute and chronic inflammation in both mouse models and clinical trials.

4.3. Fucoidan Ingestion and Atherosclerosis in Animal Models

Several studies have reported beneficial effects of fucoidan on outcome of atherosclerosis, a disease related to long-term inflammation of the arterial vessel wall.

ApoE-deficient mice are the most frequently used model for assessing atherosclerotic plaque development. In one study, sulphated polysaccharides from *Laminaria japonica* supplementation markedly reduced the thickness of the lipid-rich plaque, lipid peroxidation and foamy macrophage accumulation in the aorta via suppression of MAPKs and NF-κB signaling [112]. In line, Wang et al. found that *Laminaria japonica* fucoidan (50–100 mg/kg/day for 16 weeks) attenuated atherosclerosis by reducing inflammation and oxidative stress [113]. Furthermore, LMW fucoidan extracted from *Laminaria japonica* inhibited the formation of atherosclerotic plaques; and ameliorated the occurrence and development of atherosclerosis [114]. It decreased the production of inflammatory cytokines and prevented macrophages from developing into foam cells and diminished smooth muscle cells from migrating into the intimal layer of the aorta.

Furthermore, also in other models of atherosclerosis, fucoidan appears to have beneficial effects. In a rat allogenic aorta transplantation model ad libitum ingestion of fucoidan from *Fucus vesiculosus* mediated anti-atherosclerotic activity by inhibiting inflammation, suppressing ROS production and down-regulating LOX-1 expression in the vascular wall [115]. In a rat aorta transplantation model fucoidan (LMW) from *Laminaria japonica* (200 mg/kg/day for 35 days) decreased the number of macrophages in the vascular wall by blocking P-selectin activity thereby preventing the development of aortic aneurysms [116]. In the LDLR$^{-/-}$ mouse model of atherosclerosis *Laminaria japonica* fucoidan (50–100 mg/kg/day for 16 weeks) was shown to result in atherosclerosis attenuation by reducing inflammation and oxidative stress [113]. In conclusion, fucoidan appears to be promising in the battle against atherosclerosis by decreasing macrophage infiltration in the vascular wall, as well as by reducing inflammation and oxidative stress.

5. Phenolic Compounds and Phytosterols

5.1. Phytosterols

Phytosterols, including both sterols, stanols and oxysterols, such as fucosterol, saringosterol and 24-hydroperoxy-24-vinyl-cholesterol, are a group of functional lipid compounds. Compared to other bioactive molecules produced by brown algae, phytosterols exhibit various health-improving effects, especially neuroprotective and anti-inflammatory. Table A3 presents an overview of the significant inflammation-related outcomes in animal models after oral administration of phytosterols. Fucosterol, the most abundant sterol in brown seaweed, when administered in different animal models was found to induce a significant therapeutic effect on injury- or infection-related inflammation. Mo et al. [117] showed anti-inflammatory effects of fucosterol-pretreatment in Concanavalin A-treated mice as a model for acute liver injury. After Concanavalin A-treatment, NF-κB p65 increased markedly and the expression of a nuclear receptor in its upstream pathway, PPARγ, decreased. Both NF-κB p65 and PPARγ are closely related to the release of inflammatory factors such as TNF-α, IL-6 and IL-1β. Fucosterol pretreatment down-regulated the inflammatory response and subsequently necrosis and apoptosis by inhibiting the NF-κB pathway and activating PPARγ.

Anti-inflammatory effects of fucosterol were demonstrated using regular *Sargassum fusiforme* extracts (NH) and enzyme-modified *Sargassum fusiforme* extracts (EH) [22]. Enzyme modification significantly increased the fucosterol concentration in the extract (EH) leading to better results in decreasing pro-inflammatory cytokines as compared to the NH pretreatment group. In addition, both NH and EH reduced the production of NO without inducing any cytotoxicity and even increased cell viability in cultured RAW264.7 macrophages at a concentration of 10 μg/mL or higher. Anti-inflammatory effects of fucosterol were also observed in DNCB-induced NC/Nga mice as a model for atopic dermatitis. Oral administration of fucosterol significantly reduced [22] the scratching behavior of the mice and suppressed the production of pro-inflammatory cytokines (TNF-α and IL-4), resulting in reduced circulating IgE levels.

Bogie et al. showed that 24(S)-Saringosterol, an oxyphytosterol present in *Sargassum fusiforme*, has anti-inflammatory effects likely via activation of liver X receptor (LXR)β in a mouse model of Alzheimer's disease (AD) [118]. *Sargassum fusiforme* extract rich in 24(S)-Saringosterol activated LXRβ preferentially and to a lesser extent also LXRα. LXRβ plays a key role in the down-regulation of the expression of multiple inflammatory genes [119,120]. AD is characterized by an accumulation of extracellular amyloid-β (Aβ), intracellular neurofibrillary tangles, loss of synapses, neuroinflammation and by a gradual progression of memory loss. After 45 days of dietary supplementation with *Sargassum fusiforme* the formation of Aβ plaques which is related to cognitive decline, was found to be dramatically reduced (~80% reduction) in AD mice. The expression of the LXR-target gene APOE in the central nervous system was increased due to administration of *Sargassum fusiforme* lipid extract. Apolipoprotein (Apo)E increased the clearance of Aβ by microglial cells and suppressed the secretion of Aβ by neurons in vitro. Therefore, the anti-inflammatory effects of 24(S)-Saringosterol may be explained by activation of the LXR-ApoE axis [121]. Similar to 24(S)-Saringosterol-mediated LXRβ activation, fucosterol can activate both LXRα and LXRβ, regulating different aspects of inflammatory gene expression.

5.2. *Polyphenols*

Polyphenols are another class of bioactive compounds from brown seaweed that have attracted great interest in recent years due to their pharmaceutical and biomedical properties. Polyphenols are classified based on their structure. Phlorotannins, highly abundant in brown seaweeds, are polymerized phenolic compounds consisting of phloroglucinol monomer units. Phlorotannins identified in brown seaweed, include eckol, dieckol, phlorofucofuroeckol A. Numerous studies have demonstrated the potential of polyphenol classes as antioxidant, anti-inflammatory, antidiabetic, antitumor, antihypertensive, anti-allergic, hepato-protective and anti-cancer. Table A4 provides an overview of the reported inflammation-related outcomes after oral administration of polyphenols in different animal models and in clinical trials. (Phase 0, 1, 2 and 3).

5.2.1. Polyphenols: Effects in Steady State

The mechanisms underlying the anti-inflammatory effects of polyphenols are complex and are related to various stages of the inflammatory response that are sequential but overlapping. Disturbance of the steady state causes parenchymal tissue and immune cells to respond to injury or irritation through an innate cascade driving inflammation. Irfan et al. [122] demonstrated that phlorotannins strongly inhibit in vivo platelet aggregation in Sprague Dawley rats. In line with this in vitro, phlorotannins downregulated adenosine diphosphate-induced platelet activation (Ca-mobilization, fibrinogen binding, granule release—mediated via decreased Src, PI3K, PLCγ2, MAPK signaling). A clinical trial with 80 overweight participants showed that phlorotannins modestly decrease DNA damage [7], although no significant difference was found in acute phase proteins, anti-oxidant status or in inflammatory cytokines.

5.2.2. Polyphenols: Effects on Allergy

NF-κB is one of the transcription factors that regulates eosinophilic inflammation and IgE-mediated hyperreactivity following allergic inflammation. Oral administration of polyphenols suppressed NF-κB pathway activation, and also inhibited inhibitor kappa B (IκB) that binds to NF-κB [123,124]. Polyphenols were found to alleviate particulate matter-induced airway inflammation in an allergic asthma mouse model [124]. Polyphenol treatment was found to decrease the inflammatory cell count in blood, including eosinophils, neutrophils, basophils. The level of epithelial cytokines, including IL-25, IL-33 and IL-8 also were reduced in the polyphenol-treated mice [124]. Han et al. [123] reported that pretreatment of BALB/c mice, as a model for passive cutaneous anaphylaxis, with Eckol inhibited the production of IL-4, IL-5, IL-6 and IL-13. Moreover, Eckol-treatment suppressed levels of β-hexosaminidase, secreted during the degranulation of mast cells. In addition, oral polyphenol administration was found to reduce the levels of FcεRI on the surface of IgE/bovine serum albumin (BSA)-stimulated mouse bone marrow-derived cultured mast cells (BMCMC). Cross-linking of FcεRI and allergen-specific IgE triggers allergic reactions that may be prevented by polyphenols. These results suggested that Eckol has anti-allergic potential. In a mouse ear-edema model both oral and local administration of phlorotannin, 1-21h prior to irritant application, strongly inhibited arachidonic acid (AA), 12-O-tetradecanoylphorbol-13-acetate(TPA)and immune-mediated (oxazolone (OXA))-induced ear swelling (30–80%) [125]. This suggests that the inhibitory effects of polyphenols are comparable to those of known anti-allergic agents. It was presumed that polyphenols play an anti-inflammatory effect by inhibiting mast cell degranulation, COX-2- and LOX-, and to lesser extent PLA2 activities.

5.2.3. Polyphenols: Effects in Acute and Chronic Inflammation

Inflammation is a beneficial host response for foreign invaders and necrotic tissue with phase 1 being the first, immediate response, typified by ROS production and release of early mediators such as IL-1, TNF-α and arachidonic acid pro-inflammatory metabolites. Once detected extracellularly, ingested microbes will lead to upregulation of TLRs a family of proteins involved in the initial phase of host defense against invading pathogens. TLR4 is the most common member in inflammation phases. Excessive TLR activation, however, disrupts the immune homeostasis by sustained production of pro-inflammatory cytokines and chemokines.

Polyphenols were demonstrated to be very well capable of suppressing the increase of TLRs, including TLR4 [124,126,127], TLR2 [124,127] and TLR7 [124]. TLRs as primary sensors of microbial products activate signaling pathways that lead to the induction of immune- and inflammatory genes, such as the NF-κB pathway. Polyphenol treatment also significantly decreased NF-κB and thereby the modulation of inflammation-related signaling cascades [123,124,126–131].

ROS are a crucial factor in the inflammatory response, playing multiple roles after tissue injury, including initiation of acute inflammation, clarifying infection and necrotic tissue and mediation of various intracellular signal transduction pathways. Anti-inflammatory effects of polyphenols via antioxidant activities were demonstrated by Kang et al. [132]. They found that serum ferric reducing antioxidant power (FRAP) significantly increased 30 min after polyphenol treatment in the Sprague Dawley rat-model but declined quickly thereafter. Polyphenols showed anti-inflammatory effects by reducing the expression of ROS [131,133–135]. Administration of the polyphenol-rich fraction of *Ecklonia cava* reduced ROS and NO generation in LPS-stimulated inflammation in zebrafish [133]. ROS can activate a variety of transcription factors leading to the differential expression of genes involved in inflammatory pathways. On the other hand, excessive production of ROS can cause irreversible damage to DNA. Due to this dual effect, polyphenols often show a crucial effect in tumor models by upregulating ROS in tumor cells and at the same time downregulating ROS in healthy cells. Yang et al. [129] indicated that oral administration of phlorotannins to the SKOV3-bearing mouse model of ovarian cancer enhances cancer cell apoptosis via upregulation of the ROS pathway but protects against healthy kidney cell damage by downregulating ROS levels. Tissue damage leads to a rapid increase in ROS which stimulates PGE2 production via the activation of COX-2. Polyphenols exert anti-inflammatory effects not only by suppressing COX-2 [128,130,131,136], but also by reducing PGE2 production [130,131] which exacerbates inflammatory responses and immune diseases. The production of early inflammatory cytokines such as TNF-α, IL-6 and IL-1β, that is increased in phase 1, was significantly inhibited by oral administration of polyphenols in vivo, in different mouse, rat and zebrafish models [124,126–128,130,131,135,137–139], and also in vitro, in RAW264.7 macrophages [131,134]. Polyphenol treatment [128,138] decreased the expression of CCL2/MCP-1 and consequently may reduce the infiltration of macrophages and subsequently inflammation [126,128].

In phase 2, the regulation of inflammation is amplified via positive feedback and adaptive immunity is activated. Oral administration of polyphenols significantly affects macrophage infiltration and the balance of macrophages with an M1 or M2 phenotype. Oral administration of polyphenols decreased the expression of CD11b and CD80, markers for M1 macrophages [126,127,137,138]. The M2 type is identified by marker CD206, and can prominently expresses IL-10, a cytokine with potential anti-inflammatory effects and plays an important role in limiting the host immune response to pathogens. Polyphenol treatment was found to increase the expression of the M2 markers, CD206 [126,127,137,138] and IL-10 in acute liver injured mouse model and HFD with or without seaweed supplement mouse model [124,127,135,138]. Oral administration of polyphenols induced a decrease in the level of iNOS [130,131,134,138] and the levels of NO in both.

5.2.4. Polyphenols: Late Consequences of Inflammation and Sequels

Oral administration of polyphenols may be promising for the treatment of severe or chronic inflammation and its consequences (phase 3). Oral administration of polyphenols resulted in a reduction in food intake and in body weight [126–128,137,140], as well as in the storage of triglyceride (TG) and total cholesterol (TC) [128,137]. After oral administration of extracts rich in polyphenols, also the leptin/adiponectin ratio, an important marker for inflammation and obesity, decreased [128]. Choi et al. and Son et al. [127,137] showed that the production of receptor for advanced glycation end-products (RAGE), closely related to inflammation and visceral fat hypertrophy, and RAGE-RAGE ligand binding was reduced in obese individuals treated with polyphenols. In obesity-associated type 2 diabetes, low-grade chronic inflammation can lead to an increase of blood glucose levels and to insulin resistance. Polyphenol treatment improved insulin sensitivity [128] and suppressed the increase of blood glucose levels in high-fat diet-induced obese mice [140]. A novel derivative from phloroglucinol called Compound 21 significantly exerted protective effects on multiple sclerosis through promotion of remyelination and suppressing neuroinflammation in a cuprizone-induced mouse model for multiple sclerosis [141]. In another study these authors showed that Compound 21 reduced the population of Th1/Th17 cells and inhibited their infiltration into the CNS. These results indicated a potential neuroprotective effect of Compound 21 [142].

In conclusion, polyphenols have various therapeutic effects including anti-inflammatory, anti-obesity, anti-diabetic and antioxidant. Polyphenols may be a highly promising treatment strategy for diseases involving chronic low-grade inflammation, but further clinical studies are needed.

6. Fucoxanthin(ol) and Meroterpenoids

Fucoxanthin is a major marine carotenoid present in chloroplasts of brown seaweeds, and particularly seaweeds such as *Undaria pinnatifida, Laminaria japonica* and *Sargassum honeri* are rich in fucoxanthin [143]. Ingested fucoxanthin is metabolized to fucoxanthinol in the small intestine and then absorbed [144]. Therefore, fucoxanthinol has a higher bioavailability than fucoxanthin. Multiple potentially beneficial health effects have been ascribed to both fucoxanthin and fucoxanthinol; e.g., anti-oxidative, anti-inflammatory, anti-obesity, anti-diabetic and anti-carcinogenic properties [145,146]. Meroterpenoids are partially derived from a terpenoid pathway (mero- means partial). Tetraterpenoids, of which carotenoids are the most common representatives, belong to the terpenoids and consist of eight isoprene units [147]. Meroterpenes can be isolated from brown algae such as *Sargassum serratifolium*. The meroterpenoid-rich fraction from the ethanolic extract (MES) of this brown alga is known for its antioxidant and anti-inflammatory activities [148]. Table A5 presents an overview of the significant outcomes of studies investigating oral administration of fucoxanthin, fucoxanthinol or MES in animal models related to the inflammatory response.

6.1. *Fucoxanthin(ol) and Meroterpenoids: Effects in Acute and Chronic Inflammation*

Fucoxanthin is known for its antioxidant potential through its ability to scavenge radicals effectively and to enhance enzymatic antioxidant activity [149]. Enhanced activity of antioxidant enzymes superoxide dismutase (SOD), catalase (Cat) and glutathione peroxidase (GPx) was observed in plasma and testis of rats [150], mice [151] and hamsters [152] after treatment with fucoxanthin. Fucoxanthin reduced the increased production of ROS as a consequence of increased oxidative stress and reduced the increased malondialdehyde formation during lipid peroxidation, which in turn causes upregulation of pro-inflammatory cytokine production [153,154]. Malondialdehyde levels were reduced in plasma, sperm and/or testicular tissue of rats [150], mice [151,155] and hamsters [152] after fucoxanthin treatment. Additionally, a reduction of ROS, such as superoxide (O_2^-) and hydrogen peroxide (H_2O_2) was seen both in vivo and in vitro after oral administration of fucoxanthin [150–152,155].

Oral administration of fucoxanthin in different animal models decreased the expression of various pro-inflammatory mediators, including cytokines such as TNF-α, IL-6 and IL-1β. This was observed in white adipose tissue, plasma, testis and colonic tissue after stimulation with various inflammatory triggers [150,151,155,156]. Sugiura et al. demonstrated the anti-inflammatory and inhibitory effects of oral or percutaneous administration of fucoxanthin on mouse ear swelling induced by different irritants [157]. Fucoxanthin and fucoxanthinol were shown to inhibit the enzymatic activities of PLA_2 and COX-2, thus restraining the generation of pro-inflammatory arachidonic acid metabolites in these mice. These anti-inflammatory effects of fucoxanthin and fucoxanthinol were also confirmed in vitro using rat basophilic leukemia cells, which showed reduced mRNA expression of $sPLA_2$ and COX-2 upon treatment with both fucoxanthin and fucoxanthinol [157]. In addition, Tan et al. found that COX-2 and iNOS mRNA expression were downregulated in obese mice upon fucoxanthin administration [155]. Similarly, Maeda et al. demonstrated decreased expression of MCP-1 in white adipose tissue in mice with obesity-related inflammation upon treatment [145,156]. Since MCP-1 is a pro-inflammatory cytokine this suggests an anti-inflammatory effect of fucoxanthin on adipocytes. In mouse models of DSS-induced colitis and colitis-associated colon carcinoma, a reduction in total NO content and in NO release in colonic tissue was observed after oral administration of fucoxanthin [151]. Moreover, NO production was also reduced after treatment with fucoxanthin in cisplatin-induced testicular damage in hamsters [152]. Additionally, the meroterpenoid-rich fraction of an ethanolic extract from *Sargassum serratifolium* (MES) induced anti-inflammatory activities in high-cholesterol-fed mice. The mice demonstrated decreased serum levels of MCP-1 and keratinocyte chemoattractant, which are pro-inflammatory chemokines causing monocyte adhesion in vascular lesions. Furthermore, MES supplementation resulted in reduction of ICAM-1, VCAM-1, MCP-1, COX-2 and MMP-9 expression in aortic tissue. These results indicated that MES prevented vascular inflammation in these mice [148]. Similarly, mice fed a high-fat diet supplemented with MES, compared to un-supplemented high-fat diet showed decreased expression of macrophage markers F4/80 and MCP-1, indicating a suppression of inflammation [158].

6.2. Fucoxanthin(ol) and Meroterpenoids: Late Consequences of Inflammation and Sequels

Related to the inflammation in adipose and other tissues, oral fucoxanthin supplementation was also shown to have effects counteracting obesity and obesity-related morbidity [159]. Administration of fucoxanthin to mice fed a high-fat diet reduced gain in body weight and in weight of white adipose tissue as compared to chow-fed control mice [155,156]. Furthermore, Maeda et al. demonstrated that mice fed a high-fat diet also receiving fucoxanthin displayed significantly lower plasma levels of LDL-cholesterol and leptin compared to mice that were fed a high-fat diet only, indicative of moderated metabolic dysregulation [156]. Additionally, the obesity-related reduction in expression of beta-3 adrenergic receptor (ADRB3), responsible for lipolysis and thermogenesis [160], appeared significantly restored in mice upon addition of fucoxanthin to their high fat diet [156]. Moreover, Tan et al. showed a decrease in myeloperoxidase (MPO) activity in mice with high-fat diet-induced obesity after oral fucoxanthin administration, which implies a reduction in polymorphonuclear cell infiltration [155]. In addition, MES supplementation suppressed body weight, TG, glucose and free fatty acid concentrations in plasma of high fat diet-fed mice. In addition, the lower HDL cholesterol levels increased to comparable levels as in the control group. Moreover, increased expression of UCP-1 and ADRB3 in subcutaneous tissues demonstrates that MES supplementation causes conversion of white to beige/brite adipocytes which resembles brown adipose tissue [158]. These results suggest anti-obesity effects and inhibition of lipogenesis by MES supplementation.

In line with improvement of metabolic functions induced by fucoxanthin supplementation, anti-diabetic effects of fucoxanthin have been observed [145,150,156]. Feeding a high fat-diet containing fucoxanthin resulted in decreased plasma insulin and blood glucose levels in mice, to levels similar as in mice fed a regular diet [156]. Moreover, mRNA levels of GLUT4, encoding the insulin-sensitive glucose transporter in adipose tissue and muscle, were restored to normal levels when the high fat diet was supplemented with fucoxanthin [156]. In the diabetic KK-Ay mouse model fucoxanthin consumption was found to decrease elevated plasma blood glucose concentrations [145]. It was also shown that glucose intolerance improved by fucoxanthin dietary addition [145]. More recently, Kong et al. found that treatment of diabetic rats with fucoxanthin significantly reduced levels of plasma glucose compared to diabetic rats without any treatment [150]. Insulin concentrations and homeostatic model assessment of insulin resistance (HOMA-IR) levels were significantly reduced in these rats. Finally, fucoxanthin supplementation inhibited the expression of the suppressor of cytokine singaling-3 (SOCS-3), involved in the induction of insulin resistance [150]. Together, these results indicate that fucoxanthin possesses anti-diabetic effects by suppressing inflammation and thereby improving insulin sensitivity.

7. Concluding Remarks

In this comprehensive systematic review, we aimed to provide an overview on the modulating role of intake of complete brown seaweed, its extracts or selected compounds on the modulation on different aspects of the inflammatory immune response. This includes the impacts of seaweed consumption on steady state immune parameters, effects on allergies, the innate and adaptive immune response and also on chronic, low-grade inflammation. Brown seaweeds constitute a group of approximately 2000 species containing several common but also unique bioactive molecules with immunomodulatory functions. We therefore distinguished the impact of four different categories of compounds, i.e., polysaccharides, (poly)phenols, phytosterols and carotenoids.

We identified three common denominators across the effects of brown seaweed and constituents thereof on inflammation (represented in Figure 2). Firstly, most purified compounds, despite their diverse chemical nature, appear to inhibit similar aspects of inflammation, in particular synthesis of reactive oxygen species. This common effect is rather puzzling. Yet, a caveat of some studies may be the use of compound concentrations beyond physiological levels. In addition, purification of the compounds used in the cited studies mostly left sufficient room for concentrations of contaminants that might contribute to, or even explain, the claimed effects. In our view, this calls for rigorous comparative research of the various compounds in an identical experimental setting. Nevertheless, the suppression of reactive oxygen species provides an interesting angle to study immune-modulatory effects of brown seaweed constituents.

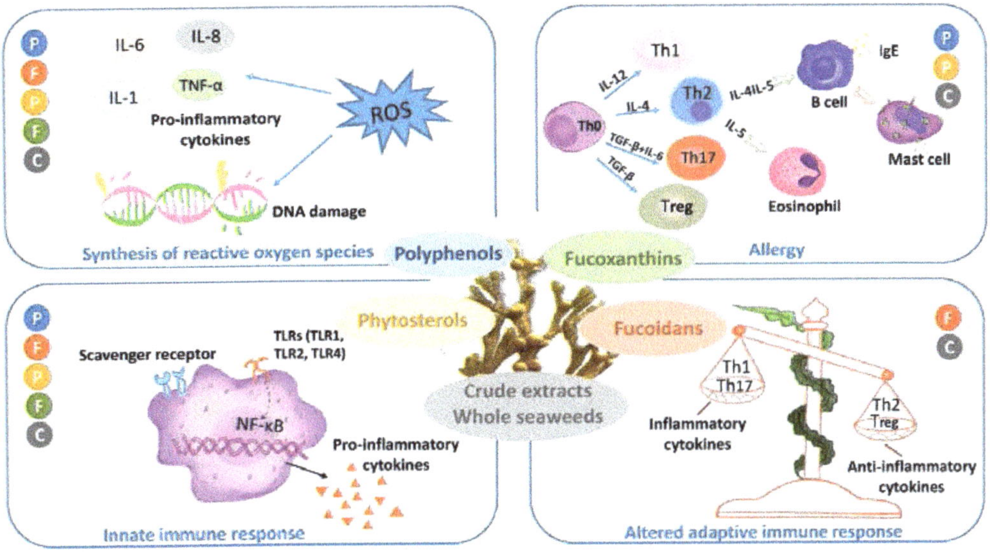

Figure 2. Overview of the general effects of crude brown seaweed or its extracts (C, grey) and of its constituents: polyphenol (P, blue), Fucoxanthin (F, green), Fucoidan (F, brown) and Phytosterols (P, yellow) on the different phases of the inflammatory processes, including allergy. The colored dots showing the crude seaweed or its extract and its constituents in the separate boxes indicate in which phases of the immune response they exert their effects.

Secondly, brown seaweeds interfere with the innate immune response on the level of TLR-induced NF-κB signaling. This route is actually linked with the previous one since oxygen radicals drive and amplify innate NF-κB-mediated activation. In conjunction, a plethora of in vitro and in vivo experiments support a suppression of IL-6, IL-1, iNOS and TNF-α upon treatment with brown seaweeds. Polyphenols, fucosterol, fucoxanthin and fucoidan all seem to be active in this. Yet, when fucoidan was applied in the pretreatment setting, it was also shown to potentiate the NF-κB axis to reduce susceptibility to infection via scavenger receptor A and TLR4 activation in an antibiotic-like fashion. These seemingly contradictory findings underscore the versatile properties of brown seaweed constituents in the modulation of the innate immune response. Therefore, interpretation should be performed with care when extrapolating in vitro findings to human applications. Nevertheless, most in vivo studies summarized in this review show unequivocal anti-inflammatory effects. The mechanistic background of brown seaweed health benefits probably goes beyond direct inhibition of inflammation. Multiple studies indicate that brown seaweed compounds interact with pathways and processes involved with energy sensing and survival, such as AKT/mTOR/AMPK and autophagy. Thus, brown seaweed compounds might also function as caloric restriction mimetics and thereby stimulate vitality.

Thirdly, the adaptive immune response displays an altered Th1/Th2 response in response to brown seaweed. Different constituents have been identified to skew the Th1/Th2 balance. Depending on the molecular weight of the fucoidan, different outcomes have been identified. Fucosterol has been shown to skew Th0 cells into Th2 cells in a model for allergy, whereas polyphenols suppress the Th1/Th17 response in an animal model for MS. The net effect of brown seaweed on Th1/Th2 skewing cannot be generalized and is largely dependent on the composition of the different constituents in the seaweed. On the level of allergy, Th2 suppression reduces IL-4 cytokine levels, decreased IgE production and suppressed mast cell activity.

Inflammation is a universal response of the body to damage, infection or otherwise disturbed homeostasis. For this review, we have restricted the search terms to those that are related to inflammation, allergy and immunity. The broad implication of the inflammatory response in the maintenance of bodily integrity, however, entails that several studies that focus on aspects only indirectly related to inflammation, were not included in the final result, whereas similar studies were, based on different choices by the respective authors for their specific wording. Nevertheless, we are confident that this review covers the main aspects of oral supplementation of brown seaweeds and their components on aspects of inflammation, allergy and immunity in a broad sense. More well-designed human studies applying individual seaweed constituents as well as whole seaweed (extracts) will provide more insight into the applicability of brown seaweed as immune-modulatory nutritional intervention strategies.

Supplementary Materials: The following are available online at https://www.mdpi.com/article/10.3390/nu13082613/s1, Supplementary Figure S1: Flowchart of systematic search and search strategies.

Author Contributions: Conceptualization, S.E.M.O., P.J.M.L. and M.T.M.; methodology, S.E.M.O., P.J.M.L. and M.T.M.; validation, S.E.M.O., X.W., B.T., P.J.M.L. and M.T.M.; investigation, S.E.M.O., X.W., B.T., S.K., P.J.M.L. and M.T.M.; data curation, S.E.M.O., B.T.; writing—original draft preparation, S.E.M.O., X.W., B.T., T.V., S.K., P.J.M.L. and M.T.M.; writing—review and editing, S.E.M.O., X.W., B.T., T.V., S.K., P.J.M.L. and M.T.M.; visualization, X.W., P.J.M.L., M.T.M.; supervision, P.J.M.L., M.T.M.; project administration, S.E.M.O., M.T.M.; funding acquisition, M.T.M. All authors have read and agreed to the published version of the manuscript.

Funding: This work was supported by Health Holland (HHINT Kickstarter #LSH17004), Alzheimer Nederland (WE.03-2018-06 AN) and NWO-TTW (#16437).

Acknowledgments: The authors wish to thank librarian Elise Krabbendam for her important contributions and support in the initial phases of the generation of this review.

Conflicts of Interest: The funders had no role in the design of the study; in the collection, analyses, or interpretation of data; in the writing of the manuscript, or in the decision to publish the results. S.E.M.O., X.W., T.V., P.J.M.L. and M.T.M. declare no conflicts of interest. S.K. and B.T. are professionally involved with seaweed farming and -products but have no commercial interests in the content of this review.

Appendix A

Table A1. Effects of oral supplementation of whole brown seaweeds or crude extracts on inflammation-related parameters.

Compound/Format	Effective Concentrations	Algal Source	Experimental Model	Inflammatory Phase [1]	Significant Findings [2]	Reference
EtOH extract	300–2000–5000 mg/kg; single dose	*Sargassum micracanthum*	BALB/c mouse; toxicity; follow-up 2 weeks	steady state	non-toxic up to 5000 mg/kg	[3]
EtOH-H_2O extracts	single dose LD50 = 500–2000 mg/kg; oxidative stress reduction: 200 mg/kg/day, 4 weeks	*Fucus vesiculosus*	Swiss mouse; Sprague Dawley rat; toxicity: acute anti-oxidant activity: 4 weeks	steady state	LD50 acute toxic > 750 mg/kg 4 week supplementation: ↓ 13% food intake in 4 week 30–50% ↓ WBC but no change in differential counts 20–25% ↑ in liver and kidney weight ↑ PON-1 activity (protects against ox-LDL) ↑ SOD	[5]
MeOH-, hexane- or chloroform-extract	>2000 mg/kg, single dose	*Cystoseira compressa (Esper)*	male and female albino mouse; toxicity	steady state	no lethality in oral testing (2000 mg/kg)	[4]
EtOH extract, polyphenol-rich ultrafiltrate	100 mg/day, 8 weeks	*Ascophyllum nodosum*	human, BMI ≥ 25	steady state	marginal ↓ lymphocyte DNA damage in only obese =CRP =cholesterol, HDL, LDL, TG anti-oxidant =in vitro LPS-/TPA-induced pro-inflammatory cytokines in monocytes, lymphocytes =total plasma peroxide	[7]
whole, dried	4.8 g/dayay, 4 weeks	*Sargassum muticum*	human	steady state	↓ oxLDL (14%); correlated to ↑ total antioxidant status ↑ NK count; =lymphocyte count ↓ fatigue ↑ liver function (↓ AST, ALT)	[6]

Table A1. *Cont.*

Compound/Format	Effective Concentrations	Algal Source	Experimental Model	Inflammatory Phase [1]	Significant Findings [2]	Reference
commercial extract	2% in feed, 3 weeks	*Ascophyllum nodosum*	Spanish and Boer × Spanish goat; stress by transport and feed withholding	steady state	=cortisol =WBC and subset counts, but ↓ Eosinophil =phagocytosis ↓ lipid peroxidation; ↓ SOD (strain diff.) ↑ glutathione peroxidase (10–15%)	[8]
commercial extract	0.5% in feed; 41 weeks	*Ascophyllum nodosum*	Lohmann LSL-Lite and Lohmann Brown-Lite hens, heat stress	steady state	short-term↓ feed intake; ↑ feed/egg efficiency strain-dependent ↑ production, ↑ feed efficiency, ↑ heat stress resistance, improved ALP, ALT, GGT liver parameters	[161]
seaweed powder + anti-bacterial peptides (cecropin)	1–5% of basal diet	*Laminaria japonica*	Arbor Acres broiler chicks	steady state	synergistic effect of seaweed + cecropin: ↑ antibodies, ↑ lymphocytes, microbiota: ↑ *Lactobacillus*, ↓ *E.coli* ↑ feed conversion ratio	[9]
hot water extract	100–500 mg/kg in feed, 12 weeks	*Sargassum oligocystum*	fish, Pangasius *(Pangasianodon hypophthalmus)*	steady state	↑ weight, daily growth rate, feed conversion ratio ↑ WBC, RBC, Hb, Hc, platelets	[10]
seaweed meal	3–6–9% in feed, 6 weeks	*Sargassum ilicifolium*	fish, Asian sea bass *(Lates calcarifer)*	steady state	↑ growth; ↑ pancreatic enzyme activities ↑ serum Ig; (alternative) complement pathway components, lysozyme ↑ Ig in skin mucus ↑ liver SOD, IL-1β mRNA	[11]

Table A1. *Cont.*

Compound/Format	Effective Concentrations	Algal Source	Experimental Model	Inflammatory Phase [1]	Significant Findings [2]	Reference
whole seaweed	10% in feed, 8 weeks	*Sargassum ilicifolium*	fish, great sturgeon (*Huso huso*)	steady state	↑ growth ↑ serum protein, lysozyme, IgM, respiratory burst, complement ↑ Hb, RBC, WBC =TG, LDH, AST, ALT ↓ blood cholesterol ↑ survival upon *Yersinia ruckeri* infection (14 days infection)	[12]
aqueous extract	400 mg/kg, 8 weeks	*Sargassum angustifolium*	fish, rainbow trout (*Oncohrynchus mykiss*)	steady state	↑ weight gain ↑ Hb, Hc, RBC, WBC, total protein, albumin↑ survival and immune response to *Yersinia ruckeri* infection	[13]
EtOH extract	200 mg/kg, 31 days	*Sargassum horneri*	male C57BL/6 mouse; in vivo anti-DNP-IgE i.v. + DNP i.n. challenge	allergy	↓ nasal rubbing ↓ mast cell degranulation Proposed mechanism: Chlorophyll-C2 →↓ PI-3K + ↓ Btk, ↓ Syk active upon FcεR trigger →↓ Ca-mobilization	[15]
EtOH extract	10, 50, 100 mg/kg, 1×/day, 3 weeks	*Sargassum horneri*	female NC/Nga mouse; house dust mite/dayNCB skin sensibilization DNCB skin challenge (2×/week)	allergy	↓ atopic dermatitis symptoms ↓ epidermal hyperplasia, hyperkeratosis, skin dryness ↓ mast cell + eosinophil skin infiltrates ↓ IL-25, IL-33 → ↓ Rantes (CCL5), Eotaxin (CCL11), TARC (CCL17) in skin ↓ IL-4, -5, -13, ↓ IL-6, -10, IFN-γ in serum ↓ spleen size increase ↓ IgG1, IgG2a increase	[16]

Table A1. *Cont.*

Compound/Format	Effective Concentrations	Algal Source	Experimental Model	Inflammatory Phase [1]	Significant Findings [2]	Reference
EtOH extract	100–300 mg/kg/day in diet, 5 weeks	*Costaria costata*	male NC/Nga mouse; DNCB-induced atopic dermatitis skin allergy	allergy	↓ inflammatory cell and mast cell infiltration, epidermal thickness, erythema, hemorrhage, dead skin cell layers, skin dehydration ↓ Eotaxin and TARC (CCL17)↓ serum IgE and histamine normalization of spleen lymphocyte proliferation and cytokine production	[17]
water extract	100–300 mg/kg/day, 4 weeks	*Laminaria japonica*	female NC/Nga mouse; DNCB-sensitized and challenged dorsal skin	allergy	↓ dermatitis severity; ↓ inflammatory mediators ↑ skin moisture in vitro: ↓ p38 MAPK,↑ ERK, ↓ STAT1 in HaCaT human keratinocyte cell line	[18]
dried powder	1–5–10%, 6 weeks	*Eisenia (=Ecklonia) arborea*	female Brown Norway rat; oral OVA immunization	allergy	↓ OVA-specific IgE, ↓ total IgE (n.s. with 10% diet) ↓ serum histamine ↑ IFN-γ in spleen and MLN ↓ IL-10 in spleen and MLN	[14]
EtOH extract	0.1–0.3 g/kg/day, 3 days prior to experiment	*Laminaria japonica*	male SD rat; carrageenan-induced paw inflammation	1,2	↓ paw swelling, leukocyte infiltration in vitro: ↓ IκB phosphorylation → ↓ iNOS, COX2, TNF-α, IL-1β, IL-6	[19]
diethylether fraction	0.1–1 mg/mouse, 2×	*Sargassum fusiforme*	ICR mouse; ear-swelling after irritant	1,2	↓ ear swelling by arachidonic acid, TPA, oxazolone in vitro: ↓ mast cell degranulation by inhibiting PLA2, COX2, LOX, HA	[20]
ethyl acetate extract	200 mg/kg, 5 weeks	*Eisenia bicyclis = Ecklonia bicyclis*	male SD rat; isolated peritoneal Mϕ	1,2	↓ LPS-induced iNOS expression and NO production ↓ NF-κB nuclear translocation with and without LPS =(LPS-induced) tumoricidal activity against B16 melanoma	[21]

Table A1. *Cont.*

Compound/Format	Effective Concentrations	Algal Source	Experimental Model	Inflammatory Phase [1]	Significant Findings [2]	Reference
EtOH extract +/− enzyme-treatment	50–200 mg/kg, 3 weeks	*Hizikia. = Sargassum fusiforme*	male C57BL mouse; LPS-stimulation of isolated peritoneal Mϕ conA-induced splenic lymphocyte proliferation	1,2	↑ LPS-stimulated IL-1β, TNF-α, IL-6 by peritoneal Mϕ ↑ conA-induced splenocyte proliferation	[22]
EtOH extract	6–24 mg/kg/day, 4 days	*Sargassum serratifolium*	male ICR mouse; LPS i.p. (2 mg/kg), blood sampling after 2 h	1	↓ TNF, IL-1β, IL-6 confirming in vitro findings	[23]
enzymatic extract	50–200 µg/mL, 1 h prior to and during exp	*Sargassum polycystum* *Chnoospora minima*	zebrafish embryos; in vivo 24 h challenge w/H_2O_2 (10 µg/mL) or LPS (10 µg/mL), monitored 5d	1,2	↓ H_2O_2-induced ROS levels and cell death ↓ LPS-induced ROS, NO and cell death	[24]
whole seaweed	0–4% feed, 40 days	*Sargassum latifolium*	Barki sheep (*Ovis aries*); i.v. LPS challenge (1.25 µg/kg), after day.28 + day.35	2	↓ body temperature, respiration rate ↓ leukocytosis, ESR, HSP70 ↑ blood anti-oxidant capacity (CAT, SOD) ↓ damage-related molecules: malondialdehyde (MDA; lipid peroxidation product); ALAT, LDH	[25]
whole seaweed	0–4% feed, 40 days	*Sargassum latifolium*	Barki sheep (*Ovis aries*); heat stress (solar experiment 8–17) vs. mild temperature without solar exposure	2,3	↓ Δ leukocytosis, Δ ESR ↓ proinflammatory cytokines, HSP70 ↑ body weight gain, kidney function, blood anti-oxidant function	[26]

Table A1. *Cont.*

Compound/Format	Effective Concentrations	Algal Source	Experimental Model	Inflammatory Phase [1]	Significant Findings [2]	Reference
commercial extract	1% feed, 27 days prior to and during exp (10 days)	*Ascophyllum nodosum*	crossbred wether lamb (*Ovis aries*); heat stress	1,2	↓ heat stress-induced reduction of phagocyte oxidative burst ↑ SOD ↓ heat stress-induced changes in GSH-peroxidase activity ↓ lipid peroxidation ↑ leukocyte phagocytosis	[27]
EtOH and organic-purified fraction	15 mg/kg/day, 6 weeks	*Turbinaria ornata*	female C57BL/6J mouse; DSS-induced colitis	1,2	↓ disease activity index ↓ histopathology incl. length reduction, neutrophil infiltrate ↓ TNF ↑ FoxP3, Treg, but = Th17 ↑ IL-10	[28]
aqueous extract (AE) + probiotic mix	100–300 mg/kg, 2×/day for 7 days	*Laminaria japonica*	male BALB/c mouse; DSS-induced colitis	1,2	AE alone: ↓ colitis, incl. ↓ colonic IL-1. IL-6 AE + probiotics: synergistic ↓ colitis, ↓ IL-1, -6, -12p40; not IFN-γ, IL-10, IL-12p70	[29]
EtOH-extract	50–200 mg/kg/day, d.28–98	*Sargassum muticum*	male DBA/1J mouse; collagen-induced arthritis	1,2,3	↓ arthritis and edema ↓ IL-6, TNF, IFN-γ in serum ↓ joint degradation, inflammatory cytokines in joints possibly explained by apo-9′ fucoxanthinone (sim. effects)	[30]
GFS = hot water extract with galactofucan sulfate	therapy: 4 × 560 mg/day, 10 days, 1–24 mo. maintenance: 2 × 560 mg/day	*Undaria pinnatifida*	human; response in patients with active or latent Herpes infection	2	15/15 patients with active disease: ↓ symptoms or full clearance of infection no side effects inhibition of relapse in pts with latent disease	[31]

Table A1. *Cont.*

Compound/Format	Effective Concentrations	Algal Source	Experimental Model	Inflammatory Phase [1]	Significant Findings [2]	Reference
hot water extract, and HCl-EtOH extract	100 mg/kg/day (heat extract), 10 mg/kg/day (HCl-EtOH extract), for 3–7 days	*Laminaria japonica*	kuruma shrimp (*Marsupenaeus japonicas*) in vivo WSSV infection in vitro hemocyte analysis	1,2	↑ survival upon WSSV infectionhemocyte fMet-Leu-Phe stimulation: ↑ chemotaxis ↑ superoxide production ↑ phenol oxidase activity ↑ phagocytosis	[32]
hot water-EtOH extract	20–200 mg/kg; 7 weeks	*Hizikia.* = *Sargassum fusiforme*	male SD rat; ligature-induced periodontitis	3	↓ alveolar bone loss due to inflammation	[33]
hot water extract of 10 different herbs incl. *S. fusiforme*	10 mg/kg; 10 weeks	*Sargassum fusiforme*	female SD rat; induced autoimmune thyroiditis by CFA-IFA thyroglobulin immunization	3	↓ AI-thyroiditis (cellular infiltrate), ↓ auto antibodies	[34]
whole seaweed, freeze-dried, powdered	5% in chow, 8 weeks	*Undaria pinnatifida*	male C57BL/6J mouse, $HFD^{+/-}$ seaweed supplement	3	↑ total plasma cholesterol (cf. HFD) fecal cholesterol excretion ↓↓ MCP-1 induction obese phenotype not prevented no glycemic improvement in i.p. glucose tolerance test microbiome composition (HFD + seaweed) is closer to LFD than to HFD	[38]

Table A1. *Cont.*

Compound/Format	Effective Concentrations	Algal Source	Experimental Model	Inflammatory Phase [1]	Significant Findings [2]	Reference
whole seaweed, freeze-dried	5% in chow, 16 weeks	*Undaria pinnatifida* (*Up*), *Laminaria japonica* (*Lj*), *Sargassum fulvellum* (*Sf*), *Hizikia* = (*Sargassum*) *fusiforme* (*Hf*)	male C57BL/6N mouse, HFD$^{+/-}$ seaweed supplement	3	body weight or adiposity: no change cf. HFD, except HFD+Up →↑ weight + ↑ subcutaneous adipose tissue adiponectin: ↓ with HFD+Up leptin: ↓ with HFD+Lj, +Sf, +Hf insulin resistance: ↓ w/HFD+Lj blood glucose: ↓ HFD+Lj or +Hf = LFD; HFD+Up = HFD crown-like structures in adipose tissue: ↓↓ with all; =LFD LPS-induced pro-inflammatory cytokines by BMDM: ↓ with all	[36]
MeOH extract + carob pod	0.1%/0.9%, 4 weeks	*Undaria pinnatifida*	male Wistar rat, MetS after 8 weeks western diet	2,3	=body weight, fat mass or muscle =food or energy intake ↓ systolic blood pressure at low dose; ↑ at high dose =CRP ↑ insulin, =glucose ↓ non-esterified fatty acids	[37]
hot aqueous extract	500 mg/kg/day started after 2 weeks, for 2 weeks	*Ecklonia cava*	female Sprague Dawley rat, letrozole-induced PCOS	2,3	↓ vaginal leukocyte infiltration restore normal estrous cycle restore normal plasma hormonal levelsnormalize expression of gonadotropin- and steroid hormone-related genes = weight gain upon PCOS induction	[39]

[1] Phases of inflammation as indicated in Table 1: 1-initiation of inflammation; 2-short term amplification and regulation; 3-short- or long-term consequences. [2] Symbols used: ↑—increase; ↓—decrease; =—no change.

Appendix B

Table A2. Effects of oral supplementation of brown seaweed derived polysaccharides on inflammation-related parameters.

Compound/Format	Effective Concentrations	Algal Source	Experimental Model	Inflammatory Phase [1]	Significant Findings [2]	Reference
Fucoidan	0.2%, 2% fucoidan chow for 8 weeks	*Cladosiphon okamuranus*	Rats	steady state	Uptake through intestinal tract	[46]
Fudoidan 7 kDa (LMW)	2000 mg/kg, Single dose	*Ascophyllum nodusum*	rats	steady state	No toxicity	[50]
Sodium-Alginate (SA)	5% SA containing chow for 6 weeks	Unknown	Male C57BL/6, Non-alcoholic steatosis model	steady state	↑ Intestinal barrier function in small intestine ↓ Hepatic lipid accumulationLiver: ↓ TNFα, ↓ Collagen-1α1 Liver: ↓ Macrophage infiltration	[162]
Fucoidan-rich extract	50–100–150 mg/kg for 14 days	*Undaria pinnatifada*	In vitro: RAW 264.7 In vivo: C57BL/6 mice	1,2	In vitro: ↑ NO, ↑ TNF-α, IL-1α, IL-1β, IL-6 In vivo: ↑ CD3+, CD4+ ↑ TNF-α, IFN-γ ↑ IgM	[163]
Fucoidan	10–100 mg/kg for 21 days	*Cladosiphon tokida*	In vitro: RAW 264.7 + S180 tumor cells In vivo: S180 tumor bearing mice	2	↑ NO-production by macrophages↑ NK-cell mediated cytotoxicity ↑ NF-κβ translocation in macrophages	[52]
Fucoidan-rich extract	50–100–150 mg/kg for 14 days	*Undaria pinnatifada*	Cyclophosmamide-immunosuppressed male C57BL/6 mice	1,2	All doses: ↑ NK cytotoxicity 100–150 mg/kg: ↑ proliferation of T cells 150 mg/kg: ↑ TNF-α, IgM and total IgG	[93]
Extracted polysaccharides (SFP)	100–200–300 mg/kg, 14 days	*Sargassum fusiforme*	Cyclophosmamide-immunosuppressed ICR mice	1,2	Compared to the cy model group: ↑ Thymus and spleen indices Liver: ↑ SOD, ↓ MDA, ↑ GSH ↑ villus height in normal mice, = in CY-treated mice ↑ intestinal epithelial lymphocyte and Goblet cells in normal and CY-treated mice	[164]

Table A2. *Cont.*

Compound/Format	Effective Concentrations	Algal Source	Experimental Model	Inflammatory Phase [1]	Significant Findings [2]	Reference
Fucoidan	Buccal LPS injection on 3 separate days followed by fucoidan ingestion *P. gingivalis* infection day 10–25, followed Fucoidan ingestion day 29–49	*Sargassum wightii*	Mouse–Buccal LPS and bacterial (*P. gingivalis*) induced inflammation In vitro: RAW 264.7—LPS induced inflammation	2	↓ TNF-α, IL-1β, IL-6 ↓ monocyte and dendritic cell recruitment =lymphocyte numbers ↓ IL-17, ↑ IL-10 ↓ inhibited antigen-specific immune response No inhibition of bacterial induced periodontitis In vitro: ↓ iNOS, COX-2 ↓ NO, PGE2 ↓ TNF-α, IL-6, IFN-γ	[55]
Sulfated polysaccharide >30 kDa	25–50–100 μg/mL	*Sargassum horneri*	Zebrafish embryos—LPS induced inflammation In vitro: RAW 264.7—LPS induced inflammation	1,2	↓ LPS-induced NO production, toxicity, cell death In vitro: ↓ p-IKβ ↓ iNOS, COX-2 ↓ NO, PGE2 ↓ TNF-α, IL-6	[56]
Fucoidan	25–50–100 μg/mL	*Fucus vesiculosus*	Zebrafish embryos—LPS induced inflammation In vitro: RAW 264.7—LPS induced inflammation	1,2	↓ NO, ROS production ↓ neutrophil and macrophage recruitment dose dependent activity In vitro: ↓ iNOS, COX-2 ↓ NO, PGE2 ↓ TNF-α, IL-1β	[103]
Fucoidan	12.5–25–50 μg/mL	*Turbinaria ornata*	Zebrafish embryos—LPS induced inflammation In vitro: RAW 264.7—LPS induced inflammation	1,2	↓ NO, ROS production ↑ Cell viability dose dependent activity In vitro: ↓ iNOS, COX-2 ↓ NO, PGE2 ↓ TNF-α, IL-1β	[53]
Fucoidan	12.5–25–50 μg/mL	*Laminaria japonica*	Zebrafish embryos—LPS induced inflammation In vitro: RAW 264.7—LPS induced inflammation	1,2	↓ NO, ROS production ↓ cell death rate dose dependent activity In vitro: ↓ MAPK, NF-κβ ↓ NO, ↓ TNF-α, IL-1β, IL-6	[54]

Table A2. *Cont.*

Compound/Format	Effective Concentrations	Algal Source	Experimental Model	Inflammatory Phase [1]	Significant Findings [2]	Reference
Laminarin, fucoidan and ash	Laminarin 1 g, fucoidan 0.8 g, day 107 of gestation until weaning (day 26)	*Laminaria*	Pregnant + Lactating sows	2	↑ Colostrum IgA ↑ Piglets serum IgG	[64]
Fucoidan	6 g/kg feed for 21 days	*Laminaria japonica*	Immunosuppressed African catfish	1,2	Macrophages: ↑ oxidative burst, ↑ phagocytic activity Lymphocytes: ↑ transformation index Serum: ↑ lysozyme, NO and bactericidal activity ↑ Survival rate in challenge test	[70]
Fucoidan-rich extract	2% inclusion, 45 days feeding trial	*Sargassum wightii*	Sutchi Catfish	1,2	Macrophages: ↑ oxidative burst, ↑ phagocytic activity ↑ Total lymphocyte count ↓ Albumin/Globulin ratio ↑ IFN-γ ↑ Survival rate in challenge test	[73]
Laminarin	0.2 g/kg/day for 21 days	*Laminaria digitata*	Rainbow trout	1,2	↑ TNF-α, IL-8	[72]
Fucoidan	0.05%, 0.1% and 0.2% per kilogram feed	*Sargassum horneri*	Yellow catfish	1,2	Macrophages: ↑ oxidative burst, ↑ phagocytic activity Serum: ↑ lysozyme ↑ CAT ↑ SOD, ↓ MDA↑ Survival rate in challenge test	[75]
Fucoidan-rich seaweed extract (FRSE) + Methionine	2% FRSE + 0.3 methionine	*Sargassum wightii*	Carp (*Labeo rohita*)	1,2	↑ respiratory burst activity, phagocytic activity, ↑ MPO activity, lysozyme activity, ↑ total immunoglobulin and TLC	[71]
Laminarin	0.5–1.0% inclusion in diet for 48 days	Commercially sourced	Grouper (*Epinephelus coioides*)	1,2	↑ IL-1β, IL-8 and TLR2 ↑ lysozyme, CAT and SOD ↑ growth rate and the feed efficiency	[76]

Table A2. *Cont.*

Compound/Format	Effective Concentrations	Algal Source	Experimental Model	Inflammatory Phase [1]	Significant Findings [2]	Reference
Fucoidan	250–500 mg/kg feed	*Sargassum*, *Padina* and *Turbinaria*	White shrimp (*Litopenaeus vannamei*)	1,2	↑ THC, PA and SOD ↑ LGBP, Toll and lectin	[74]
Laminarin, fucoidan and ash	Lam 1 g, Fuc 0.8 g, d 83 of gestation until weaning (day 28)	*Laminaria*	Pregnant + Lactating sows	1,2	↑ Piglets villus height in the jejunum and ileum	[67]
Extract	D 83 of gestation until weaning (day 28)	*Ascophyllum nodosum* and *Fucus*	Pregnant + Lactating sows	1,2	↑ Piglets CD4+ and CD8+ T cells	[66]
Laminarin	1 g/day Pre-weaning (0–62 days) and post-weaning (63–93 days)	*Laminaria*	Holstein Friesian bull calves	1,2	↓ Growth ↑ serum haptoglobin ↓ lymphocyte levels ↓ stimulated IFN-γ (in vitro challenges)	[68]
Sodium alginate oligosaccharides	0, 0.04 and 0.2% of diet	Unknown	broiler chickens Salmonella Enteritis challenge	1,2	in unchallenged animals: ↑ IFN-γ, IL-10, IL-1β	[69]
Highly viscous polysaccharide extract (HVPE)	10–30–100 mg/kg/day for 14 days	*Kjellmania crassifolia*	C57BL/6 mice	1,2	ConA-stimulated spleen cells: ↑ IFN-γ, IL-12, IL-6, IgA secretion ConA-stimulated Peyer's patch cells: ↑ IgA ↑ Peritoneal macrophage phagocytic activity	[77]
Fucoidan (LMWF)	200–400–1000 mg/kg/day (6 weeks)	*Laminaria japonica*	Sprague Dawley rats, Mycoplasma pneumoniae antigen stimulation	1,2	↑ Spleen weight, splenocyte proliferation potential ↑ NK cell activity, ↑ Phagocytic activity ↑ IFN-γ, IL-2 IL-4 ↑ IgG, IgA, ↓ IgE ↑ Antigen-specific antibodies	[61]
Fucoidan containing product (88.3% purity)	410–1025 mg/kg	*Cladosiphon okamuramus*	Balb/c mice, OVA-immunized	1,2	↑ Immune cell proliferation, IL-2, ↑ macrophage phagocytic activity ↑ serum IgM, IgG, IgA ↓ IgE ↓ IL-4, -5	[62]

Table A2. *Cont.*

Compound/Format	Effective Concentrations	Algal Source	Experimental Model	Inflammatory Phase [1]	Significant Findings [2]	Reference
Fucoidan (LMWF, MMWF, HMWF)	Unclear	*Cladosiphon okamuranus*	Pathogen free Male Balb/c mice	1,2	HMWF: ↑ Proportion CD8+ T-cells in spleen, ↓ CD4/CD8 ratio	[78]
water-soluble polysaccharides	50–100–200 for 10 days	*Sargassum fusiforme*	Cyclophosmamide-immunusuppressed ICR mice	1,2	Compared to the cy model group: ↑ spleen index ↑ splenic lymphocyte proliferation potential ↑ splenocyte produced IL-2, Il-6, IFN-γPeritoneal macrophages: ↑ phagocytic activity, ↑ produced IL-2, Il-6, IFN-γ	[92]
Fucoidan	150 mg/kg daily for 2 weeks	*Fucus vesiculosus*	mice	1,2	increased cytolytic activity of NK cells	[89]
Laminarin	500–1000 mg/kg/day for 10 days	Unknown, commercially sourced	Cyclophosmamide-immunusuppressed male Balb/c mice	1,2	Compared to the cy model group: ↑ cytotoxicity of NK cells ↑ serum IL-12 and IFN-γ	[94]
Fucoidan	0.1–0.5 mg/day for 14 days	*Undaria pinnatifada*	HSV-1 infected mice/5 fluorouracil immunosuppressed mice	2	Protection against herpes infection ↑ Macrophage NO production ↑ CTL activity ↑ B-cell blastogenesis↑ Antibody titers	[82]
Fucoidan	25–250 mg/kg/day, 3 times weekly for 4 weeks	Unknown	Leishmania infection, Balb/c mice	3	In vitro: 93% reduction Amastigote multiplication in macrophagesIn vivo: complete elimination parasite in liver and spleen Th2 ◊ Th1 response Splenocytes: ↑ superoxide and NO production	[79]
Fucoidan	500 mg/kg every 2 days, for 6 weeks	*Fucus vesiculosus*	Schistosomiasis Japonica infeced C57BL/6 mice	3	↓ Granuloma size, ↓ hepatic inflammation↓ IL-6, IL-12, TNF-α ↑ IL-4, IL-13 ↑ IL-10, TGF-β ↑ Th2 response ↑ Treg infiltration in liver	[106]

Table A2. *Cont.*

Compound/Format	Effective Concentrations	Algal Source	Experimental Model	Inflammatory Phase [1]	Significant Findings [2]	Reference
Fucoidan	0.034 g/mouse/day 10 days before, 40 days after tumor inoculation	*Undaria pinnatifada*	A20-tumor bearing Male Balb/c mice/Do-11-10-Tg mice (transgenic for TCR)	3	↑ NK cell activity ↑ CTL activity ↑ IFN-γ, IL-12	[84]
Ascophyllan	500 mg/kg. 4 days before–10 days after tumor implantation	*Ascophyllum nodosum*	S-180 sarcoma bearing SPF male ddY mice	3	Oral route stronger anti-tumor effects than intraperitoneal injection, effects via interaction with intestinal immune system ↓ 69% tumor size reduction ↑ TNF-α, IL-12	[91]
Polysaccharides	100–200 mg/kg for 28 days	*Sargassum fusiforme*	A549 adenolungcarcinoma-bearing Balb/c nuce mice	3	↑ Serum TNF-α ↑ Peritoneal macrophage production of TNF-α, IL-1β ↑ Splenocyte lymphocyte proliferation Liver and Kidney: ↑ SOD, ↓ MDA	[87]
Fucoidan (LMWF/IMWF/HMWF)	5 g/kg/day	*Cladosiphon okamuramus*	Colon 26 tumor bearing mice; Myd-88$^{-/-}$ mice	3	No intestinal absorption Effects via interaction with intestinal immune system All: ↑ Median survival timeIMWF ↓ Tumor weight, ↓ Cell divisions, ↑ Apoptosis HMWF: ↑ Splenic NK cell numbers	[85]
Polysaccharide	100–200–400 mg/kg for 28 days	*Sargassum fusiforme*	HepG2-tumor-bearing mice	3	↑ Serum TNF-α, NO, IL-1β, IgM ↑ Peritoneal macrophage production of TNF-α, IL-1β ↑ Apoptosis in HepG2-tumor cells	[86]
Polysaccharides	50–100–200 mg/kg	*Sargassum fusiforme*	CNE- tumor- bearing mice	3	↑ Serum NO, IL-1β, IgM ↑ Peritoneal macrophage production of NO, IL-1β↑ Splenocyte lymphocyte proliferation ↑ IgM production Activity is, in part, mediated via TLR2 and TLR4	[88]

Table A2. *Cont.*

Compound/Format	Effective Concentrations	Algal Source	Experimental Model	Inflammatory Phase [1]	Significant Findings [2]	Reference
Fucoidan	200–400 mg/kg (6×/week) for 4 months	*Fucus vesiculosus*	DMBA-induced mammary carcinogenesis in female Sprague Dawley rats	3	↓ Tumor incidence, ↓ Tumor weight ↑ Serum IL-6, IL-12p40, IFN-γ ↓ IL-10 TGF-β ↓ MRNA expression levels of FOXP3, TGF-β, ↑ IFN-γ in tumors ↓ Foxp3, PD1, =PDL1/PDL2 protein levels in tumors ↓ p-PI3K, p-AKT protein levels in tumors	[90]
Fucoidan	5 mg/day for 10 days (start direct after infection)	*Undaria pinnatifada*	Influenza infected mice	3	↑ Survival ↑ Inhibition of viral replication ↑ Mucosal antibody levels (IgA) ↑ Serum antibody titer (IgM, IgG)	[80]
Fucoidan	6 g/day for 6–13 months	Unknown, commercially obtained	Human, HTLV-infected patients	3	↓ HTLV proviral load =$CD4^+$, $CD8^+$ frequencies =NK-cell, mDC and pDC frequencies	[81]
Oligo Fucoidan	4400 mg/day for 48 weeks	Unknown, commercially sourced	Patients with chronic Hepatitis B virus 2193	3	vitamin D dependent activity ↓ HBV DNA ↑ $CD4^+CD45RO^+$ and $CD8^+CD45RO^+$	[109]
Fucoidan	3 g/day for 6 months	*Cladosiphon okamuranus*	Human, male cancer survivors	3	↑ NK-cell activity	[165]
Fucoidan	100–1000 mg/day for 4 weeks	*Fucus vesiculosus* (85%), *Macrocystis pyrifera* (10%), *Laminaria japonica* (5%)	Human volunteers	3	↑ Cytotoxic T cell numbers ↑ Phagocytic capacity monocytes ↓ IL-6No dose response	[110]
Fudoidan (75% pure)	3 g/day for 12 days	*Undaria pinnatifada*	Human volunteers	3	↑ CD34+ cells, CD34+ cells expressing CXCR4 (45% ◊ 90%)↑ serum SDF-1 and IFN-γ	[166]
Fucoidan	300 mg/day for 4 weeks	*Mekabu fucoidan*	Human, Influenza vaccinated elderly	3	↑ Antibody titers	[167]

Table A2. *Cont.*

Compound/Format	Effective Concentrations	Algal Source	Experimental Model	Inflammatory Phase [1]	Significant Findings [2]	Reference
Sulphated polysaccharide	20–40–80 mg/kg/day for 5 days	*Sargassum hemiphyllum*	Mouse–arachidonic acid induced ear-inflammation	2	dose dependent activity ↓ Ear swelling and erythema ↓ TNF-α, IL-1β, IL-6 ↓ neutrophil infiltration	[95]
Combined preparation including fucoidan	Fucoidan: 18–54 mg/kg/day for 7 days	*Laminaria japonica*	Mouse–Carrageenan induced pouch inflammation	2	↓ NO, PGE2 ↓ neutrophil and macrophage recruitment	[96]
Fucoidan	0.02 g/kg for 14 days	Commercially sourced from Sigma	Mouse–aspirin induced ulcer	2,3	Protection against ulceration ↓ AST, ALT ↓ PGE2 ↓ IFN-γ, IL-6, IL-10 ↓ stomach glycogen	[97]
Fucoidan	0.05% *w*/*w* in chow	*Cladosiphon okamuranus*	Mouse–DSS-induced colitis	2	↓ IFN-γ, IL-6 ↑ IL-10, TGF-β	[98]
Fucoidan & Fucoidan-polyphenol complex (Synergy)	10 mg/mouse, 400 mg/kg/day	*Fucus vesiculosus*	Mouse–DSS-induced colitis	2,3	Oral synergy best results ↑ colon length ↓ colon weight/body weight, spleen weight ↓ Histology damage score↓ TNF-α, IL-12 IP fucoidan actually worsened colitis symptoms	[99]
Fucoidan	200 mg/kg	*Laminaria japonica*	Mouse–myocardial ischemia-reperfusion injury	2,3	↓ myocardial infarct size ↓ HMGB1, p-IκB-a and NF-κB ↓ TNF-a and IL-6, ↑ IL-10 ↓ infiltration PMNs, ↓ MPO activity ↓ histopathological damages in myocardium	[104]
Fucoidan	200 mg/kg/day for 14 days	*Sargassum hemiphyllum*	C57BL/6 mice, irradiation induced pneumonitis and lung fibrosis	3	↓ Lung fibrosis ↓ Neutrophil and macrophage infiltration ↓ TIMP-1, CXCL-1, MCP-1, MIP-2, IL-Ra ↓ Procollagen-1α	[105]

Table A2. *Cont.*

Compound/Format	Effective Concentrations	Algal Source	Experimental Model	Inflammatory Phase [1]	Significant Findings [2]	Reference
Fucoidan (LMWF: 1 ± 0.2 kDa MMWF: 3.5 ± 0.3 kDa HMWF: 100 ± 4 kDa)	300 mg/kg dissolved in water, one day after booster immunization until the end of the experiment (day 22–47)	*Undaria pinnatifada*	Male DBA/1J mice, Collagen-induced arthritis	2,3	LMWF: ↓ reduced severity of inflammation↓ IFN-γ, TNF-α ↓ IgG2a, HMWF: ↑ Increased severity of inflammation↑ IFN-γ, TNF-α ↑ IgG2a ↑ adhesion molecules and migration potential of macrophages	[102]
Fucoidan	4 g/day for 4 at least weeks	*Cladosiphon novae Caledoniae*	Advanced cancer patients	3	Responsiveness of IL-1β independent prognostic factor for QOL-scores ↓ TNF-α, IL-1β, IL-6 = QOL-scores	[111]
Fucoidan	400 mg/kg for 22 weeks	*Fucus vesiculosus*	Mice, colorectal carcinogenesis model	2,3	↑ Beneficial Microbiome modulation ↓ IL-17 and IL-23 ↑ IFN-γ, IL-4 and IL-10 ↑ NK cells, CD4+ T cells in blood	[101]
Fucoidan	300–600 mg/kg/day for 5 weeks	*Fucus vesiculosus*	Non-obese diabetic mice (autoimmune diabetes model)	2	↑ serum insulin levels Delayed onset and ↓ incidence of diabetes ↓ Th1 cytokines ↑ Th2 cytokines Dendritic cells: ↓ MHC-II, ↓ CD86 Pancreatic cells: ↓ TLR4 expression Microbiota: ↑ Akkermansia, ↑ Lactobaccillus, ↓ Bacteroides	[108]
Fucoidan	400 mg/kg	*Ascophyllum nodosum*	Colonic inflammation-induced SPF-C57BL/6 mice (antibiotics treated)	3	↓ Colonic inflammation↓ Dysbiosis ↓ Infiltration of inflammatory cells ↓ Space between mucosa and submucosa ↑ Crypt depth ↓ TNF-α, IL-1β, IL-6, ↑ IL-10	[100]
Fucoidan	200 mg/kg	*Laminaria japonica/Ascophyllum nodosum*	SPF C57BL/6 J mice, High fat diet	3	↓ Body weight gain, ↓ Fat mass ↓ Insulin resistance ↓ Endotoxemia (↓ LBP) ↓ Systemic inflammation (↓ TNF-α, IL-1β, MCP-1) Microbiome: ↑ Akkermansia, Alloprevotella, Blautia, Bacteroides	[47]

Table A2. *Cont.*

Compound/Format	Effective Concentrations	Algal Source	Experimental Model	Inflammatory Phase [1]	Significant Findings [2]	Reference
Fucoidan and Fucoxanthin	275 mg LMF	Unknown, commercially obtained	Clinical trial, NAFLD patients		↓ AST =adiponectin	[168]
Fucoidan	50 mg/kg for 14 days	*Fucus vesiculosus*	Balb/C mice, ConA-induced acute liver injury	3	↓ AST, ALT ↓ Histopathological changes ↓ IFN-γ, TNF-α ↓ Apoptosis inhibition	[107]
Fucoidan	1 or 5% inclusion in diet (12 weeks)	*Cladosiphon okamuranus*	Diet-induced dyslipidemia in $ApoE^{-/-}$ mice	3	↓ Tissue weight (liver and WAT), hepatic steatosis ↓ Blood lipids (TC, TG, non-HDL-C)↓ Blood glucose ↑ Plasma LPL activity, HDL-C ↑ Insulin-sensitivity ↑ Pparα, ↓ Srebf1	[169]
LJP12	50–100–200 mg/kg/day	*Laminaria japonica*	Atherogenic diet fed $LDLR^{-/-}$ mice	3	Dose-dependent activity ↓ Atherosclerotic plaque formation ↓ Plasma lipids (TC, TG, LDL-C), ↑ HDL-C/LDL-C ↓ Systemic inflammation (↓ TNF-α, IL-1β, IL-6, MCP-1, ↑ IL-10) ↑ SOD, ↓ MDA ↓ p-p 65, p-I$\kappa\beta$, p-ERK, p-JNK, p-P38	[112]
Fucoidan (LMWF)	200 mg/kg/day	*Laminaria japonica*	Diet-induced dyslipidemia in $ApoE^{-/-}$ mice	3	↓ TG, OX-LDL ↓ p-JNK, cyclin-D1 ↓ IL-6, ↑ IL-10 ↓ Macrophages ◊ foam cells ↓ Migration of SMCs into intimal layer	[114]
Fucoidan	Unknown	*Fucus vesiculosus*	Rats	3	↓ Endothelial hyperplasia, ↓ vascular modulation ↓ α-actin^{+} cells ↓ Vessel inflammation↓ Macrophage infiltration ↓ Apoptosis in vessel wall	[115]

Table A2. *Cont.*

Compound/Format	Effective Concentrations	Algal Source	Experimental Model	Inflammatory Phase [1]	Significant Findings [2]	Reference
Fucoidan (LMWF)	200 mg/kg/day	*Laminaria japonica*	Diet-induced dyslipidemia in ApoE$^{-/-}$ mice	3	↓ Inflammatory infiltration ↓ Limit enlargement of AAA ↓ Maximal aortic diameter Preserved elastin ↓ TNF-α, IL-1β, MCP-1 ↓ MMP	[116]
Fucoidan	50–100 mg/kg/day	*Laminaria japonica*	In vitro: oxLDL treatment RAW264.7 In vivo: Atherogenic diet fed LDLR$^{-/-}$ mice	3	In vitro: ↓ oxLDL-induced LOX-1, ↓ TNF-α, IL-1β, IL-6 ↓ ICAM-1, VCAM-1 ↓ ROS In vivo: ↓ Atherosclerotic plaque formation ↑ Plaque stability↓ Macrophage infiltration ↓ Plasma lipids (TC, TG, non-HDL-C) ↓ ROS	[113]
Fucoidan (LMWF) <3 kDa	5 mg/day. With or without added probiotics	*Undaria pinnatifada*	Male Balb/c mice.	2	Fucoidan enhances effects of probiotic strain ↑ IFN-γ, TNF-α, IL-6, IL-12 OVA-immunization: ↓ lgE, IL-4, ↑ IFN-γ Desulphated Fucoidan no effects	[170]
Extracted polysaccharide (not fucoidan)	50 mg/kg/day for 2 weeks	*Laminaria japonica*	SPF Kunming mice, OVA-immunized asthma model	2,3	↓ Eosinophils in BALF ↓ lung inflammation↓ Serum IgE ↑ IL-12, ↓ TGF-β, IL-13 in BALF and lung	Lin et al. 2015. Multidiscip Respir Med. PMID: 26110056 [58]

Table A2. *Cont.*

Compound/Format	Effective Concentrations	Algal Source	Experimental Model	Inflammatory Phase [1]	Significant Findings [2]	Reference
Fucoidans	100–400 mg/kg/day	*Undaria pinnatifada*	OVA-immunized female Balb/c mice, exacerbated allergic astma	2,3	↓ Infiltration of macrophages, neutrophils, $CD4^+$ lymphocytes ↓ Lipid peroxidation ↓ IL-4, ↓ IgE ↓ Mast cell activation, degranulation ↓ Mucus hypersecretion, ↓ Goblet cell hyperplasia	Herath et al. 2020. Molecules. PMID: 32580518 [57]
Fucoidan	100–400 µg/mouse/day for 4 days	*Laminaria japonica*	Female Balb/c mice, passive cutaneous anaphylaxis model	2,3	↓ allergic symptoms (Edema) Oral fucoidan anti-allergy effects, IP-fucoidan no effects =Serum IgE, IgG1 ↑ Galectin-9 secretion by intestinal epithelial cells	[59]
Fucoidan	60 µg/mouse/day for 17 days after OVA-immunization	*Laminaria japonica*	Female Balb/c mice, OVA-immunized	2,3	↓ allergic symptoms (Rectal temperature reduction) =Serum IgE ↑ Serum galectin-9 ↓ Degranulation of mast cells ↓ IgE-attachment on mast cells	[60]
Alginate, 108 kDa	2 mg alginate one day before every OVA-challenge	*Laminaria japonica*	OVA-immunized female Balb/c mice, food allergy model	2,3	↓ IL-4, ↓ IgE ↓ Mast cell degranulation, ↓ Histamine ↑ Number of Tregs in spleen ↓ Th0 ◊ Th2	[63]

[1] Phases of inflammation as indicated in Table 1: 1-initiation of inflammation; 2-short term amplification and regulation; 3-short- or long-term consequences. [2] Symbols used: ↑—increase; ↓—decrease; =—no change.

Appendix C

Table A3. Effects of oral supplementation of brown seaweed derived phytosterols on inflammation-related parameters.

Compound/Format	Effective Concentrations	Algal Source	Experimental Model	Inflammatory Phase [1]	Significant Findings [2]	Reference
Seaweed extract containing fucosterol Enzyme-modified extract	NH: 200 mg/kg EH: 50, 100 and 200 mg/kg; 3 weeks	*Sargassum fusiforme* = *Hijiki*	Male C57BL/6 mice; ConA activation induced splenocyte proliferation	1	↑ splenocyte proliferation In peritoneal Macrophages:↓ IL-6 ↓ IL-1β ↓ TNF-α (EH > NH)	[22]
Lipid extract containing 24(S)-Saringo-sterol	50% (*w*/*w*), 10 weeks	*Sargassum fusiforme*	APPswePS1ΔE9 Mice, Alzheimer's disease +/− seaweed supplement	3	↑ LXR-responsive gene expression in CNS ↑ ApoE	[118]
Fucosterol extract	25, 50 or 100 mg/kg; 3 days	*Eisenia bicyclis*	Male BALB/c mice, ConA- induced acute liver injury after Fucosterol pretreatment	1	↓ TNF-a ↓ IL-6 ↓ IL-1β ↓ NF-κB p65 ↑ PPARγ expression ↑ p-P38 MAPK levels	[117]
ethyl acetate extract => Fucosterol	200 mg/kg 2 weeks	*Sargassum fusiforme*	NC/N ga male mice, DNCB induced AD-like dermatitis	allergy	↓ Scratching ↓ epidermal thickness of dorsal skin ↓ mast cells ↓ serum level of IgE In cultured splenocytes: ↓ IL-4 ↓ TNF-α ↑ IFN-γ	[171]

[1] Phases of inflammation as indicated in Table 1: 1-initiation of inflammation; 2-short term amplification and regulation; 3-short- or long-term consequences; [2] Symbols used: ↑—increase; ↓—decrease; =—no change.

Appendix D

Table A4. Effects of oral supplementation of brown seaweed derived phenolic compounds on inflammation-related parameters.

Compound/Format	Effective Concentrations	Algal Source	Experimental Model	Inflammatory Phase [1]	Significant Findings [2]	Reference
(poly)phenol-rich seaweed extract	400 mg; 8 weeks	*Ascophyllum nodosum*	80 participants; 30–65 years old with a body mass index (in kg/m^2) $\geq$ 25.	Steady state	$\downarrow$ DNA damage in obese people No significant differences in CRP, antioxidant status and inflammatory cytokines	[7]
MeOH–EtAc extract => phlorotannin	100–200 mg/kg/day 3 days before experiment	*Eisenia bicyclis* = *Ecklonia bicyclis*	Sprague Dawley rat, platelet aggregation + activation	Steady state	Eisenia bicyclis extract strongly inhibits in vivo platelet aggregation. Mechanism in vitro: downregulated ADP-induced platelet activation (Ca-mobilization, fibrinogen binding, granule release—mediated via decreased Src, PI3K, PLCgamma2, MAPK signaling)	[122]
MeOH extract=> Eckol	25, 50, 100 µg/mL, pretreatment 2 h prior to anaphylaxis induction 25, 50, 100 µg/mL, 24 h	*Ecklonia cava*	BALB/c mice; anti-DNP-IgE induced PCA C57BL/6 male mice; anti-DNP-IgE and IgE-BSA induced allergic	allergy	$\uparrow$ binding of IgE to FcϵRI $\downarrow$ FcϵRI expression $\downarrow$ mRNA level of IL-1β, IL-6, TNF-α, IFN-γ $\downarrow$ NF-κB nuclear translocation $\downarrow$ IκB degradation $\downarrow$ secretion of inflammatory mediators, such as histamine, β-hexosaminidase, leukotrienes and prostaglandins $\downarrow$ production of Th2-type cytokines, such as IL-4, IL-5, IL-13 (not 25 ug/mL)	[123]
MeOH-ethyl acetate extract (crude (72–74%) phlorotannins = phloroglucinol oligomer);	oral 1$\times$ or 2$\times$; 0.1, 1 mg/ear	*Ecklonia kurome* *Ecklonia arborea*	male ICR mouse; AA induced ear oedema	allergy	inhibitory effects are comparable to known anti-allergic agents presumed mechanisms: inhibition of MC degranulation + inhibition of COX-2 and LOX, and to lesser extent PLA2, activities	[125]

Table A4. *Cont.*

Compound/Format	Effective Concentrations	Algal Source	Experimental Model	Inflammatory Phase [1]	Significant Findings [2]	Reference
(poly)phenol-rich brown algae extract	100 mg/kg 500 mg/kg; 12 weeks	*Ecklonia cava*	C57BL/6 male mice; HFD+/− seaweed supplement	1,3	↓ MCP1 ↓ TNF-α ↓ IL-1 ↓ COX-2 (slightly decreased) protein level ↓ NF-κB protein level↓ Body weight ↓ adipose tissue weight ↓ deposition in the liver ↓ TG and TC levels ↓ leptin and adiponectin ratio ↓ Glucose Tolerance Test ↓ Insulin Resistance ↑ protein levels of P-AMPK ↑ SIRT1 protein levels ↑ PGC1α protein levels in the nucleus	[128]
A single compound Pyrogallol-Phloroglucinol-6,6-Bieckol from the brown algae ethanoic extract	2 mg/kg; 4 weeks	*Ecklonia cava*	C57Bl/6N mice; HFD $^{+/-}$ seaweed supplement	1,2,3	In visceral fat: ↓ TNF-a mRNA levels ↓ IL-1β mRNA levels ↓ Macrophage infiltration ↓ M1/M2 ratio (CD86 and CD80 lower, CD163 and CD206 higher) ↓ RAGE ↓ RAGE-RAGE Ligand Bonding ↓ Body weight ↓ size of visceral adipocytes ↓ fat mass ↓ serum TG and TC ↓ AGEs, HMGB1 and S100beta secret by adipocytes	[137]

Table A4. *Cont.*

Compound/Format	Effective Concentrations	Algal Source	Experimental Model	Inflammatory Phase [1]	Significant Findings [2]	Reference
Polyphenols-rich Ethanol extract	200, 400 mg/kg; 7 days	*Sargassum horneri*	BALB/c female mice; PM induced airway inflammation in allergic asthma	1,2	↓ granulocyte infiltration ↓ macrophage infiltration ↓ TLRs expression in lungs ↓ NF-κB pathway activation ↓ pro-inflammatory cytokines IL-1β, TNF-α, IL-6 expression in lung↑ IL-10 expression in lung ↓ IL-25, IL-33, IL-8, TGF-β in lung ↓ cytokines (IL-25, IL-33) in serum	[124]
phlorotannins-rich extract	PREC (75 and 150) or dieckol (50 and 100 mg/kg; 4 weeks Combination cisplatin (1, 3 or 5 μM) with PREC (35, 50, 75 or 100 μg/mL); 48 h	*Ecklonia cava*	BALB/c athymic female nude mice; SKOV3 cells induced ovarian carcinoma	1	PREC: ↓ Akt in ovarian cancer ↓ NF-κB in ovarian cancer ↑ intracellular ROS in ovarian cancer ↓ cisplatin-induced ROS in normal HEK293 ↑ ROS production in SKOV3 cells the combined treatment of PREC and cisplatin: ↓ ROS production in in SKOV3 cells and normal HEK 293cells Dieckol: ↑pathwayscisplatin-induced ROS production Dieckol: ↓pathwayscisplatin-induced ROS production (slightly)	[129]

Table A4. *Cont.*

Compound/Format	Effective Concentrations	Algal Source	Experimental Model	Inflammatory Phase [1]	Significant Findings [2]	Reference
phlorotannin derived from the brown alga	0.5 and 1.0 mg/kg/day; Pretreatment for 7 days	*Ecklonia stolonifera*	Male Kunming mice; CCl4 induced acute liver injury	1,2	↓ TNF-a ↓ IL-1β ↓ IL-6 ↓ ROS ↑ IL-10 ↑ CD11c+	[135]
Ethyl acetate fraction containing diphlorethohydroxy-carmalol	6.25, 12.5, 25 μg/mL	*Ishige okamurae*	zebrafish embryo; FD induced ROS, NO production and cell death	1,2	↓ ROS production ↓ NO production	[134]
polyphenol-rich brown algae extract	25, 50 and 100 μg/mL; pretreatment for 1 h	*Ecklonia cava*	Zebrafish embryos	1,2	↓ ROS expression ↓ NO expression ↓ iNOS	[133]
chloroform–methanol extract Apo-9′-fucoxanthinone (AF)	25, 50 and 100 μg/mL; pretreatment for 1 h	*Sargassum muticum*	Zebrafish embryos	1,2	↓ IL-1β, ↓ TNF-α ↓ ROS ↓ COX-2 ↓ NF-κB ↓ MAPK ↓ iNOS ↓ NO	[131]
Six purified phlorotannins ((eckol; 6,60-bieckol; 6,80-bieckol; 8,80-bieckol; phlorofucofuroeckol (PFF)-A and PFF-B)	10, 100, 200 μM	*Eisenia arborea*	ICR mice; AA, TPA and OXA induced ear swelling	1	↓ COX-2 mRNA expression ↓ COX-2 enzymatic activity	[136]

Table A4. *Cont.*

Compound/Format	Effective Concentrations	Algal Source	Experimental Model	Inflammatory Phase [1]	Significant Findings [2]	Reference
Ethanol extract including Dieckol, 2,7-phloroglucinol-6,6-bieckol (PHB), PFF-A and pyrogallol-phloroglucinol-6,6′-bieckol (PPB)	2.5 mg/kg; 4 weeks	*Ecklonia cava*	57BL/6N mice; $HFD^{+/-}$ seaweed supplement	1,2,3	↓ TNF-α ↓ IL-6 ↓ TLR4 expression ↓ NF-kB expression ↓ CD11b ↓ CD86 ↑ CD206 ↓ body weight ↓ food intake ↓ fat mass ↓ leptin resistance ↑ leptin sensitivity ↓ PPAR expression ↓ CEBP expression ↓ FAS expression ↓ ACC expression ↓ ER	[126]
Ethanolic extract including phlorotannin	70 mg/kg; 4 weeks	*Ecklonia cava*	male 57BL/6N mice, $HFD^{+/-}$ seaweed supplement	1,2,3	↓ NF-kB ↓ TNF-α ↓ IL-6 ↓ TLR4 expression ↓ CD4 and CD8a ↓ CD11b ↑ CD206 ↓ CD80 ↑ IL-10 ↑ TGF-β mRNA ↓ Body weight ↓ Food intake ↓ Adipocyte size ↓ Leptin resistance ↓ RAGE	[127]

Table A4. *Cont.*

Compound/Format	Effective Concentrations	Algal Source	Experimental Model	Inflammatory Phase [1]	Significant Findings [2]	Reference
phlorotannins-rich ethanolic extract	100, 200, 400 mg/kg/day; 8 weeks	*Ecklonia cava*	Male Sprague Dawley rats; Experimental periodontitis induced by placing a sterile 4-0 silk ligature around the gingival cervix of the right mandibular second molar teeth	1,2	↓ COX-2 activity ↓ PGE2 ↓ NO ↓ iNOS	[130]
Ethanolic extract and PPB	ECE: 70 mg/kg PPB: 2.5 mg/kg; 4 weeks	*Ecklonia cava*	Male C57BL/6N mice; $HFD^{+/-}$ seaweed supplement	1,2,3	↓ MCP-1 ↓ TNF-α ↓ IL-6 ↓ IL-10 ↓ iNOS ↓ CD80 ↑ CD206 ↓ ER stress ↓ size of white adipocytes ↑ PPAR expression	[138]

Table A4. *Cont.*

Compound/Format	Effective Concentrations	Algal Source	Experimental Model	Inflammatory Phase [1]	Significant Findings [2]	Reference
baicalein, luteolin, rosmarinic acid and cis-2-decenoic acid (10-HAD)	Test diet for mice:(baicalein (650 mg/kg diet), luteolin (300 mg/kg diet), rosmarinic acid (500 mg/kg diet), monolaurin (500 mg/kg diet), 10-HAD (500 mg/kg diet)); 4 weeks Capsule for human: (baicalein 250 mg/day, luteolin 75 mg/day, rosmarinic acid 100 mg/day, monolaurin 250 mg/day, 10-HAD 100 mg/day and iodine 0.15 mg/day); 3 times/day 6 months	Unknown	C3H/HeN inbred female mice; LD infection $^{+/-}$ test diet Adult human; with a history of acute LD	1,2	Animal study: ↓ IL-6 ↓ TNF-α ↓ INF-γ Human study: By administration of the composition, 17.7% had slight physical and psychological improvement, and 17.7% were none responsive ↓ IL-17	[139]
commercial (30%) polyphenolic fraction (VNP)oral admin	Rat: 1 dose 2000 mg/kg Human: 2400 mg/day 8 weeks	*Sargassum fusiformi*, *Ecklonia kurome*, *Ecklonia Stolonifera*, *Eisenia bicyclis* and *Ecklonia cava*	Sprague Dawley rat human (men)	1	rat: serum ferric reducing antioxidant power (FRAP) significantly elevated 30 min after treatment, but declined quickly thereafterhuman: ↑ erectile function (significantly) =usefulness of these polyphenolic compounds as chemo preventive agents against vascular risk factors originating from oxidative stress	[132]

Table A4. *Cont.*

Compound/Format	Effective Concentrations	Algal Source	Experimental Model	Inflammatory Phase [1]	Significant Findings [2]	Reference
Compound 21	35 mg/kg; 7 days	Synthesized by the department of pharmaceutical chemistry	Female Lewis rats; MBP-induced EAE	3	Attenuates neurological deficits, immune infiltration and demyelination in EAE rats Reduces the population of Th1/Th17 cells and inhibits their infiltration into the CNS	[142]
phloroglucinol derivative Compound 21	1, 5, 10, 20, 50 mg/kg; 4 weeks	Unknown	Cuprizone induced intoxication in mice,	3	significantly improved neurological dysfunction and motor coordination impairment decreased microglia and astrocytes activities and the subsequent neuro-inflammatory response	[141]
EtOAc Fraction of seaweed Crude Extract	200 mg/kg; 8 weeks	*Ecklonia* cava from Jeju or Gijang	C57BL/6 mice; HFD+/− seaweed supplement	1,3	↓ TNF-a mRNA levels ↓ IL-1ß mRNA levels↓ body weights ↓ weight gain ↓ fat tissue mass ↓ Plasma ALT and cholesterol levels ↑ PPARγ2 mRNA expression ↑ C/EBPα, mRNA ↓ Blood glucose levels	[140]

[1] Phases of inflammation as indicated in Table 1: 1-initiation of inflammation; 2-short term amplification and regulation; 3-short- or long-term consequences; [2] Symbols used: ↑—increase; ↓—decrease; =—no change.

Appendix E

Table A5. Effects of oral supplementation of brown seaweed derived Fucoxanthin(ol) and Meroterpenoid on inflammation-related parameters.

Compound/Format	Effective Concentrations	Algal Source	Experimental Model	Inflammatory Phase [1]	Significant Findings [2]	Reference
Fucoxanthin-containing extract	13, 26 and 65 mg/kg Fx for 4 weeks	*Laminaria japonica*	Male Sprague Dawley rats (5 weeks old; ±200 g), diabetes induced by i.p. administration of streptozotocin followed by nicotinamide	1,3	↓ TNF-α mRNA expression in plasma and testis ↓ IL-6 mRNA expression in plasma and testis ↑ CAT, SOD and GPx activity in plasma ↑ SOD activity in testis ↓ H_2O_2 and O_2^- levels ↓ MDA level in plasma, testis and sperm ↓ Plasma glucose level ↓ Insulin level ↓ HOMA-IR ↓ SOCS-3 mRNA expression in hypothalamus	[150]
Fucoxanthin-rich brown algae extract	Colitis: 1, 2 or 5 g/kg bw/day for 7 days; CACC: 0.5, 1 or 2.5 g/kg bw/day for 11 weeks	*Sargassum muticum*	Male BALB/c mice (6–8 weeks old), 3% DSS-induced colitis for 14 days or CACC induction by a single i.p. injection of azoxymethane + 2% DSS for 7 consecutive days at weeks 3, 6 and 9	1,2	DSS-induced colitis: ↓ MDA level in colonic tissue=MDA level in plasma ↑ SOD levels in plasma ↓ TNF-α in colonic tissue and plasma ↓ IL-6 in plasma =IL-6 in colonic tissue ↓ Total NO content in colonic tissue↓ NO release in colonic tissueCACC: ↓ MDA level in colonic tissue=MDA level in plasma ↑ SOD levels in plasma ↓ TNF-α in colonic tissue and plasma =IL-6 in colonic tissue and plasma ↑ proliferation T and B cells ↓ Total NO content in colonic tissue↓ NO release in colonic tissue	[151]

Table A5. *Cont.*

Compound/Format	Effective Concentrations	Algal Source	Experimental Model	Inflammatory Phase [1]	Significant Findings [2]	Reference
Fucoxanthin-rich extract	100, 200 or 500 mg/kg Fx, dissolved in olive oil and administered for 5 days	*Sargassum glaucescens*	Male Syrian hamsters (7 weeks old; 80–90 g), Cisplatin induced testicular damage group with CP i.p. injection before Fx treatment	1,2,3	↑ CAT and GPx activity in testis ↑ SOD activity in plasma ↓ MDA level in testicular tissue, sperm and plasma ↓ NO levels =Body weight	[152]
Fucoxanthin	0.4% and 0.6% Fx for 5 weeks	Unknown	Kunming strain mice (20–22 g) fed regular chow or HFD for 9 weeks + 5 weeks with Fx added to diet	1,2,3	↓ Mammary gland inflammation↓ MPO activity ↓ IL-1β in the blood ↓ TNF-α in the blood ↓ MDA level ↓ COX-2 and iNOS mRNA expression ↓ Body weight gain	[155]
Fucoxanthin-rich wakame lipids (WLs)	WLs with 1.06% Fx; WLs with 2.22% Fx, administered for 5 weeks	*Undaria pinnatifida*	Male C57BL/6J mice (8 weeks old), normal-fat or HFD for 10 weeks followed by normal-fat or HFD with Fx for 5 weeks	2,3	↓ MCP-1 mRNA expression in WAT =TNF-alfa mRNA expression in WAT ↓ Body weight gain =Food intake ↓ WAT weight gain ↓ LDL cholesterol ↓ Plasma leptin ↓ Plasma insulin ↓ blood glucose ↓ Leptin mRNA expression in WAT ↑ ADRB3 mRNA expression in WAT ↑ GLUT4 mRNA expression in skeletal muscle	[156]
Fucoxanthin and fucoxanthinol	150 nmol/mouse, before induction of ear swelling	Unknown	Male ICR-strain mice (4 weeks old), AA-, TPA- or OXA-induced ear swelling	1	↓ Inflammation in mouse ear swelling ↓ PLA2 and COX-2 enzymatic activities	[157]

Table A5. *Cont.*

Compound/Format	Effective Concentrations	Algal Source	Experimental Model	Inflammatory Phase [1]	Significant Findings [2]	Reference
Lipid extracts containing fucoxanthin	0.10% Fx, 1 g Fx/kg of diet for 27 days	*Undaria pinnatifida*	Male KK-Ay mice (3 weeks old), fed an experimental diet with or without Fx	1,2,3	↓ TNF-α mRNA expression in WAT ↓ MCP-1 mRNA expression in WAT ↓ Blood glucose level ↑ Glucose intolerance	[145]
Meroterpenoid	90 and 180 mg MES/kg BW for 10 weeks	*Sargassum serratifolium*	Male C57BL/6J mice (6-weeks old), HCD	2	↓ MCP-1, KC concentration in serum ↓ COX-2, ICAM-1, VCAM-1, MMP-9, MCP-1, beta-actin expression in aortic tissues	[148]
Meroterpenoid	60 and 120 mg MES/kg BW/day for 8 weeks	*Sargassum serratifolium*	Male C57BL/6J mice (7-weeks old), HFD-induced obesity	2,3	↓ F4/80, MCP-1 expression in epididymal tissue ↓ Body, liver and epididymal issue weight ↓ ALT, AST ↓ TG, glucose, free fatty acid in plasma ↑ HDL cholesterol ↑ UCP-1, ADRB3 expression in subcutaneous fat	[158]
Fucoxanthin and fucoidan	275 mg LMF and 275 mg HSFx in 1 capsule, 3 capsules twice daily for 12 weeks	Unknown	Patients with nonalcoholic fatty liver disease (20–75 y/o)	3	↓ BMI ↓ ALT =liver steatosis (=CAP) =adiponectin =fasting insulin =insulin resistance	[159]

[1] Phases of inflammation as indicated in Table 1: 1-initiation of inflammation; 2-short term amplification and regulation; 3-short- or long-term consequences; [2] Symbols used: ↑—increase; ↓—decrease; =—no change.

References

1. Cherry, P.; O'Hara, C.; Magee, P.; McSorley, E.M.; Allsopp, P.J. Risks and benefits of consuming edible seaweeds. *Nutr. Rev.* **2019**, *77*, 307–329. [CrossRef] [PubMed]
2. Bramer, W.; Milic, J.; Mast, F. Reviewing retrieved references for inclusion in systematic reviews using EndNote. *J. Med. Libr. Assoc.* **2017**, *105*, 84–87. [CrossRef]
3. Jeong, D.-H.; Kim, K.-B.; Kim, M.-J.; Kang, B.-K.; Ahn, D.-H. Anti-inflammatory activity of ethanolic extract of *Sargassum micracanthum*. *J. Microbiol. Biotechnol.* **2013**, *23*, 1691–1698. [CrossRef]
4. Güner, A.; Köksal, C.; Erel, B.; Kayalar, H.; Nalbantsoy, A.; Sukatar, A.; Yavaşoğlu, N.K. Antimicrobial and antioxidant activities with acute toxicity, cytotoxicity and mutagenicity of *Cystoseira compressa* (Esper) Gerloff & Nizamuddin from the coast of Urla (Izmir, Turkey). *Cytotechnology* **2013**, *67*, 135–143. [CrossRef] [PubMed]
5. Zaragozá, M.C.; López, D.; Sáiz, M.P.; Poquet, M.; Pérez, J.; Puig-Parellada, P.; Màrmol, F.; Simonetti, P.; Gardana, C.; Lerat, Y.; et al. Toxicity and Antioxidant Activity in Vitro and in Vivo of Two Fucus vesiculosus Extracts. *J. Agric. Food Chem.* **2008**, *56*, 7773–7780. [CrossRef] [PubMed]
6. Park, S.Y.; Seo, I.S.; Lee, S.J.; Lee, S.P. Study on the Health Benefits of Brown Algae (*Sargassum muticum*) in Volunteers. *J. Food Nutr. Res.* **2015**, *3*, 126–130. [CrossRef]
7. Baldrick, F.R.; McFadden, K.; Ibars, M.; Sung, C.; Moffatt, T.; Megarry, K.; Thomas, K.; Mitchell, P.; Wallace, J.M.W.; Pourshahidi, L.K.; et al. Impact of a (poly)phenol-rich extract from the brown algae *Ascophyllum nodosum* on DNA damage and antioxidant activity in an overweight or obese population: A randomized controlled trial. *Am. J. Clin. Nutr.* **2018**, *108*, 688–700. [CrossRef]
8. Kannan, G.; Saker, K.; Terrill, T.; Kouakou, B.; Galipalli, S.; Gelaye, S. Effect of seaweed extract supplementation in goats exposed to simulated preslaughter stress. *Small Rumin. Res.* **2007**, *73*, 221–227. [CrossRef]
9. Bai, J.; Wang, R.; Yan, L.; Feng, J. Co-Supplementation of Dietary Seaweed Powder and Antibacterial Peptides Improves Broiler Growth Performance and Immune Function. *Braz. J. Poult. Sci.* **2019**, *21*. [CrossRef]
10. Baleta, F.N.; Bolanos, J.M. Growth and immune response of *Pangasius hypophthalmus* fed diets containing seaweed extracts as immunostimulant. *Braz. Arch. Biol. Technol.* **2019**, *62*. [CrossRef]
11. Zeynali, M.; Bahabadi, M.N.; Morshedi, V.; Ghasemi, A.; Mozanzadeh, M.T. Replacement of dietary fishmeal with *Sargassum ilicifolium* meal on growth, innate immunity and immune gene mRNA transcript abundance in *Lates calcarifer* juveniles. *Aquacult. Nutr.* **2020**, *26*, 1657–1668. [CrossRef]
12. Yeganeh, S.; Adel, M. Effects of dietary algae (*Sargassum ilicifolium*) as immunomodulator and growth promoter of juvenile great sturgeon (*Huso huso* Linnaeus, 1758). *J. Appl. Phycol.* **2018**, *31*, 2093–2102. [CrossRef]
13. Zeraatpisheh, F.; Firouzbakhsh, F.; Khalili, K.J. Effects of the macroalga *Sargassum angustifolium* hot water extract on hematological parameters and immune responses in rainbow trout (*Oncohrynchus mykiss*) infected with Yersinia rukeri. *Environ. Boil. Fishes* **2018**, *30*, 2029–2037. [CrossRef]
14. Sugiura, Y.; Matsuda, K.; Okamoto, T.; Kakinuma, M.; Amano, H. Anti-allergic effects of the brown alga *Eisenia arborea* on Brown Norway rats. *Fish. Sci.* **2008**, *74*, 180–186. [CrossRef]
15. Yoshioka, H.; Ishida, M.; Nishi, K.; Oda, H.; Toyohara, H.; Sugahara, T. Studies on anti-allergic activity of *Sargassum horneri* extract. *J. Funct. Foods* **2014**, *10*, 154–160. [CrossRef]
16. Han, E.J.; Fernando, I.P.S.; Kim, H.-S.; Jeon, Y.-J.; Madusanka, D.M.D.; Dias, M.K.H.M.; Jee, Y.; Ahn, G. Oral Administration of *Sargassum horneri* Improves the HDM/DNCB-Induced Atopic Dermatitis in NC/Nga Mice. *Nutr.* **2020**, *12*, 2482. [CrossRef]
17. Kim, O.-K.; Lee, M.; Kwon, H.O.; Lee, D.; Park, J.; Kim, E.; You, Y.; Lim, Y.T.; Jun, W.; Lee, J. Costaria costata Extract Suppresses Development of Atopic Dermatitis in chloro-2,4-dinitrobenzene-treated NC/Nga Mice. *Skin Pharmacol. Physiol.* **2018**, *31*, 212–219. [CrossRef] [PubMed]
18. Hwang, Y.-H.; Song, H.-K.; Lee, A.; Ha, H.; Kim, T. *Laminaria japonica* Suppresses the Atopic Dermatitis-Like Responses in NC/Nga Mice and Inflamed HaCaT Keratinocytes via the Downregulation of STAT1. *Nutrients* **2020**, *12*, 3238. [CrossRef] [PubMed]
19. Park, S.K.; Park, S.J.; Park, S.M.; Cho, I.J.; Park, C.I.; Kim, Y.W.; Kim, S.C. Inhibition of acute phase inflammation by *Laminaria japonica* through regulation of inos-Nf- B pathway. *Evid. Based Complement Altern. Med.* **2013**, *2013*. [CrossRef] [PubMed]
20. Sugiura, Y.; Kinoshita, Y.; Abe, M.; Murase, N.; Tanaka, R.; Matsushita, T.; Usui, M.; Hanaoka, K.-I.; Miyata, M. Suppressive effects of the diethyl ether fraction from a brown alga *Sargassum fusiforme* on allergic and inflammatory reactions. *Fish. Sci.* **2016**, *82*, 369–377. [CrossRef]
21. Namkoong, S.; Kang, S.-C.; Do, H.; Jang, K.-H.; Jang, S.-A.; Choung, M.-G.; Sohn, E.-H. Immunomodulatory Effects of Supplementation with Extracts from the Marine Brown Alga *Eisenia bicyclis* on Macrophages. *Korean J. Plant Resour.* **2011**, *24*, 298–303. [CrossRef]
22. Park, S.Y.; Hwang, E.; Shin, Y.-K.; Lee, D.-G.; Yang, J.-E.; Park, J.-H.; Yi, T.-H. Immunostimulatory effect of enzyme-modified *Hizikia fusiforme* in a mouse model in vitro and ex vivo. *Mar. Biotechnol.* **2017**, *19*, 65–75. [CrossRef]
23. Joung, E.J.; Gwon, W.-G.; Shin, T.-S.; Jung, B.-M.; Choi, J.; Kim, H.-R. Anti-inflammatory action of the ethanolic extract from *Sargassum serratifolium* on lipopolysaccharide-stimulated mouse peritoneal macrophages and identification of active components. *J. Appl. Phycol.* **2017**, *29*, 563–573. [CrossRef]

24. Fernando, I.P.S.; Sanjeewa, K.K.A.; Samarakoon, K.W.; Lee, W.W.; Kim, H.-S.; Ranasinghe, P.; Gunasekara, U.K.D.S.S.; Jeon, Y.-J. Antioxidant and anti-inflammatory functionality of ten Sri Lankan seaweed extracts obtained by carbohydrase assisted extraction. *Food Sci. Biotechnol.* **2018**, *27*, 1761–1769. [CrossRef] [PubMed]
25. Ramadan, G.; Fouda, W.A.; Ellamie, A.M.; Ibrahim, W.M. Dietary supplementation of *Sargassum latifolium* modulates thermo-respiratory response, inflammation, and oxidative stress in bacterial endotoxin-challenged male Barki sheep. *Environ. Sci. Pollut. Res.* **2020**, *27*, 33863–33871. [CrossRef] [PubMed]
26. Ellamie, A.M.; Fouda, W.A.; Ibrahim, W.M.; Ramadan, G. Dietary supplementation of brown seaweed (*Sargassum latifolium*) alleviates the environmental heat stress-induced toxicity in male Barki sheep (*Ovis aries*). *J. Therm. Biol.* **2020**, *89*, 102561. [CrossRef]
27. Saker, K.E.; Fike, J.H.; Veit, H.; Ward, D.L. Brown seaweed- (TascoTM) treated conserved forage enhances antioxidant status and immune function in heat-stressed wether lambs. *J. Anim. Physiol. Anim. Nutr.* **2004**, *88*, 122–130. [CrossRef] [PubMed]
28. Kim, N.-H.; Lee, S.M.; Na Kim, Y.; Jeon, Y.-J.; Heo, J.-D.; Jeong, E.J.; Rho, J.-R. Standardized Fraction of *Turbinaria ornata* Alleviates Dextran Sulfate Sodium-Induced Chronic Colitis in C57BL/6 Mice via Upregulation of FOXP3$^+$ Regulatory T Cells. *Biomolecules* **2020**, *10*, 1463. [CrossRef]
29. Ko, S.J.; Bu, Y.; Bae, J.; Bang, Y.-m.; Kim, J.; Lee, H.; Lee, B.-J.; Hyun, Y.H.; Park, J.-W. Protective effect of Laminaria japonica with probiotics on murine colitis. *Mediat. Inflamm.* **2014**, *2014*, 417814. [CrossRef]
30. Jeon, H.; Yoon, W.-J.; Ham, Y.-M.; Yoon, S.-A.; Kang, S.C. Anti-Arthritis Effect through the Anti-Inflammatory Effect of Sargassum muticum Extract in Collagen-Induced Arthritic (CIA) Mice. *Molecules* **2019**, *24*, 276. [CrossRef]
31. Cooper, R.; Dragar, C.; Elliot, K.; Fitton, J.; Godwin, J.; Thompson, K. GFS, a preparation of Tasmanian *Undaria pinnatifida* is associated with healing and inhibition of reactivation of Herpes. *BMC Complement. Altern. Med.* **2002**, *2*, 11. [CrossRef]
32. Imai, T.; Takahashi, Y. Chemotaxis Assay for *Marsupenaeus japonicas* Hemocytes and Application for the Development of an Oral Immunostimulant Against White Spot Syndrome Virus. *Front. Cell Dev. Biol.* **2020**, *8*, 46. [CrossRef] [PubMed]
33. Lee, D.-G.; Shin, Y.-K.; Park, J.-H.; Park, S.-Y.; Hwang, E.; Yang, J.-E.; Jo, H.; Kim, K.-Y.; Mavlonov, G.T.; Yi, T.-H. Alveolar Bone Protective Effect of Hiziki Extracts on the Progression of Periodontitis. *Mar. Biotechnol.* **2018**, *20*, 313–323. [CrossRef]
34. Song, X.-H.; Zan, R.-Z.; Yu, C.-H.; Wang, F. Effects of modified Haizao Yuhu Decoction in experimental autoimmune thyroiditis rats. *J. Ethnopharmacol.* **2011**, *135*, 321–324. [CrossRef] [PubMed]
35. Wu, H.; Ballantyne, C.M. Metabolic Inflammation and Insulin Resistance in Obesity. *Circ. Res.* **2020**, *126*, 1549–1564. [CrossRef]
36. Oh, J.-H.; Kim, J.; Lee, Y. Anti-inflammatory and anti-diabetic effects of brown seaweeds in high-fat diet-induced obese mice. *Nutr. Res. Pract.* **2016**, *10*, 42–48. [CrossRef] [PubMed]
37. Martínez-Villaluenga, C.; Peñas, E.; Rico, D.; Martin-Diana, A.B.; Portillo, M.P.; Macarulla, M.T.; De Luis, D.A.; Miranda, J. Potential Usefulness of a Wakame/Carob Functional Snack for the Treatment of Several Aspects of Metabolic Syndrome: From In Vitro to In Vivo Studies. *Mar. Drugs* **2018**, *16*, 512. [CrossRef]
38. Mendez, R.L.; Miranda, C.; Armour, C.R.; Sharpton, T.J.; Stevens, J.F.; Kwon, J.Y. Supplementation with Sea Vegetables *Palmaria mollis* and *Undaria pinnatifida* Exerts Metabolic Benefits in Diet-Induced Obesity in Mice. *Curr. Dev. Nutr.* **2020**, *4*, nzaa072. [CrossRef] [PubMed]
39. Yang, H.; Lee, S.Y.; Lee, S.R.; Pyun, B.-J.; Kim, H.J.; Lee, Y.H.; Kwon, S.W.; Suh, D.H.; Lee, C.H.; Hong, E.-J.; et al. Therapeutic Effect of Ecklonia cava Extract in Letrozole-Induced Polycystic Ovary Syndrome Rats. *Front. Pharmacol.* **2018**, *9*. [CrossRef]
40. Berteau, O.; Mulloy, B. Sulfated fucans, fresh perspectives: Structures, functions, and biological properties of sulfated fucans and an overview of enzymes active toward this class of polysaccharide. *Glycobiology* **2003**, *13*, 29R–40R. [CrossRef]
41. Holdt, S.L.; Kraan, S. Bioactive compounds in seaweed: Functional food applications and legislation. *Environ. Boil. Fishes* **2011**, *23*, 543–597. [CrossRef]
42. Van Weelden, G.; Bobinski, M.; Okla, K.; van Weelden, W.J.; Romano, A.; Pijnenborg, A.M.A. Fucoidan Structure and Activity in Relation to Anti-Cancer Mechanisms. *Mar. Drugs* **2019**, *17*, 32. [CrossRef] [PubMed]
43. Laurienzo, P. Marine Polysaccharides in Pharmaceutical Applications: An Overview. *Mar. Drugs* **2010**, *8*, 2435–2465. [CrossRef]
44. Zhao, X.; Xue, C.-H.; Li, B.-F. Study of antioxidant activities of sulfated polysaccharides from *Laminaria japonica*. *J. Appl. Phycol.* **2007**, *20*, 431–436. [CrossRef]
45. Fitton, J.H.; Stringer, D.N.; Karpiniec, S.S. Therapies from Fucoidan: An Update. *Mar. Drugs* **2015**, *13*, 5920–5946. [CrossRef] [PubMed]
46. Nagamine, T.; Nakazato, K.; Tomioka, S.; Iha, M.; Nakajima, K. Intestinal Absorption of Fucoidan Extracted from the Brown Seaweed, *Cladosiphon okamuranus*. *Mar. Drugs* **2014**, *13*, 48–64. [CrossRef]
47. Shang, Q.; Song, G.; Zhang, M.; Shi, J.; Xu, C.; Hao, J.; Li, G.; Yu, G. Dietary fucoidan improves metabolic syndrome in association with increased *Akkermansia* population in the gut microbiota of high-fat diet-fed mice. *J. Funct. Foods* **2017**, *28*, 138–146. [CrossRef]
48. Kadena, K.; Tomori, M.; Iha, M.; Nagamine, T. Absorption Study of Mozuku Fucoidan in Japanese Volunteers. *Mar. Drugs* **2018**, *16*, 254. [CrossRef]
49. Iraha, A. Fucoidan enhances intestinal barrier function by upregulating the expression of claudin-1. *World J. Gastroenterol.* **2013**, *19*, 5500–5507. [CrossRef]
50. Chauvierre, C.; Aid-Launais, R.; Aerts, J.; Chaubet, F.; Maire, M.; Chollet, L.; Rolland, L.; Bonafé, R.; Rossi, S.; Bussi, S.; et al. Pharmaceutical Development and Safety Evaluation of a GMP-Grade Fucoidan for Molecular Diagnosis of Cardiovascular Diseases. *Mar. Drugs* **2019**, *17*, 699. [CrossRef]

51. Li, N.; Zhang, Q.; Song, J. Toxicological evaluation of fucoidan extracted from *Laminaria japonica* in Wistar rats. *Food Chem. Toxicol.* **2005**, *43*, 421–426. [CrossRef] [PubMed]
52. Mori, N.; Takeda, K.; Tomimori, K.; Kimura, R.; Ishikawa, C.; Nowling, T.K. Anti-tumor activity of fucoidan is mediated by nitric oxide released from macrophages. *Int. J. Oncol.* **2011**, *40*, 251–260. [CrossRef]
53. Jayawardena, T.U.; Fernando, I.P.S.; Lee, W.W.; Sanjeewa, K.K.A.; Kim, H.-S.; Lee, D.-S.; Jeon, Y.-J. Isolation and purification of fucoidan fraction in *Turbinaria ornata* from the Maldives; Inflammation inhibitory potential under LPS stimulated conditions in in-vitro and in-vivo models. *Int. J. Biol. Macromol.* **2019**, *131*, 614–623. [CrossRef] [PubMed]
54. Ni, L.; Wang, L.; Fu, X.; Duan, D.; Jeon, Y.-J.; Xu, J.; Gao, X. In vitro and in vivo anti-inflammatory activities of a fucose-rich fucoidan isolated from *Saccharina japonica*. *Int. J. Biol. Macromol.* **2020**, *156*, 717–729. [CrossRef]
55. Park, J.; Cha, J.-D.; Choi, K.-M.; Lee, K.-Y.; Han, K.M.; Jang, Y.-S. Fucoidan inhibits LPS-induced inflammation in vitro and during the acute response in vivo. *Int. Immunopharmacol.* **2017**, *43*, 91–98. [CrossRef]
56. Sanjeewa, K.K.A.; Fernando, I.; Kim, S.-Y.; Kim, H.-S.; Ahn, G.; Jee, Y.; Jeon, Y.-J. In vitro and in vivo anti-inflammatory activities of high molecular weight sulfated polysaccharide; containing fucose separated from *Sargassum horneri*: Short communication. *Int. J. Biol. Macromol.* **2018**, *107*, 803–807. [CrossRef] [PubMed]
57. Herath, K.H.I.N.M.; Kim, H.J.; Kim, A.; Sook, C.E.; Lee, B.-Y.; Jee, Y. The Role of Fucoidans Isolated from the Sporophylls of *Undaria pinnatifida* against Particulate-Matter-Induced Allergic Airway Inflammation: Evidence of the Attenuation of Oxidative Stress and Inflammatory Responses. *Molecules* **2020**, *25*, 2869. [CrossRef]
58. Lin, R.; Liu, X.; Meng, Y.; Xu, M.; Guo, J. Effects of *Laminaria japonica* polysaccharides on airway inflammation of lungs in an asthma mouse model. *Multidiscip. Respir. Med.* **2015**, *10*, 20. [CrossRef] [PubMed]
59. Tanino, Y.; Hashimoto, T.; Ojima, T.; Mizuno, M. F-fucoidan from *Saccharina japonica* is a novel inducer of galectin-9 and exhibits anti-allergic activity. *J. Clin. Biochem. Nutr.* **2016**, *59*, 25–30. [CrossRef]
60. Mizuno, M.; Sakaguchi, K.; Sakane, I. Oral Administration of Fucoidan Can Exert Anti-Allergic Activity after Allergen Sensitization by Enhancement of Galectin-9 Secretion in Blood. *Biomolecules* **2020**, *10*, 258. [CrossRef]
61. Hwang, P.-A.; Lin, H.-T.V.; Lin, H.-Y.; Lo, S.-K. Dietary Supplementation with Low-Molecular-Weight Fucoidan Enhances Innate and Adaptive Immune Responses and Protects against *Mycoplasma pneumoniae* Antigen Stimulation. *Mar. Drugs* **2019**, *17*, 175. [CrossRef] [PubMed]
62. Tomori, M.; Nagamine, T.; Miyamoto, T.; Iha, M. Evaluation of the Immunomodulatory Effects of Fucoidan Derived from *Cladosiphon Okamuranus* Tokida in Mice. *Mar. Drugs* **2019**, *17*, 547. [CrossRef]
63. Yu, B.; Bi, D.; Yao, L.; Li, T.; Gu, L.; Xu, H.; Li, X.; Li, H.; Hu, Z.-L.; Xu, X. The inhibitory activity of alginate against allergic reactions in an ovalbumin-induced mouse model. *Food Funct.* **2020**, *11*, 2704–2713. [CrossRef] [PubMed]
64. Leonard, S.G.; Sweeney, T.; Bahar, B.; O'Doherty, J.V. Effect of maternal seaweed extract supplementation on suckling piglet growth, humoral immunity, selected microflora, and immune response after an ex vivo lipopolysaccharide challenge1. *J. Anim. Sci.* **2012**, *90*, 505–514. [CrossRef]
65. Walsh, A.M.; Sweeney, T.; O'Shea, C.J.; Doyle, D.N.; O'Doherty, J.V. Effects of supplementing dietary laminarin and fucoidan on intestinal morphology and the immune gene expression in the weaned pig. *J. Anim. Sci.* **2012**, *90*, 284–286. [CrossRef] [PubMed]
66. Azizi, A.F.N.; Miyazaki, R.; Yumito, T.; Ohashi, Y.; Uno, S.; Miyajima, U.; Kumamoto, M.; Uchiyama, S.; Yasuda, M. Effect of maternal supplementation with seaweed powder on immune status of liver and lymphoid organs of piglets. *J. Vet. Med. Sci.* **2018**, *80*, 8–12. [CrossRef] [PubMed]
67. Heim, G.; O'Doherty, J.V.; O'Shea, C.; Doyle, D.N.; Egan, A.M.; Thornton, K.; Sweeney, T. Maternal supplementation of seaweed-derived polysaccharides improves intestinal health and immune status of suckling piglets. *J. Nutr. Sci.* **2015**, *4*, e27. [CrossRef] [PubMed]
68. McDonnell, R.P.; Doherty, J.V.O.; Earley, B.; Clarke, A.M.; Kenny, D.A. Effect of supplementation with n-3 polyunsaturated fatty acids and/or β-glucans on performance, feeding behaviour and immune status of Holstein Friesian bull calves during the pre- and post-weaning periods. *J. Anim. Sci. Biotechnol.* **2019**, *10*, 1–17. [CrossRef]
69. Yan, G.L.; Guo, Y.M.; Yuan, J.M.; Liu, D.; Zhang, B.K. Sodium alginate oligosaccharides from brown algae inhibit *Salmonella* Enteritidis colonization in broiler chickens. *Poult. Sci.* **2011**, *90*, 1441–1448. [CrossRef] [PubMed]
70. El-Boshy, M.; El-Ashram, A.; Risha, E.; Abdelhamid, F.; Zahran, E.; Gab-Alla, A. Dietary fucoidan enhance the non-specific immune response and disease resistance in African catfish, *Clarias gariepinus*, immunosuppressed by cadmium chloride. *Vet. Immunol. Immunopathol.* **2014**, *162*, 168–173. [CrossRef]
71. Mir, I.N.; Sahu, N.P.; Pal, A.K.; Makesh, M. Synergistic effect of l-methionine and fucoidan rich extract in eliciting growth and non-specific immune re-sponse of *Labeo rohita* fingerlings against *Aeromonas hydrophila*. *Aquaculture* **2017**, *479*, 396–403. [CrossRef]
72. Morales-Lange, B.; Bethke, J.; Schmitt, P.; Mercado, L. Phenotypical parameters as a tool to evaluate the immunostimulatory effects of laminarin in *Oncorhynchus mykiss*. *Aquac. Res.* **2014**, *46*, 2707–2715. [CrossRef]
73. Prabu, D.L.; Sahu, N.P.; Pal, A.K.; Dasgupta, S.; Narendra, A. Immunomodulation and interferon gamma gene expression in sutchi cat fish, *Pangasianodon hypophthalmus*: Effect of dietary fucoidan rich seaweed extract (FRSE) on pre and post challenge period. *Aquac. Res.* **2016**, *47*, 199–218. [CrossRef]
74. Setyawan, A.; Isnansetyo, A.; Murwantoko, M.; Soedarmanto, I.; Handayani, C.R. Comparative immune response of dietary fucoidan from three indonesian brown algae in white shrimp *Litopenaeus vannamei*. *AACL Bioflux* **2018**, *11*, 1707–1723.

75. Yang, Q.; Yang, R.; Li, M.; Zhou, Q.; Liang, X.; Elmada, Z.C. Effects of dietary fucoidan on the blood constituents, anti-oxidation and innate immunity of juvenile yellow catfish (*Pelteobagrus fulvidraco*). *Fish Shellfish Immunol.* **2014**, *41*, 264–270. [CrossRef] [PubMed]
76. Yin, G.; Li, W.; Lin, Q.; Lin, X.; Lin, J.; Zhu, Q.; Jiang, H.; Huang, Z. Dietary administration of laminarin improves the growth performance and immune responses in *Epinephelus coioides*. *Fish Shellfish Immunol.* **2014**, *41*, 402–406. [CrossRef]
77. Katayama, S.; Nishio, T.; Kishimura, H.; Saeki, H. Immunomodulatory Properties of Highly Viscous Polysaccharide Extract from the Gagome Alga (*Kjellmaniella crassifolia*). *Plant Foods Hum. Nutr.* **2012**, *67*, 76–81. [CrossRef] [PubMed]
78. Shimizu, J.; Wada-Funada, U.; Mano, H.; Matahira, Y.; Kawaguchi, M.; Wada, M. Proportion of Murine Cytotoxic T Cells is Increased by High Molecular-Weight Fucoidan Extracted from *Okinawa mozuku* (*Cladosiphon okamuranus*). *J. Health Sci.* **2005**, *51* 394–397. [CrossRef]
79. Kar, S.; Sharma, G.; Das, P.K. Fucoidan cures infection with both antimony-susceptible and -resistant strains of *Leishmania donovani* through Th1 response and macrophage-derived oxidants. *J. Antimicrob. Chemother.* **2011**, *66*, 618–625. [CrossRef]
80. Hayashi, T.; Hayashi, K.; Kanekiyo, K.; Ohta, Y.; Lee, J.-B.; Hashimoto, M.; Nakano, T. Promising Antiviral Glyco-Molecules from an Edible Alga. In *Combating the Threat of Pandemic Influenza: Drug Discovery Approaches*; John Wiley & Sons: Hoboken, NJ, USA, 2007; pp. 166–182. [CrossRef]
81. Araya, N.; Takahashi, K.; Sato, T.; Nakamura, T.; Sawa, C.; Hasegawa, D.; Ando, H.; Aratani, S.; Yagishita, N.; Fujii, R.; et al. Fucoidan therapy decreases the proviral load in patients with human T-lymphotropic virus type-1-associated neurological disease. *Antivir. Ther.* **2011**, *16*, 89–98. [CrossRef]
82. Hayashi, K.; Nakano, T.; Hashimoto, M.; Kanekiyo, K.; Hayashi, T. Defensive effects of a fucoidan from brown alga *Undaria pinnatifida* against herpes simplex virus infection. *Int. Immunopharmacol.* **2008**, *8*, 109–116. [CrossRef] [PubMed]
83. Hayashi, K.; Lee, J.-B.; Nakano, T.; Hayashi, T. Anti-influenza A virus characteristics of a fucoidan from sporophyll of *Undaria pinnatifida* in mice with normal and compromised immunity. *Microbes Infect.* **2013**, *15*, 302–309. [CrossRef] [PubMed]
84. Maruyama, H.; Tamauchi, H.; Iizuka, M.; Nakano, T. The Role of NK cells in Antitumor Activity of Dietary Fucoidan from *Undaria pinnatifida* Sporophylls (Mekabu). *Planta Med.* **2006**, *72*, 1415–1417. [CrossRef] [PubMed]
85. Azuma, K.; Ishihara, T.; Nakamoto, H.; Amaha, T.; Osaki, T.; Tsuka, T.; Imagawa, T.; Minami, S.; Takashima, O.; Ifuku, S.; et al. Effects of Oral Administration of Fucoidan Extracted from *Cladosiphon okamuranus* on Tumor Growth and Survival Time in a Tumor-Bearing Mouse Model. *Mar. Drugs* **2012**, *10*, 2337–2348. [CrossRef] [PubMed]
86. Fan, S.; Zhang, J.; Nie, W.; Zhou, W.; Jin, L.; Chen, X.; Lu, J. Antitumor effects of polysaccharide from *Sargassum fusiforme* against human hepatocellular carcinoma HepG2 cells. *Food Chem. Toxicol.* **2017**, *102*, 53–62. [CrossRef]
87. Chen, X.; Nie, W.; Yu, G.; Li, Y.; Hu, Y.; Lu, J.; Jin, L. Antitumor and immunomodulatory activity of polysaccharides from *Sargassum fusiforme*. *Food Chem. Toxicol.* **2012**, *50*, 695–700. [CrossRef] [PubMed]
88. Fan, S.; Yu, G.; Nie, W.; Jin, J.; Chen, L.; Chen, X. Antitumor activity and underlying mechanism of *Sargassum fusiforme* polysaccharides in CNE-bearing mice. *Int. J. Biol. Macromol.* **2018**, *112*, 516–522. [CrossRef]
89. Atashrazm, F.; Lowenthal, R.M.; Woods, G.; Holloway, A.; Karpiniec, S.S.; Dickinson, J.L. Fucoidan Suppresses the Growth of Human Acute Promyelocytic Leukemia Cells In Vitro and In Vivo. *J. Cell. Physiol.* **2015**, *231*, 688–697. [CrossRef]
90. Xue, M.; Liang, H.; Tang, Q.; Xue, C.; He, X.; Zhang, L.; Zhang, Z.; Liang, Z.; Bian, K.; Zhang, L.; et al. The Protective and Immunomodulatory Effects of Fucoidan Against 7,12-Dimethyl benz[a]anthracene-Induced Experimental Mammary Carcinogenesis Through the PD1/PDL1 Signaling Pathway in Rats. *Nutr. Cancer* **2017**, *69*, 1234–1244. [CrossRef]
91. Jiang, Z.; Abu, R.; Isaka, S.; Nakazono, S.; Ueno, M.; Okimura, T.; Yamaguchi, K.; Oda, T. Inhibitory effect of orally-administered sulfated polysaccharide ascophyllan isolated from *ascophyllum nodosum* on the growth of sarcoma-180 solid tumor in mice. *Anticancer Res.* **2014**, *34*, 1663–1671. [PubMed]
92. Chen, X.; Nie, W.; Fan, S.; Zhang, J.; Wang, Y.; Lu, J.; Jin, L. A polysaccharide from *Sargassum fusiforme* protects against immunosuppression in cyclophospha-mide-treated mice. *Carbohydr. Polym.* **2012**, *90*, 1114–1119. [CrossRef]
93. Lee, H.H.; Cho, Y.; Kim, G.-H.; Cho, H. *Undaria pinnatifida* Fucoidan-Rich Extract Recovers Immunity of Immunosuppressed Mice. *J. Microbiol. Biotechnol.* **2020**, *30*, 439–447. [CrossRef]
94. Zhu, X.; Zhu, R.; Jian, Z.; Yu, H. Laminarin enhances the activity of natural killer cells in immunosuppressed mice. *Cent. Eur. J. Immunol.* **2019**, *44*, 357–363. [CrossRef]
95. Hwang, P.-A.; Hung, Y.-L.; Chien, S.-Y. Inhibitory activity of *Sargassum hemiphyllum* sulfated polysaccharide in arachidonic acid-induced animal models of inflammation. *J. Food Drug Anal.* **2015**, *23*, 49–56. [CrossRef] [PubMed]
96. Kyung, J.; Kim, D.; Park, D.; Yang, Y.-H.; Choi, E.-K.; Lee, S.-P.; Kim, T.-S.; Lee, Y.-B.; Kim, Y.-B. Synergistic anti-inflammatory effects of *Laminaria japonica* fucoidan and *Cistanche tubulosa* extract. *Lab. Anim. Res.* **2012**, *28*, 91–97. [CrossRef]
97. Choi, J.-I.; Raghavendran, H.R.B.; Sung, N.-Y.; Kim, J.-H.; Chun, B.S.; Ahn, D.H.; Choi, H.-S.; Kang, K.-W.; Lee, J.-W. Effect of fucoidan on aspirin-induced stomach ulceration in rats. *Chem. Interact.* **2010**, *183*, 249–254. [CrossRef]
98. Matsumoto, S.; Nagaoka, M.; Hara, T.; Kimura-Takagi, I.; Mistuyama, K.; Ueyama, S. Fucoidan derived from *Cladosiphon okamuranus* Tokida ameliorates murine chronic colitis through the down-regulation of interleukin-6 production on colonic epithelial cells. *Clin. Exp. Immunol.* **2004**, *136*, 432–439. [CrossRef]
99. Lean, Q.Y.; Eri, R.D.; Fitton, J.H.; Patel, R.P.; Gueven, N. Fucoidan Extracts Ameliorate Acute Colitis. *PLoS ONE* **2015**, *10*, e0128453. [CrossRef] [PubMed]

100. Wang, L.; Ai, C.; Wen, C.; Qin, Y.; Liu, Z.; Wang, L.; Gong, Y.; Su, C.; Wang, Z.; Song, S. Fucoidan isolated from *Ascophyllum nodosum* alleviates gut microbiota dysbiosis and colonic inflammation in antibiotic-treated mice. *Food Funct.* **2020**, *11*, 5595–5606. [CrossRef] [PubMed]
101. Xue, M.; Liang, H.; Ji, X.; Zhou, Z.; Liu, Y.; Sun, T.; Zhang, L. Effects of fucoidan on gut flora and tumor prevention in 1,2-dimethylhydrazine-induced colorectal carcino-genesis. *J. Nutr. Biochem.* **2020**, *82*, 108396. [CrossRef]
102. Park, S.-B.; Chun, K.-R.; Kim, J.-K.; Suk, K.; Jung, Y.-M.; Lee, W.-H. The differential effect of high and low molecular weight fucoidans on the severity of collagen-induced arthritis in mice. *Phytother. Res.* **2010**, *24*, 1384–1391. [CrossRef] [PubMed]
103. Jeong, J.-W.; Hwang, S.J.; Han, M.H.; Lee, D.-S.; Yoo, J.S.; Choi, I.-W.; Cha, H.-J.; Kim, S.; Kim, H.-S.; Kim, G.-Y.; et al. Fucoidan inhibits lipopolysaccharide-induced inflammatory responses in RAW 264.7 macrophages and zebrafish larvae. *Mol. Cell. Toxicol.* **2017**, *13*, 405–417. [CrossRef]
104. Li, C.; Gao, Y.; Xing, Y.; Zhu, H.; Shen, J.; Tian, J. Fucoidan, a sulfated polysaccharide from brown algae, against myocardial ischemia-reperfusion injury in rats via regulating the inflammation response. *Food Chem. Toxicol.* **2011**, *49*, 2090–2095. [CrossRef] [PubMed]
105. Yu, H.-H.; Chengchuan Ko, E.; Chang, C.-L.; Yuan, K.S.-P.; Wu, A.T.H.; Shan, Y.-S.; Wu, S.-Y. Fucoidan Inhibits Radiation-Induced Pneumonitis and Lung Fibrosis by Reducing Inflammatory Cytokine Expression in Lung Tissues. *Mar. Drugs* **2018**, *16*, 392. [CrossRef]
106. Bai, X.; Li, M.; Wang, X.; Chang, H.; Ni, Y.; Li, C.; He, K.; Wang, H.; Yang, Y.; Tian, T.; et al. Therapeutic potential of fucoidan in the reduction of hepatic pathology in murine *schistosomiasis japonica*. *Parasites Vectors* **2020**, *13*, 1–14. [CrossRef]
107. Li, J.; Chen, K.; Li, S.; Liu, T.; Wang, F.; Xia, Y.; Lu, J.; Zhou, Y.; Guo, C. Pretreatment with Fucoidan from *Fucus vesiculosus* Protected against ConA-Induced Acute Liver Injury by Inhibiting Both Intrinsic and Extrinsic Apoptosis. *PLoS ONE* **2016**, *11*, e0152570. [CrossRef] [PubMed]
108. Xue, M.; Liang, H.; Ji, X.; Liu, Y.; Ge, Y.; Hou, L.; Sun, T. Fucoidan prevent murine autoimmune diabetes via suppression TLR4-signaling pathways, regulation DC/Treg induced immune tolerance and improving gut microecology. *Nutr. Metab.* **2019**, *16*, 1–15. [CrossRef]
109. Ko, W.-S.; Shen, F.-P.; Shih, C.-J.; Chiou, Y.-L. The 25(OH)Vitamin D Status Affected the Effectiveness of Oligo Fucoidan in Patients with Chronic Hepatitis B Virus Infection with Immune Tolerance Phase. *Nutrients* **2020**, *12*, 321. [CrossRef]
110. Myers, S.P.; O'Connor, J.; Fitton, J.H.; Brooks, L.; Rolfe, M.; Connellan, P.; Wohlmuth, H.; Cheras, P.A.; Morris, C. A combined phase I and II open label study on the effects of a seaweed extract nutrient complex on oste-oarthritis. *Biol. Targets Ther.* **2010**, *4*, 33–44. [CrossRef]
111. Takahashi, H.; Kawaguchi, M.; Kitamura, K.; Narumiya, S.; Kawamura, M.; Tengan, I.; Nishimoto, S.; Hanamure, Y.; Majima, Y.; Tsubura, S.; et al. An Exploratory Study on the Anti-inflammatory Effects of Fucoidan in Relation to Quality of Life in Advanced Cancer Patients. *Integr. Cancer Ther.* **2017**, *17*, 282–291. [CrossRef]
112. Peng, F.-H.; Zha, X.-Q.; Cui, S.-H.; Asghar, M.-N.; Pan, L.-H.; Wang, J.-H.; Luo, J.-P. Purification, structure features and anti-atherosclerosis activity of a Laminaria japonica polysaccharide. *Int. J. Biol. Macromol.* **2015**, *81*, 926–935. [CrossRef] [PubMed]
113. Wang, X. Fucoidan attenuates atherosclerosis in LDLR−/− mice through inhibition of inflammation and oxidative stress. *Int. J. Clin. Exp. Pathol.* **2016**, *9*, 6896–6904.
114. Xu, Y.; Xu, J.; Ge, K.; Tian, Q.; Zhao, P.; Guo, Y. Anti-inflammatory effect of low molecular weight fucoidan from *Saccharina japonica* on atherosclerosis in apoE-knockout mice. *Int. J. Biol. Macromol.* **2018**, *118*, 365–374. [CrossRef]
115. Soin, J.; Kurzejamska, E.; Gaciong, Z.; Henrykowska, G.; Bojakowski, K. Fucoidan Inhibits Vascular Remodeling in Transplant Vasculopathy in Rat. *Funct. Foods Health Dis.* **2018**, *8*, 323. [CrossRef]
116. Zhou, M.; Ding, Y.; Cai, L.; Wang, Y.; Lin, C.; Shi, Z. Low molecular weight fucoidan attenuates experimental abdominal aortic aneurysm through interfering the leukocyte-endothelial cells interaction. *Mol. Med. Rep.* **2018**, *17*, 7089–7096. [CrossRef]
117. Mo, W.; Wang, C.; Li, J.; Chen, K.; Xia, Y.; Li, S.; Xu, L.; Lu, X.; Wang, W.; Guo, C. Fucosterol Protects against Concanavalin A-Induced Acute Liver Injury: Focus on P38 MAPK/NF-kappaB Pathway Activity. *Gastroenterol. Res. Pract.* **2018**, *2018*, 2824139. [CrossRef] [PubMed]
118. Bogie, J.; Hoeks, C.; Schepers, M.; Tiane, A.; Cuypers, A.; Leijten, F.; Chintapakorn, Y.; Suttiyut, T.; Pornpakakul, S.; Struik, D.; et al. Dietary Sargassum fusiforme improves memory and reduces amyloid plaque load in an Alzheimer's disease mouse model. *Sci. Rep.* **2019**, *9*, 1–16. [CrossRef]
119. Lacy, M.; Atzler, D.; Liu, R.; de Winther, M.; Weber, C.; Lutgens, E. Interactions between dyslipidemia and the immune system and their relevance as putative therapeutic targets in atherosclerosis. *Pharmacol. Ther.* **2019**, *193*, 50–62. [CrossRef]
120. Schulman, I.G. Liver X receptors link lipid metabolism and inflammation. *FEBS Lett.* **2017**, *591*, 2978–2991. [CrossRef]
121. Gonzalez, N.A.; Castrillo, A. Liver X receptors as regulators of macrophage inflammatory and metabolic pathways. *Biochim. Biophys. Acta.* **2011**, *1812*, 982–994. [CrossRef]
122. Irfan, M.; Kwon, T.-H.; Yun, B.-S.; Park, N.-H.; Rhee, M.H. *Eisenia bicyclis* (brown alga) modulates platelet function and inhibits thrombus formation via impaired P 2 Y 12 receptor signaling pathway. *Phytomedicine* **2018**, *40*, 79–87. [CrossRef]
123. Han, E.J.; Kim, H.S.; Sanjeewa, K.K.A.; Herath, K.; Jeon, Y.J.; Jee, Y.; Lee, J.; Kim, T.; Shim, S.Y.; Ahn, G. Eckol from *Ecklonia cava* Suppresses Immunoglobulin E-mediated Mast Cell Activation and Passive Cutaneous Anaphylaxis in Mice. *Nutrients* **2020**, *12*, 1361. [CrossRef] [PubMed]

124. Herath, K.H.I.N.M.; Mihindukulasooriya, S.P.; Kim, H.J.; Kim, A.; Kim, H.J.; Jeon, Y.J.; Jee, Y. Oral administration of polyphenol-rich *Sargassum horneri* suppresses particulate matter exacer-bated airway inflammation in murine allergic asthma: Relevance to the TLR mediated NF-kappa B pathway inhibition. *J. Funct. Foods* **2020**, *71*, 103991. [CrossRef]
125. Sugiura, Y.; Nagayama, K.; Kinoshita, Y.; Tanaka, R.; Matsushita, T. The anti-allergic effect of the ethyl acetate fraction from an *Ecklonia kurome* extract. *Food Agric. Immunol.* **2015**, *26*, 181–193. [CrossRef]
126. Son, M.; Oh, S.; Choi, J.; Jang, J.T.; Choi, C.H.; Park, K.Y.; Son, K.H.; Byun, K. Attenuation of Inflammation and Leptin Resistance by Pyrogallol-Phloroglucinol-6,6-Bieckol on in the Brain of Obese Animal Models. *Nutrients* **2019**, *11*, 2773. [CrossRef] [PubMed]
127. Son, M.; Oh, S.; Choi, J.; Jang, J.T.; Choi, C.H.; Park, K.Y.; Son, K.H.; Byun, K. The Phlorotannin-Rich Fraction of *Ecklonia cava* Extract Attenuated the Expressions of the Markers Related with Inflammation and Leptin Resistance in Adipose Tissue. *Int. J. Endocrinol.* **2020**, *2020*, 1–11. [CrossRef]
128. Eo, H.; Jeon, Y.-J.; Lee, M.; Lim, Y. Brown Alga *Ecklonia cava* Polyphenol Extract Ameliorates Hepatic Lipogenesis, Oxidative Stress, and Inflammation by Activation of AMPK and SIRT1 in High-Fat Diet-Induced Obese Mice. *J. Agric. Food Chem.* **2014**, *63*, 349–359. [CrossRef] [PubMed]
129. Yang, Y.-I.; Ahn, J.-H.; Choi, Y.S.; Choi, J.-H. Brown algae phlorotannins enhance the tumoricidal effect of cisplatin and ameliorate cisplatin nephrotoxicity. *Gynecol. Oncol.* **2015**, *136*, 355–364. [CrossRef] [PubMed]
130. Kim, S.; Choi, S.-I.; Kim, G.-H.; Imm, J.-Y. Anti-Inflammatory Effect of *Ecklonia cava* Extract on Porphyromonas gingivalis Lipopolysaccharide-Stimulated Macrophages and a Periodontitis Rat Model. *Nutrients* **2019**, *11*, 1143. [CrossRef]
131. Kim, E.A.; Kim, S.Y.; Ye, B.R.; Kim, J.; Ko, S.C.; Lee, W.W.; Kim, K.N.; Choi, I.W.; Jung, W.K.; Heo, S.J. Anti-inflammatory effect of Apo-9′-fucoxanthinone via inhibition of MAPKs and NF-kB signaling pathway in LPS-stimulated RAW 264.7 macrophages and zebrafish model. *Int. Immunopharmacol.* **2018**, *59*, 339–346. [CrossRef]
132. Kang, K.; Park, Y.; Hwang, H.J.; Kim, S.H.; Lee, J.G.; Shin, H.-C. Antioxidative properties of brown algae polyphenolics and their perspectives as chemopreventive agents against vascular risk factors. *Arch. Pharmacal. Res.* **2003**, *26*, 286–293. [CrossRef] [PubMed]
133. Kim, S.Y.; Kim, E.A.; Kang, M.C.; Lee, J.H.; Yang, H.W.; Lee, J.S.; Lim, T.I.; Jeon, Y.J. Polyphenol-rich fraction from *Ecklonia cava* (a brown alga) processing by-product reduces LPS-induced in-flammation in vitro and in vivo in a zebrafish model. *Algae* **2014**, *29*, 165–174. [CrossRef]
134. Fernando, I.P.S.; Kim, H.S.; Sanjeewa, K.K.A.; Oh, J.Y.; Jeon, Y.J.; Lee, W.W. Inhibition of inflammatory responses elicited by urban fine dust particles in keratinocytes and macro-phages by diphlorethohydroxy-carmalol isolated from a brown alga *Ishige okamurae*. *Algae* **2017**, *32*, 261–273. [CrossRef]
135. Li, S.; Liu, J.; Zhang, M.; Chen, Y.; Zhu, T.; Wang, J. Protective Effect of Eckol against Acute Hepatic Injury Induced by Carbon Tetrachloride in Mice. *Mar. Drugs* **2018**, *16*, 300. [CrossRef]
136. Sugiura, Y.; Usui, M.; Katsuzaki, H.; Imai, K.; Kakinuma, M.; Amano, H.; Miyata, M. Orally Administered Phlorotannins from *Eisenia arborea* Suppress Chemical Mediator Release and Cy-clooxygenase-2 Signaling to Alleviate Mouse Ear Swelling. *Mar. Drugs* **2018**, *16*, 267. [CrossRef] [PubMed]
137. Choi, J.; Oh, S.; Son, M.; Byun, K. Pyrogallol-Phloroglucinol-6,6-Bieckol Alleviates Obesity and Systemic Inflammation in a Mouse Model by Reducing Expression of RAGE and RAGE Ligands. *Mar. Drugs* **2019**, *17*, 612. [CrossRef] [PubMed]
138. Son, M.; Oh, S.; Lee, H.S.; Chung, D.-M.; Jang, J.T.; Jeon, Y.-J.; Choi, C.H.; Park, K.Y.; Son, K.H.; Byun, K. *Ecklonia Cava* Extract Attenuates Endothelial Cell Dysfunction by Modulation of Inflammation and Brown Adipocyte Function in Perivascular Fat Tissue. *Nutrients* **2019**, *11*, 2795. [CrossRef]
139. Goc, A.; Gehring, G.; Baltin, H.; Niedzwiecki, A.; Rath, M. Specific composition of polyphenolic compounds with fatty acids as an approach in helping to reduce spirochete burden in Lyme disease: In vivo and human observational study. *Ther. Adv. Chronic Dis.* **2020**, *11*. [CrossRef]
140. Park, E.Y.; Kim, E.H.; Kim, M.H.; Seo, Y.W.; Lee, J.I.; Jun, H.S. Polyphenol-Rich Fraction of Brown Alga *Ecklonia cava* Collected from Gijang, Korea, Reduces Obesity and Glucose Levels in High-Fat Diet-Induced Obese Mice. *Evid. Based Complement Altern. Med.* **2012**, *2012*, 418912. [CrossRef]
141. Zhao, Z.; Bao, X.Q.; Zhang, Z.; Liu, H.; Zhang, D. Phloroglucinol derivative compound 21 attenuates cuprizone-induced multiple sclerosis mice through pro-moting remyelination and inhibiting neuroinflammation. *Sci. China Life Sci.* **2020**, *63*, 905–914. [CrossRef]
142. Zhao, Z.; Bao, X.-Q.; Zhang, Z.; Li, F.; Liu, H.; Zhang, D. Novel phloroglucinol derivative Compound 21 protects experimental autoimmune encephalomyelitis rats via inhibiting Th1/Th17 cell infiltration. *Brain Behav. Immun.* **2020**, *87*, 751–764. [CrossRef]
143. Miyashita, K.; Hosokawa, M. Carotenoids as a Nutraceutical Therapy for Visceral Obesity. In *Nutrition in the Prevention and Treatment of Abdominal Obesity*; Watson, R.R., Ed.; Academic Press: San Diego, CA, USA, 2014; pp. 329–340.
144. Sugawara, T.; Baskaran, V.; Tsuzuki, W.; Nagao, A. Brown Algae Fucoxanthin Is Hydrolyzed to Fucoxanthinol during Absorption by Caco-2 Human Intestinal Cells and Mice. *J. Nutr.* **2002**, *132*, 946–951. [CrossRef]
145. Maeda, H.; Kanno, S.; Kodate, M.; Hosokawa, M.; Miyashita, K. Fucoxanthinol, Metabolite of Fucoxanthin, Improves Obesity-Induced Inflammation in Adipocyte Cells. *Mar. Drugs* **2015**, *13*, 4799–4813. [CrossRef] [PubMed]
146. Martin, L.J. Fucoxanthin and Its Metabolite Fucoxanthinol in Cancer Prevention and Treatment. *Mar. Drugs* **2015**, *13*, 4784–4798. [CrossRef]

147. Menna, M.; Imperatore, C.; D'Aniello, F.; Aiello, A. Meroterpenes from Marine Invertebrates: Structures, Occurrence, and Ecological Implications. *Mar. Drugs* **2013**, *11*, 1602–1643. [CrossRef]
148. Gwon, W.-G.; Joung, E.-J.; Shin, T.; Utsuki, T.; Wakamatsu, N.; Kim, H.-R. Meroterpinoid-rich fraction of the ethanol extract from *Sargassum serratifolium* suppresses TNF-α-induced monocytes adhesion to vascular endothelium and vascular inflammation in high cholesterol-fed C57BL/6J mice. *J. Funct. Foods* **2018**, *46*, 384–393. [CrossRef]
149. Zhang, H.; Tang, Y.; Zhang, Y.; Zhang, S.; Qu, J.; Wang, X.; Kong, R.; Han, C.; Liu, Z. Fucoxanthin: A Promising Medicinal and Nutritional Ingredient. *Evid. Based Complement. Altern. Med.* **2015**, *2015*, 1–10. [CrossRef]
150. Kong, Z.-L.; Sudirman, S.; Hsu, Y.-C.; Su, C.-Y.; Kuo, H.-P. Fucoxanthin-Rich Brown Algae Extract Improves Male Reproductive Function on Streptozoto-cin-Nicotinamide-Induced Diabetic Rat Model. *Int. J. Mol. Sci.* **2019**, *20*, 4485. [CrossRef] [PubMed]
151. Kong, Z.-L.; Kao, N.-J.; Hu, J.-Y.; Wu, C.-S. Fucoxanthin-Rich Brown Algae Extract Decreases Inflammation and Attenuates Colitis-associated Colon Cancer in Mice. *J. Food Nutr. Res.* **2016**, *4*, 137–147.
152. Wang, P.-T.; Sudirman, S.; Hsieh, M.-C.; Hu, J.-Y.; Kong, Z.-L. Oral supplementation of fucoxanthin-rich brown algae extract ameliorates cisplatin-induced testicular damage in hamsters. *Biomed. Pharmacother.* **2020**, *125*, 109992. [CrossRef] [PubMed]
153. Chen, L.; Deng, H.; Cui, H.; Fang, J.; Zuo, Z.; Deng, J.; Li, Y.; Wang, X.; Zhao, L. Inflammatory responses and inflammation-associated diseases in organs. *Oncotarget* **2017**, *9*, 7204–7218. [CrossRef]
154. Ayala, A.; Muñoz, M.F.; Argüelles, S. Lipid Peroxidation: Production, Metabolism, and Signaling Mechanisms of Malondialdehyde and 4-Hydroxy-2-Nonenal. *Oxid. Med. Cell. Longev.* **2014**, *2014*, 360438. [CrossRef]
155. Tan, C.-P.; Hou, Y.-H. First Evidence for the Anti-inflammatory Activity of Fucoxanthin in High-Fat-Diet-Induced Obesity in Mice and the Antioxidant Functions in PC12 Cells. *Inflammation* **2013**, *37*, 443–450. [CrossRef]
156. Maeda, H.; Hosokawa, M.; Sashima, T.; Murakami-Funayama, K.; Miyashita, K. Anti-obesity and anti-diabetic effects of fucoxanthin on diet-induced obesity conditions in a murine model. *Mol. Med. Rep.* **2009**, *2*, 897–902. [CrossRef] [PubMed]
157. Sugiura, Y.; Kinoshita, Y.; Usui, M.; Tanaka, R.; Matsushita, T.; Miyata, M. The Suppressive Effect of a Marine Carotenoid, Fucoxanthin, on Mouse Ear Swelling through Regulation of Activities and mRNA Expression of Inflammation-associated Enzymes. *Food Sci. Technol. Res.* **2016**, *22*, 227–234. [CrossRef]
158. Kwon, M.; Lim, S.-J.; Joung, E.-J.; Lee, B.; Oh, C.-W.; Kim, H.-R. Meroterpenoid-rich fraction of an ethanolic extract from *Sargassum serratifolium* alleviates obesity and non-alcoholic fatty liver disease in high fat-fed C57BL/6J mice. *J. Funct. Foods* **2018**, *47*, 288–298. [CrossRef]
159. Cheng, I.C.; Weng, S.-Y.; Wu, M.-S.; Suk, F.-M.; Lien, G.-S.; Chen, C.-N. Low-molecular-weight fucoidan and high-stability fucoxanthin decrease serum alanine transaminase in patients with nonalcoholic fatty liver disease—A double-blind, randomized controlled trial. *Adv. Dig. Med.* **2019**, *6*, 116–122. [CrossRef]
160. Kurylowicz, A.; Jonas, M.; Lisik, W.; Jonas, M.; Wicik, Z.A.; Wierzbicki, Z.; Chmura, A.; Puzianowska-Kuznicka, M. Obesity is associated with a decrease in expression but not with the hypermethylation of thermogene-sis-related genes in adipose tissues. *J. Transl. Med.* **2015**, *13*, 31. [CrossRef]
161. Borzouie, S.; Rathgeber, B.M.; Stupart, C.M.; MacIsaac, J.; MacLaren, L.A. Effects of Dietary Inclusion of Seaweed, Heat Stress and Genetic Strain on Performance, Plasma Biochemical and Hematological Parameters in Laying Hens. *Animals* **2020**, *10*, 1570. [CrossRef]
162. Kawauchi, S.; Horibe, S.; Sasaki, N.; Tanahashi, T.; Mizuno, S.; Hamaguchi, T.; Rikitake, Y. Inhibitory Effects of Sodium Alginate on Hepatic Steatosis in Mice Induced by a Methionine- and Choline-deficient Diet. *Mar. Drugs* **2019**, *17*, 104. [CrossRef] [PubMed]
163. Lee, H.H.; Cho, Y.J.; Yu, D.; Chung, D.; Kim, G.-H.; Kang, H.; Cho, H. Undaria pinnatifida Fucoidan-Rich Extract Induces Both Innate and Adaptive Immune Responses. *Nat. Prod. Commun.* **2019**, *14*. [CrossRef]
164. Wang, W.; Lu, J.-B.; Wang, C.-S.; Zhang, H.-H.; Li, C.-Y.; Qian, G.-Y. Effects of *Sargassum fusiforme* polysaccharides on antioxidant activities and intestinal functions in mice. *Int. J. Biol. Macromol.* **2013**, *58*, 127–132. [CrossRef] [PubMed]
165. Nagamine, T.; Kadena, K.; Tomori, M.; Nakajima, K.; Iha, M. Activation of NK cells in male cancer survivors by fucoidan extracted from *Cladosiphon okamuranus*. *Mol. Clin. Oncol.* **2019**, *12*, 81–88. [CrossRef] [PubMed]
166. Irhimeh, M.R.; Fitton, J.H.; Lowenthal, R.M. Fucoidan ingestion increases the expression of CXCR4 on human CD34+ cells. *Exp. Hematol.* **2007**, *35*, 989–994. [CrossRef] [PubMed]
167. Negishi, H.; Mori, M.; Mori, H.; Yamori, Y. Supplementation of Elderly Japanese Men and Women with Fucoidan from Seaweed Increases Immune Responses to Seasonal Influenza Vaccination. *J. Nutr.* **2013**, *143*, 1794–1798. [CrossRef] [PubMed]
168. Chen, Y.-C.; Cheng, C.-Y.; Liu, C.-T.; Sue, Y.-M.; Chen, T.-H.; Hsu, Y.-H.; Huang, N.-J.; Chen, C.-H. Combined protective effects of oligo-fucoidan, fucoxanthin, and L-carnitine on the kidneys of chronic kidney disease mice. *Eur. J. Pharmacol.* **2021**, *892*, 173708. [CrossRef] [PubMed]
169. Yokota, T.; Nomura, K.; Nagashima, M.; Kamimura, N. Fucoidan alleviates high-fat diet-induced dyslipidemia and atherosclerosis in ApoE(shl) mice deficient in apolipoprotein E expression. *J. Nutr. Biochem.* **2016**, *32*, 46–54. [CrossRef] [PubMed]
170. Kawashima, T.; Murakami, K.; Nishimura, I.; Nakano, T.; Obata, A. A sulfated polysaccharide, fucoidan, enhances the immunomodulatory effects of lactic acid bacteria. *Int. J. Mol. Med.* **2012**, *29*, 447–453. [CrossRef]
171. Hwang, E.; Park, S.-Y.; Shin, H.-S.; Lee, D.-G.; Yi, T.H. Effect of oral administration of fucosterol from *Hizikia fusiformis* on DNCB-induced atopic dermatitis in NC/Nga mice. *Food Sci. Biotechnol.* **2014**, *23*, 593–599. [CrossRef]

MDPI
St. Alban-Anlage 66
4052 Basel
Switzerland
Tel. +41 61 683 77 34
Fax +41 61 302 89 18
www.mdpi.com

Nutrients Editorial Office
E-mail: nutrients@mdpi.com
www.mdpi.com/journal/nutrients

www.ingramcontent.com/pod-product-compliance
Lightning Source LLC
Chambersburg PA
CBHW061513210926
51850CB00008B/1902
9783036553474